Sediment Diagenesis

NATO ASI Series

Advanced Science Institutes Series

A series presenting the results of activities sponsored by the NATO Science Committee, which aims at the dissemination of advanced scientific and technological knowledge, with a view to strengthening links between scientific communities.

The series is published by an international board of publishers in conjunction with the NATO Scientific Affairs Division

A	Life Sciences	Plenum Publishing Corporation
B	Physics	London and New York
C	Mathematical and Physical Sciences	D. Reidel Publishing Company Dordrecht, Boston and Lancaster
D	Behavioural and Social Sciences	Martinus Nijhoff Publishers
E	Engineering and Materials Sciences	The Hague, Boston and Lancaster
F	Computer and Systems Sciences	Springer-Verlag
G	Ecological Sciences	Berlin, Heidelberg, New York and Tokyo

Sediment Diagenesis

edited by

A. Parker

and

B. W. Sellwood

Sedimentology Research Laboratory,
Department of Geology, University of Reading, U.K.

D. Reidel Publishing Company

Dordrecht / Boston / Lancaster

Published in cooperation with NATO Scientific Affairs Division

Proceedings of the NATO Advanced Study Institute on
Sediment Diagenesis
Reading, U.K.
July 12-25, 1981

Library of Congress Cataloging in Publication Data

NATO Advanced Study Institute on Sediment Diagenesis 1981 : Reading,
    Berkshire)
    Sediment diagenesis.

    (NATO ASI series. Series C, Mathematical and physical sciences ; Vol. 115)
    "Proceedings of the NATO Advanced Study Institute on Sediment
Diagenesis"—T.p. verso.
    Includes index.
    1. Sediments (Geology)—Congresses.   2.   Diagenesis—Congresses.
I. Parker, A. (Andrew), 1941-   .   II.   Sellwood, B. W. (Bruce William),
1946-   .   III.   Title.   IV.   Series.
QE471.N393   1981   551.3   83-19073
ISBN 90-277-1677-3

Published by D. Reidel Publishing Company
P.O. Box 17, 3300 AA Dordrecht, Holland

Sold and distributed in the U.S.A. and Canada
by Kluwer Academic Publishers,
190 Old Derby Street, Hingham, MA 02043, U.S.A.

In all other countries, sold and distributed
by Kluwer Academic Publishers Group,
P.O. Box 322, 3300 AH Dordrecht, Holland

D. Reidel Publishing Company is a member of the Kluwer Academic Publishers Group

Printed in The Netherlands

TABLE OF CONTENTS

FOREWORD

The topic of sediment diagenesis is of fundamental importance
to industry in the evaluation of hydrocarbon and water reservoir
rocks. Detailed knowledge of the diagenetic textures, fabrics, and
minerals, and a prediction of the regional diagenetic response,
partly controls hydrocarbon recovery programmes. In other words,
knowledge of the diagenesis can aid (or even control) conservation
policy. Similarly, facies and diagenetic trends within basins can
influence exploration policy.

This volume incorporates the majority of the principal
contributions given to the NATO Advanced Study Institute held in
the University of Reading, U.K., from July 12th-25th, 1981, at
which the major themes of carbonate and terrigenous clastic
sediments were treated sequentially from deposition to deep burial.

Eighty selected scientists from twelve NATO and three other
countries participated in the Institute. The keynote addresses
which acted as the touchstones for discussion are presented here
in the expectation that they will stimulate a still wider audience.

We gratefully acknowledge the award of a grant from the
Scientific Affairs Division of NATO to run the Institute, and also
the cooperation of the University of Reading. Mrs. D. M. Powell
helped in many ways with the organisation, and also retyped the
entire manuscript of this book.

A. Parker

B. W. Sellwood

FACIES, SEQUENCES AND SAND-BODIES OF THE PRINCIPAL CLASTIC
DEPOSITIONAL ENVIRONMENTS

T.Elliott
Department of Geology
University College of Swansea
Singleton Park,
Swansea SA2 8PP
Wales, U.K.

INTRODUCTION

One of the many controls on the diagenesis of clastic sediments is the depositional environment in which the sediments
accumulated (Hayes, 1979; Nagtegaal, 1980). This control
operates in a number of ways:

1. The depositional environment dictates the facies pattern
including the size, form and overall context of initially porous
and permeable sand-bodies. These sand-bodies influence the flow
of diagenetic pore fluids through the sediment body and may
ultimately serve as hydrocarbon reservoirs.

2. The physical and bilogical processes of the depositional
environment influence the texture of the sediment in terms of
grain size, sorting and packing. As the texture defines the
initial porosity and permeability characteristics of the sediment
it is therefore crucial to the course of diagenesis.

3. To some extent the physical processes of the depositional
environment also affect the initial composition of the sediment,
though provenance and tectonic setting are also important in this
respect.

4. The depositional environment naturally dictates early, shallow
diagenesis which is important in its own right and can also influence the course of later, deeper burial diagenesis. Important
factors include the composition of the pore waters (whether
saline, fresh or mixed), oxidation-reduction reactions, early

1

A. Parker and B. W. Sellwood (eds.), Sediment Diagenesis, 1–56.
© *1983 by D. Reidel Publishing Company.*

calcite leaching or precipitation, and the developments of evaporitic or soil-related minerals.

Thus the depositional environment can influence the early
diagenetic processes, the initial raw materials on which the diagenetic processes operate, and the small- and large-scale
permeability of the sediments which controls the passage of the
diagenetic fluids.   Recognition of the depositional environments
of clastic sediments is therefore an important part of unravelling
their diagenetic history.   What follows is a review of the main
clastic depositional environments (excluding lake and glacial
settings) in terms of their processes, facies and sand-bodies.
The emphasis is towards the larger scale, sandy-body aspect with
the intention of reviewing the diversity of sand-bodies which can
be produced in different depositional environments and hence providing an environmental context for the diagenesis of sand-bodies.
Details of compositional attributes and textures are discussed
briefly but not treated at length, though in many cases textural
information can be deduced from descriptions of the facies and
their sedimentary structures.   The emphasis on diversity of sand-
bodies is important as these days it is no longer sufficient to
merely recognise the environment of deposition in sedimentary
successions.   Instead, once the environment has been recognised
with some degree of confidence it is necessary to qualify this
interpretation.   For example, in a fluvial succession, what
types of channels are involved, do they change spatially or
temporally, and what is the overall composition or 'architecture'
of the succession?   In the case of deltaic successions, which
part of the delta is represented in a given section, and again
what type of delta is it?   These points have added significance
where the sand-bodies under discussion contain hydrocarbons since
the precise nature of the environment of deposition also contains
predictions on the geometry, trend and extent of the sand-body.
Rather than presenting single, rather idealised views of environments and their products, the text strives to convey an impression
of the variety exhibited by each of the environments.

The brevity of this review inevitably means that the discussion is rather selective and curtailed in places.   For more
extensive coverage the interested reader is referred particularly
to the compilations of Reading (1978), Walker (1979), and various
Special Publications of I.A.S. and S.E.P.M.

CONTINENTAL ENVIRONMENTS AND SAND-BODIES

Alluvial fans
     Alluvial fans are localised features which tend to develop
where high relief, rapidly weathering areas are juxtaposed against
lower relief depositional plains or basins.   Commonly a fault-

defined scarp separates these areas and sediment is transferred
from the source area by slope failure and fluvial processes into
gullies or canyons incised into the fault scarp.   Expansion and
deceleration of the sediment-laden flows at the mouth of the
canyons causes appreciable sediment deposition and forms alluvial
fans which spread across the margin of the depositional basin.
The majority of alluvial fans have been described from interior
basins in semi-arid to arid settings where they enter into playa
lakes (Blissenbach, 1954; Bluck, 1964; Denny, 1967).   However,
they also occur in lakes, alluvial or deltaic plains, or marine
coastal settings ('fan deltas'), and form in arctic or tropical
humid settings (Gole and Chitale, 1966; Legget et al., 1966).

    The range of depositional processes which operate on alluvial
fans is controlled largely by the climate which influences the
weathering processes, the type(s) of sediment produced and the
water discharge responsible for transporting sediment to the fan.
Bull (1972, 1977) recognises four main depositional processes on
fans: stream channel processes, sheet-floods, waning sieve flows
and debris flows (or mud-flows).   Arid fans are characterised by
episodic, flood-related depositional events, involving debris flows
in the proximal region (assuming that clay is available), extensive
sheet-floods in the mid- to distal fan region and late-stage reworking by stream channels as the flood discharge wanes.   Humid
fans tend to be dominated by more continuous stream channel processes, aided perhaps by sheet-floods in the distal region of the
fans, and occasional debris flows in the proximal region.   It is
common for only part of a fan to be active at any time (particularly
in arid fans), and stagnant areas of the fan therefore experience
post-depositional, modifying processes which act on previously
deposited sediment (Denny, 1967; Walker, 1967).   Unstable clasts
deposited on the fan break down in situ to form finer-grained
matrix which may filter into the underlying gravels.   In addition,
soils frequently form, salts are precipitated, and in more humid
settings plant colonisation and peat accumulation can occur on the
distal fan.

    Alluvial fan successions are generally thick, coarse-
grained successions localised at the margins of depositional
basins (Fig. 1).   Both the distance and duration of sediment
transport in alluvial fans are short, perhaps involving a single,
very brief flood-induced event acting over several kilometres.
The deposits of alluvial fans are often therefore poorly sorted,
compositionally immature conglomerates/breccias and sandstones
which closely reflect the bedrock geology of the particular
drainage basin supplying the fan.   A general down-fan decrease
in grain size has been demonstrated in several examples, accompanied by changes in process, surface morphology and facies.
These changes may be expected to impart a high degree of order to
the facies patterns of alluvial fans, but isolated individual fans

Fig. 1.   Thick alluvial fan conglomerate fringing the
          northern margin of the Ebro basin in the
          Spanish Pyrenees; the Rolden Conglomerate,
          Lower Miocene.   Photograph courtesy of C.D.
          Atkinson.

are comparatively rare as adjacent fans tend to coalesce laterally
into a continuous, locally lobate front of over-lapping fans (a
bahada), thus complicating the overall facies pattern.   Typical
facies include matrix-supported conglomerates (deposited by debris
flows), horizontally bedded clast-supported conglomerates (stream
channel and sieve processes), cross-bedded pebble-bearing sandstones (stream channel or possibly sheet-flood processes) and
parallel- and ripple-laminated fine to medium sandstones (sheet-
flood processes).

    The most common basin-types in which alluvial fans occur are
grabens, half-grabens and strike-slip or oblique-slip basins,
which have abrupt fault-defined margins.   Successions of alluvial
fan deposits several kilometres thick are commonly produced in
these basins as the faults provide an in-built mechanism for subsidence.   Rigorous facies analysis of thick alluvial fan
successions often reveals sequences of facies on a variety of
scales (Steel, 1974; Heward, 1978).   Particularly common are

fining-upwards and coarsening upwards sequences on a scale of
10's to 100's m which reflect moderate to long-term fan behaviour
which is often controlled by tectonic events.   Fining-upwards
sequences are felt to reflect fan retreat due to scarp recession
and/or source-area plnation during tectonically quiet periods.
Coarsening-upwards sequences result from fan progradation,
possibly triggered by tectonically-induced rejuvenation of the
source area.   In the Devonian Hornelen Basin in Norway, alluvial
fan deposition involves more than 150 basin-wide coarsening-upwards sequences with an average thickness of 100 m.   Each sequence is considered to have formed by tectonic lowering of the
basin floor followed by fan progradation (Steel, 1976; Steel et
al., 1977).

Alluvial Channels and Floodplains
     Modern alluvial channels vary greatly, but by combining the
overall nature of the sediment load (mud/silt-, sand- and gravel-
dominated) and the channel pattern type (braided, meandering,
straight and anastomosed) it is possible to sub-divide the broad
spectrum of alluvial channel types (Miall, 1977; Collinson,
1978a; Schumm, 1981) (Fig. 2).   Of these, the most important

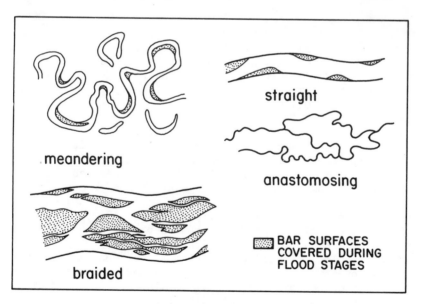

Fig. 2.   Principal types of river channel patterns,
             after Miall, 1977.

(or perhaps most studied) types are gravel-dominated braided
rivers, and sand-dominated braided and meandering rivers.
Recently, the importance of anastomosed river systems, both
modern and ancient, has been stressed by several researchers

(Putnam and Oliver, 1980; Smith and Putnam, 1980).

Gravel-dominated braided rivers are common on alluvial fans
(particularly humid fans) but also constitute extensive reaches
of major river systems where there is an abundant supply of gravel,
a relatively steep slope and high peak discharge values.   The
channel is split into a series of sub-channels by longitudinal
bars which are extensively exposed at low river stage and may
remain partially exposed during high stage periods (Bluck, 1974;
Boothroyd and Ashley, 1975).   During high stage periods imbricated
gravels accrete vertically on the shallow bar top areas whilst the
front of the bar migrates downcurrent.   The floors and flanks of
the sub-channels are generally sites of erosion and sediment
transport rather than deposition, but deposition occurs as high
stage conditions where, giving rise to cross-bedded, gravel-
bearing sands on the channel floors and distinctive sand aprons
on the channel flanks.   Some channels are temporarily abandoned
during low stage periods and silt/mud-dominated facies may
accumulate in near-stagnant waters.   Sedimentation in these
rivers is relatively disordered, involving the waxing and waning
of individual sub-channels and the migration of both sub-channels
and bars.

Sand-dominated braided rivers occur where there is an abundant supply of sand-grade material, moderate slopes and fluctuating
discharge.   They are wide, shallow channels floored by fields of
transverse or linguoid bars which split the flow during low discharge periods thus producing a braided channel pattern (Smith,
1970; Collinson, 1970; Cant and Walker, 1978).   During high stage
periods, slipfaces at the front of the bars migrate downstream
and generate laterally extensive sets of planar tabular cross-
bedding.   As river stage falls the bars are dissected, and minor
channels wander between the fragmented bars (Smith, 1971).   Inactive slipfaces are eroded slightly either by these channels or
by wind-driven waves.   These low-stage processes produce reactivation surfaces and units of lower energy facies which locally
interrupt the sets of cross-bedding (Collinson, 1970).   Sedimentation in these rivers produces erosive-based sequences of
planar tabular cross-bedded sands with subordinate trough cross-
bedding and ripple lamination.   In view of the constant shifting
of bars and low-stage sub-channels these sequences are unlikely
to show a high degree of order and it is also improbable that
their facies patterns can be summarised in a single vertical sequence.   Instead a family of sequences is required (see Cant and
Walker, 1978).

Most of the meandering rivers which have been studied are
sand-dominated, though some gravel-dominated and mud/silt-domina-
ted examples have also been researched (e.g. Gustavson, 1978;

Woodyer et al., 1979). Meandering rivers are characterised by single, undivided channels in which deposition is concentrated on point bars located on the inner side of meanders, whilst erosion occurs on the opposite, outer bank. A helicoidal flow pattern around the meander causes bed shear stress to decrease upwards across the point bar surface, resulting in a fining of sediment and decrease in transport rate as reflected by bedforms and internal structures. This pattern of flow around meanders was considered to produce erosive-based sand members which became finer-grained upwards (Allen, 1970a), but recently the extent to which this flow pattern is developed around meanders has been questioned (Jackson, 1975). One consequence of this criticism of the helicoidal flow concept is that the deposits of a meandering channel, although perhaps relatively ordered, again require several sequences to summarise their variability rather than a single sequence. Another important characteristic of meandering rivers is that they tend to exhibit a pattern of moderate to long-term channel behaviour which influences the nature of the sand-body produced by channel sedimentation (Fig. 3). The concentration of sedimentation on point bars causes these features to migrate laterally with a slight downstream component. An erosive-based, tabular sand-body is produced as the channel migrates in this fashion, but the width of the sand-body is limited as neck- and chute cut-off, and the clay plugs produced by these channel abandonment processes, confine the lateral extent of channel migration. As channel sedimentation is concentrated within a meander belt, this area soon becomes elevated above the general level of the floodplain and large-scale channel abandonment, or avulsion, involving a major shift in channel position can occur. Tabular sand bodies produced by lateral channel migration are therefore enclosed within fine-grained floodplain facies as the channels switch position in this manner (Allen, 1965a).

Ancient alluvial successions generally comprise erosive-based coarse members which represent the former channel belt, and finer grained members which accumulated on the adjacent floodplains (Fig. 4). Vertical profiles often reveal a fairly regular alternation between coarse and fine members on a scale of several metres to 10's of metres. Coarse members are composed of conglomerates and/or sandstones which are generally moderately to poorly sorted and compositionally immature. A wide range of sedimentary structures record the operation of unidirectional traction currents of fluctuating strength and imply that sediment transport and deposition within the channel took place via a series of downstream migrating bedforms and/or bars. The fine members generally show evidence of emergence or near-emergence (desiccation cracks, palaeosols, coal seams, footprints, etc.). Though floodplain sedimentation is dominated by deposition of fine-grained sediments by overbank flooding, thin sheet-like sandstone beds can be deposited by crevasse-splay events if the

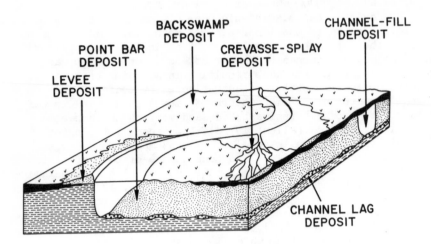

Fig. 3. Typical sub-environments and sediment bodies
of a fluvial meandering channel stressing the
erosive-based, tabular sand-body produced by
lateral migration of the point bar; after
Allen (1970a).   Compare with Fig. 6.

channel bank is locally breached during flood periods (see Fig.
4).   These beds can be important in connecting coarse channel
members which were otherwise isolated in fine-grained, relatively
impermeable floodplain facies (see later comments on alluvial
architecture).   Studies of palaeosols in floodplain fine members
in recent years have provided numerous insights into alluvial
sedimentation.   They reflect prolonged periods of minimal sed-
mentation on the floodplains and are often laterally extensive,
thus providing useful correlation horizons in alluvial successions.   They also reflect the climate at the time of deposition
which has implications on the nature of river discharge and therefore permits more refined interpretations of the channel members.
The most extensively studied palaeosols are the calcretes or
caliches which, by analogy with modern soils, are considered to
form in a semi-arid climate (Allen, 1974; Leeder, 1975).
Nodules of calcium carbonate develop by displacement and to a
lesser extent by replacement of the host sediment and can produce ordered, vertical soil profiles.   Development of these
profiles is considered to be largely controlled by time available
and they have therefore been used as a measure of floodplain
accretion (Leeder, 1975), though the rate of calcium carbonate
supply must also be important in this respect.

The recognition of channel pattern types in ancient alluvial

Fig. 4. Sandstone-dominated fluvial channel with a
high relief erosion surface cutting into a
?levee sequence dominated by thin, planar
erosive-based sandstone-siltstone couplets
interpreted as crevasse splay deposits. The
high relief of the erosion surface is considered
to reflect substantial fluctuations in river
discharge with major floods causing the channel
base to step-down appreciably. Tertiary,
Campodarbe Fm., near Campodarbe, Spanish
Pyrenees.

successions is difficult. Successions dominated by erosive-based
sequences of clast-supported conglomerates are often interpreted
as pebbly braided river deposits simply because of the predominance of coarse bedload material (Fig. 5). In cases where flat
bedded, imbricated conglomerates are intimately associated with
cross-bedded conglomerates, sandstone wedges and localised
channel fill sequences this interpretation may be correct (e.g.
Eynon and Walker, 1974), otherwise the interpretation may be over-
simplistic. Sandstone-dominated coarse members are also difficult
to interpret in terms of palaeo-sinuosity. Erosive-based fining
upwards sequences in which the coarse member shows a progressive
decrease in grain size accompanied by a waning flow sequence of

Fig. 5.   Thick, channelised unit of pebble-supported
          conglomerates (ca. 15 m) interpreted as a
          pebbly braided river deposit.   Note the basal
          erosion surface just above the road level, the
          gently dipping bedding surfaces within the
          conglomerate (straight arrows) which may
          reflect the presence of large bars in the
          channel fill, and the slight low stage sand
          apron at the upper right of the photograph
          (curved arrow).   Tertiary, Campodarbe Fm.,
          near Campodarbe, Spanish Pyrenees.

internal structures were originally interpreted as the product of
laterally migrating meandering rivers (Allen, 1970a, b), but the
validity of this has recently been challenged and the problem is
now regarded as being more complex (see earlier).   Only two
reliable criteria exist for the identification of meandering
channels in ancient successions: i) the presence in the coarse
member of gently dipping lateral accretion surfaces (or epsilon
cross-bedding) aligned normal to the direction of channel flow
and recording successive positions of the point bar surface
(Fig. 6); ii) the preservation of curved ridges on the upper bedding plane surface of the coarse member which can be interpreted
as the scroll bar topography of meander loops (Puigdefabregas,

1973; Nami, 1976). In the absence of these somewhat rare criteria the low versus high sinuosity argument for sandstone-dominated channel-fills can be pursued using the types of internal structures, the vertical sequence of structures and their relative orientations, the coarse member/fine member ratio and coarse member width (Moody-Stuart, 1966; Cant and Walker, 1976; Collinson, 1978b).

Fig. 6. Erosive-based fluvial channel sandstone in centre of photograph exhibits low-angle surfaces dipping through the sandstone member towards the left. Palaeoflow is transverse to the dip direction of these surfaces (towards the observer) and there is also an upwards decrease in grain size along the surfaces. The surfaces are therefore interpreted as lateral accretion surfaces recording successive positions of a point bar in a meandering channels as it migrated laterally towards the left. Other channel sandstones are visible at the base and top of the cliff, and the remainder of the section comprises fine-grained overbank facies with palaeosol horizons. Tertiary, Montllobat Fm., near Montanana, Spanish Pyrenees; see Puigdefabregas and van Vliet, 1978.

Another facet of recent research into alluvial sediments
involves consideration of the way in which rivers operating under
a wide range of controls deposited a large body of sediment (Fig.
7). The term 'alluvial architecture' has been coined to describe
this work, and one important aspect of the architecture of an
alluvial succession which bears on diagenetic studies of these
rocks is the extent to which coarse channel members are either
connected or isolated in fine member overbank facies (Leeder,
1978; Allen, 1978; Bridge and Leeder, 1979). Early speculation
on this problem used different modes of moderate to long-term

Fig. 7.   General view of an alluvial succession with
          major fluvial channel sandstones in the fore-
          ground and upper distance (1 and 3) and fine-
          dominated overbank sediments with sheet-like
          crevasse-splay units in the lower distance
          (2). Detailed section measuring and lateral
          correlation in such well exposed alluvial
          successions can reveal the architecture of the
          succession. Consideration of the geological
          setting then permits speculation on the con-
          trols of alluvial architecture, providing a
          text of simulated models (see text for full
          discussion).

river behaviour to predict the probable shape and arrangement of channel sandstone bodies (Allen, 1965a; 1974). For example, if rivers swept continuously across their floodplain, or skipped across in either short or long avulsive steps, or were incised into the floodplain, it was argued that these differences would be reflected in the architecture of the succession. Current work on this problem commenced with Leeder (1978) and involves numerical and computer simulation of rivers of unspecified channel pattern which are considered to shift their positions only by avulsion. The most sophisticated analysis to date is by Bridge and Leeder (1979) which models a single, avulsing river within a confined floodplain and accounts for compaction during sedimentation. The effects of channel size, aggradation rate and frequency of avulsion are modelled, and graben and half-graben settings for alluvial sedimentation are also considered. The extent to which channel sands are connected is viewed in two-dimensional, transverse to flow profiles through the successions and can be quantified. The implications of these studies to problems involving the passage of fluids (diagenetic pore waters, hydrocarbons, ground water) through the rocks are clear and the results of these studies are beginning to be tested in sub-surface and exposed fluvial successions (Galloway, 1981).

Aeolian sand seas

The transport and deposition of sand by wind is important in a number of environments but assumes particular significance in tropical deserts where extensive aeolian sand seas or 'ergs' are sometimes formed (Glennie, 1970; Wilson, 1973; Breed et al., 1979). Sand seas tend to form in structural depressions and can extend over areas of 10,000-500,000 km . Thicknesses of sand are usually less than 100 m, but can be as great as several hundred metres. The marginal areas of sand seas comprise exposed rocks, alluvial fans, ephemeral stream courses and inland sabkhas, whereas the central areas are dominated by fields of large-scale aeolian bedforms separated by gravel-strewn bedrock surfaces and occasional ephemeral lakes. A diverse suite of bedforms have been described from modern sand seas including transverse dunes, barchans, longitudinal or sief dunes, dome-shaped dunes and pyramidal or star-shaped dunes. Zonation of these bedforms has been identified in some sand seas, with barchan dunes occupying the sand-starved fringe areas and longitudinal and pyramidal dunes in the more central areas (Mainguet and Callot, 1974). Relatively little is known of the internal structures of large-scale aeolian bedforms in view of the logistical problems involved in trenching these bedforms. The limited evidence available suggests that large-scale, steeply dipping foresets composed of grain flow and grain fall layers are important. Dip directions of foresets are variable, and possibly bimodal in seif dunes (Glennie, 1970). Large-scale reactivation surfaces reflecting fluctuations in wind strength and direction seem common. The sands of which these bedforms are composed are extremely well sorted, well rounded and relatively fine-grained (0.1-1.0 mm) in view of the low viscosity of the transporting medium. The clay and silt fraction is transported in suspension and deposited as loess beyond the sand sea, and coarse-grained particles cannot be entrained and therefore form lags on deflation surfaces. Sand grains transported by wind suffer considerable abrasion and rounding due to saltative impact and air-borne grain collisions, sometimes forming perfectly rounded 'millet-seed' grains.

The recognition of ancient sand sea deposits is based on a wide range of criteria, none of which is unambiguous. However, an abundance of large-scale cross-bedding in texturally mature, generally unfossiliferous sandstones associated with sediments displaying independent evidence of arid, exposed conditions would provide a convincing combination of the main criteria. The classic major aeolian sand bodies of southwestern U.S.A. comprise 100-700 m thick successions of large-scale cross-bedded sandstones. The Navajo Sandstone for example comprises up to 600 m of fine-grained, quartz-dominated sandstones with a preserved extent of 965 km by 400 km and thus represents a major aeolian sand sea. Planar-tabular and planar-wedge sets of steeply dipping cross-bedding commonly exceeding 15 m in set thickness dominate the succession (McKee, 1979). In Europe, Permian aeolian dune sandstones (the Rotliegendes) constitute the reservoir rocks for major gas fields in the southern North Sea (Glennie, 1972). The aeolian dune facies occur between alluvial fan/wadi at the margins of the desert basin and salt-bearing playa lake deposits in the centre of the basin. Early, environment-related diagenesis in these sands includes halite and gypsum cementation (the latter subsequently converted to anhydrite), and the production of iron oxide rims to sand grains by alternation of ferro-magnesian minerals (Glennie et al., 1978).

SHORELINE ENVIRONMENTS AND SAND-BODIES

Clastic shorelines are sub-divided into deltaic shorelines formed where a major river supplies large volumes of sediment to a limited area of the basin margin, and 'non-deltaic' shorelines such as beach, beach-barrier, chenier, tidal flat and estuarine complexes produced principally by nearshore wave, wave-related and tidal processes. There is of course a great deal of overlap between these broad classes since all of the latter group are known to occur within deltas which are substantially influenced by wave and/or tidal processes. Deltas are however distinctive in producing thick successions which are localised in depositional centres around the mouths of major rivers. The ability of deltaic and non-deltaic shorelines to produce major sandstone bodies is well established and the following account

intends to review the range of sandstone bodies which can be produced by these systems.

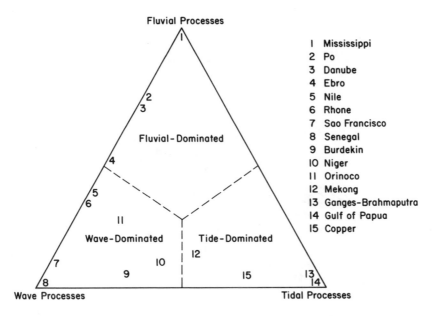

Fig. 8.   Ternary diagram of delta types on modern
          deltas; modified after Galloway, 1975.

Deltaic shorelines
      Deltas require a mature, well organised drainage basin which
culminates in a major alluvial trunk stream capable of transporting abundant mud, silt and sand to the receiving basin.   At the
river mouth the sediment-laden river flow expands, decelerates,
and mixes with the basinal waters.   As a result of these radical
hydrodynamic changes much of the sediment load of the river is
deposited in the immediate vicinity of the river mouth.   The
introduction of this sediment is however contested by basinal
processes, particularly nearshore wave and tidal processes, and
for a delta to form it is essential that the river supplies more
sediment than the basinal processes are capable of reworking
(Wright and Coleman, 1973; Coleman and Wright, 1975; Elliott,
1978).   Research into modern deltas in recent years has been
largely involved with critically comparing different deltas in
terms of their process regime.   This has led to the definition
of river-, wave- and tide-dominated classes of deltas (Fig. 8),
and in the course of this work it has also become clear that
facies patterns differ dramatically between these classes.   For
example, the river-dominated Mississippi delta is characterised
by a series of radiating 'bar finger sands' which immediately
underlie the positions of present-day distributary channels and

thus reflect linear, seaward progradation of the channels and
their mouth bar deposits (Fisk, 1961; Fig. 9).   In contrast,

BRANCHING PATTERN AND THICKNESS

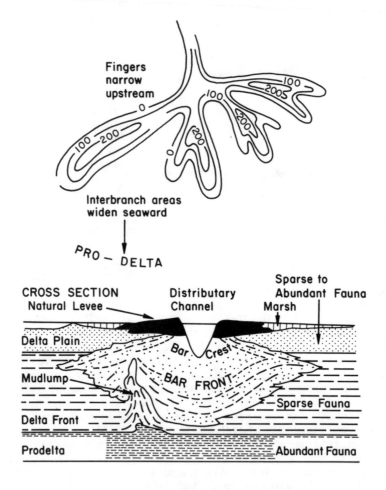

Fig. 9.   Elongate 'bar finger' sands of the modern
          Mississippi delta; isopachs in feet; after
          Fisk, 1961.   Recently it has been demonstrated
          that mud diapirism has substantially modi-
          fied sand distribution within the bar fingers,
          see Coleman et al., 1974.

the more wave-influenced Rhone delta comprises a slightly lobate

sheet-sand of coastal barrier sands which is locally cut by
linear channel sand bodies underlying present and abandoned distributary channel courses (Oomkens, 1967; Fig. 10).   The most

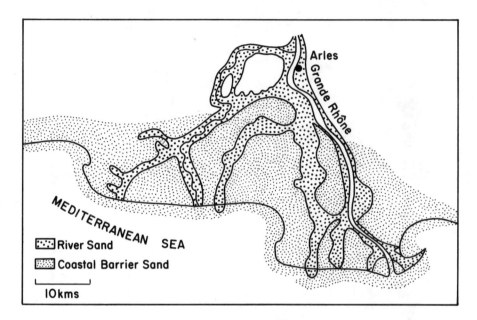

Fig. 10.   Sand-body distribution in the Rhone delta
           illustrating a slightly lobate, coast-
           parallel sheet-sand of coastal barrier
           deposits dissected by linear distributary
           channel sand bodies; after Oomkens, 1967.

comprehensive study to date of the variability of modern deltas
is a wide-ranging analysis of information from thirtyfour deltas
(Coleman and Wright, 1975).   Multivariate statistical analysis
of this information produced six discrete delta models which were
intially typified by their sand distribution pattern and then
qualified by considering a single modern delta regarded as representative of a given class (Fig. 11).   Though few of the deltas
used in this study have been thoroughly described in terms of
their facies patterns, the study provides an impression of the
range of sand-body patterns which are likely to exist in deltaic
successions.

        Common to all deltas is a simple sub-division into two main

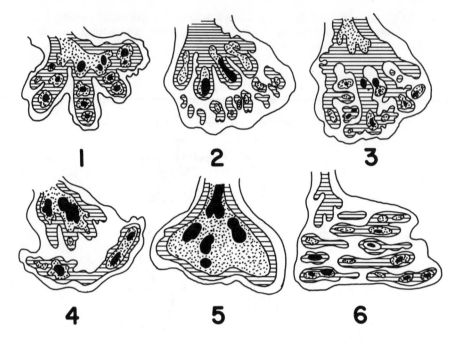

Fig. 11.   Models derived by multivariate analysis of
data from modern deltas and depicted by sand
distribution patterns.   Increasing density
of tone indicates increasing sand thickness.
Representative modern deltas for each model
are 1) Mississippi; 2) Ganges-Brahmaputra,
Ord; 3) Burdekin, Mekong, Irrawaddy; 4) Brazos;
5) Sao Francisco, Grijalva; 6) Senegal; after
Coleman and Wright, 1975.

components: the delta front which includes the river mouth and
shoreline areas and their offshore extension to the prodelta;
and the delta plain which comprises the distributary channels and
intervening bays, lakes, levees, marshes, swamps and in some cases
tidal flats.

   Delta front facies assocation. In the delta front the coarsest
sediment is deposited at the river mouth and shoreline, whilst
progressively finer sediment is transported further into the basin
and deposited in deeper water settings.   Progradation of the delta
front therefore produces a relatively large-scale coarsening-
upwards sequence, the details of which vary according to the process regime of the delta front (Oomkens, 1967; Coleman and Wright,
1975).   Locally, the upper part of this sequence is eroded by the
distributary channels as progradation continues (see Oomkens, 1967).

Delta plain facies association. Delta plain areas are generally dominated by fluvial processes, though in areas of moderate
to high tidal range the lower part of the delta plain may be tide-
dominated (e.g. the Niger delta; Allen, 1965b; Oomkens, 1974).
Fluvial-dominated delta plains generally include several fluvial-
distributary channels which divide the total discharge of the
alluvial trunk stream.   Channel avulsion is a common occurrence
on delta plains and results in numerous abandoned distributary
courses traversing the delta plain.   The facies, sequences and
sand-bodies of fluvial-distributary channels resemble those of
alluvial channels, though the frequent avulsion of the distributary channels causes the sand-bodies to have lower width to depth
ratios than alluvial channel sand-bodies in the same system
(Oomkens, 1974).   The inter-distributary areas of fluvial-
dominated delta plains are generally enclosed bays and lakes which
receive fine-grained sediment during periods of overbank flooding
of the distributary channels.   If the channel banks are breached
locally during a flood event, sand is often supplied to the bays
via crevasse channels, crevasse splay lobes and in some cases as
small-scale prograding sub-deltas (Coleman et al., 1964; Gagliano
and van Beek, 1975).   Collectively these features rapidly infill
large areas of the bays, producing a wide range of small-scale,
bay-fill coarsening-upwards sequences and thin, sheet-like sand
bodies.

Tide-dominated delta plains comprise estuarine distributary
channels which are flanked by extensive areas of supratidal and
intertidal flats dissected by tidal creeks and gullies.   In the
Niger delta, tidal-distributary channels deposit sand-dominated
sequences which resemble estuarine and tidal inlet sequences.
Tidal distributary channels are less prone to avulsion, but can
migrate laterally in the direction of longshore drift and produce
erosive-based, tabular sand-bodies in a similar manner to tidal
inlets in mesotidal barrier islands (see later).

Ancient deltaic facies and sand-bodies. Ancient deltaic
successions comprise thick but relatively localised successions
which are often divisible into repeated motifs or 'cycles'.   Each
motif involves a broadly coarsening-upwards succession which is
considered to reflect infilling of the receiving basin by delta
advance.   Within each motif it is possible to recognise numerous
facies sequences which reflect different sub-environments of the
delta, thus permitting the pattern of the delta and in some cases
the delta type to be interpreted.

The delta front is represented by relatively large-scale
coarsening-upwards sequences which record a passage from fine-
grained offshore or prodelta facies upwards into shoreline facies
which are usually sandstone-dominated.   These sequences result
from progradation of the delta front and are often truncated by

fluvial- or tidal-distributary channel sequences (Fig. 12). So

Fig. 12. Mid- to upper part (80 m) of a large-scale
(120 m) delta front coarsening-upwards sequence
truncated by a planar erosion surface which
marks the base of a 10 m minor distributary
channel sandstone. The sheet-like appearance of the channel sandstone is at least
partly related to the fact that it is viewed
parallel to flow (towards the right). Upper
Carboniferous, Co. Clare, Eire.

far mainly fluvial-dominated delta front sequences have been
recognised in the geological record (Fisher et al., 1969; Elliott,
1976a; Horne and Ferm, 1976).   These sequences commence with a
thick, uniform interval of mudstones or fine- to medium-grained
siltstones deposited from suspension at the base of the delta
front and beyond.   This facies may appear massive, but often
exhibits diffuse banding defined by slight variations in grain
size which result from fluctuations in suspended sediment supply.
In several examples of this facies, thin erosive-based coarse
siltstone-fine sandstone beds occur within the mudstones-siltstones.   Structures in these beds indicate that they are the
product of waning currents, and they often resemble thin turbidites.   These beds may reflect deposition by river-generated
density currents where the density of the sediment-laden flow
exceeded that of the basin waters, or they may have resulted

from slump events higher on the delta front.   The intermediate
parts of these sequences comprise fine-grained background sediment
into which coarse siltstone and sandstone beds are repeatedly
intercalated.   Initially the background sediment is a direct
continuation from below, but as the sequence is ascended thin
siltstone laminae, small-scale ripple laminations and ripple form
sets appear, possibly reflecting the inception of wave base.
Most of the intercalated coarse beds have planar erosive bases
and exhibit current ripple laminations.   Upper bedding surfaces
are often sharply defined and may exhibit wave ripples reflecting
post-flood wave reworking.   The sandstone-dominated member at the
top of the sequence is generally characterised by laterally continuous sandstone beds with a wide range of structures indicative
of high rates of sediment transport and deposition.   Broad
lenticular units are interpreted as minor subaqueous extensions
of the distributary channel locally dissecting the upper delta
front.   In some cases the sand-bodies at the top of these delta
front sequences are considered to have a shape analogous to the
bar fingers of the modern Mississippi delta (Collinson and Banks,
1975; Elliott, 1976a).

In wave-influenced delta front sequences the fine member at
the base of the sequence is generally thinner as suspended sediment is widely distributed beyond the immediate delta area
(Fisher et al., 1969).   The intermediate and upper parts of the
sequence may comprise well sorted sandstones with a predominance
of wave-produced structures, (e.g. the Cody Shale - Parkman Sandstone delta system in the Cretaceous of the Western Interior,
U.S.A.; Hubert et al., 1972).   Tidally-influenced delta front
sequences have not been widely described from the geological
record.   Recognition is likely to depend on the characteristics
of the sandstone member at the top of the sequence and to a lesser
extent the intermediate part of the sequence.

Fluvial-dominated examples of the delta plain facies association comprise large-scale fluvial-distributary channels, smaller
scale crevasse channels, and a.wide range of small-scale coarsening-
upwards sequences which reflect infilling of shallow water interdistributary areas (Ferm and Cavaroc, 1968; Elliott, 1974; Horne
and Ferm, 1976).

The small-scale coarsening-upwards sequences are usually
4-10 m thick and commence with finely laminated or bioturbated
mudstones-siltstones deposited from suspension across the entire
interdistributary area during river flood periods.   Plant debris
is often abundant, along with brackish or fresh-water fauna.
Sideritic ironstone nodules are common this facies and are considered to be early diagenetic features which form during shallow
burial of the organic rich muds (see Ho and Coleman, 1969 for
present-day analogue).   This facies may constitute the entire

bay-fill sequence, but more commonly the sequences terminate in
a thin sandstone member.   Facies details of the sandstones vary
depending on whether they result from levee construction by overbank flooding, crevasse splay lobes, minor mouth bar-crevasse
channel couplets or wave rewored sand spits (Elliott, 1974; Fig.13).

Fig. 13.   Small-scale coarsening-upwards sequence
           (10 m) interpreted as delta plain bay- or
           lake-fill sequence by crevasse-induced
           processes.   Initial facies (above the ridge
           at beach level) is black, organic-rich
           shales which abruptly grade into gray, lam-
           inated siltstones (crevasse initiation?).
           Sequence then coarsens-upwards with evidence
           of increasing current energy into a sandstone
           facies which is a leached (light grey)
           palaeosol overlain by coal.   Upper
           Carboniferous, Coal Measures, South-West
           Wales.

Each produces a relatively thin and impersistent sandstone body,
but they may coalesce to produce thin sheet sandstones which infill large expanses of the interdistributary area and may link
distributary channel sandstones.   Palaeosols frequently develop
towards the top of the sequences, and peat-coal may accumulate.

Crevasse channel sequences are generally 1-4 m thick and exhibit numerous reactivation surfaces, clay drapes and indications of temporary bedform emergence. These features reflect ephemeral flow in the channels resulting from healing or "stranding" during low river stage periods. Distributary channel sequences are larger scale and generally reflect more continuous discharge conditions, though still with stage fluctuations. The facies and sequences are similar to those of fluvial channels. A high sinuosity pattern is inferred in one example from the presence of lateral accretion surfaces (Elliott, 1976b), whereas in a separate example giant cross-bed sets up to 40 m thick are interpreted as side-attached alternate bars in a low sinuosity channel (McCabe, 1977). Composite, multistorey channel sandstones are commonly preserved and channel abandonment due to avulsion often produces an overall fining-upwards trend.

Tidally influenced channel sequences have been inferred from biomodal current patterns in Cretaceous delta plain associations in the Western Interior, U.S.A. (van de Graff, 1972, Hubert et al., 1972).

Fig. 14. Mud diapir affecting sand-body thickness in a
coarsening-upwards sequence; note the thin
sandstone interval above the diapir and the
substantially thicker sandstone interval on
the flanks of the diapir; height of cliff ca.
20 m. Upper Carboniferous, Co.Clare, Eire.

The major sand-bodies in deltaic successions are therefore
produced in the distributary channel and delta front areas.
Characteristics of the sand-bodies including detailed textures,
facies types, sequences, and overall sand-body shape and alignment
are controlled mainly by the delta regime and sediment load of
the system.   There is however growing recognition of the fact
that numerous deltas, both modern and ancient, experience syn-
sedimentary deformation by processes related either to the surface instability of rapidly deposited muds, or the early compactional diagenesis of these muds (Coleman et al., 1974; Weber and
Daukoru, 1975).   The latter involves the development of over-
pressured/under-compacted conditions during burial which often
leads to deep-seated, mass flowage of the clays.   Growth faults
and mud diapirs can form in response to these conditions and can
substantially modify the sand-body distribution of the deltaic
succession (Bruce, 1973; Edwards, 1976; Rider, 1978; Fig. 14).

Non-deltaic shorelines
    The range of 'non-deltaic' clastic shorelines can be sub-
divided according to the relative importance of wave and tidal
processes into i) wave-dominated beaches and barrier islands, ii)
wave- and tide-influenced barrier island/tidal inlet systems; and
iii) tidal flat/estuarine systems.

    Wave-dominated beaches and barrier islands. These are long,
narrow, coast-parallel sand-belts which tend to form on moderately
stable coastal plains with a steady supply of sand.   Both beaches
and barrier islands tend to be dominated by a seaward-dipping
beach face capped by an aeolian dune ridge, with few (if any),
tidal inlets in the barrier islands (e.g. Costa de Nayarit, Padre
Island).   The main processes are nearshore wave processes and
related wave-induced currents such as longshore currents and rip
currents.   The intensity of these processes varies substantially
between fairweather and storm periods, and is crucial to understanding the facies which accumulate on the beach face.

    Offshore parts of the beach face are often below mean storm
wave base and therefore accumulate muds and fine silts from suspension.   As the shore is approached across the shallowing beach
face profile, storm wave base is encountered.   Finely interlaminated muds, silts and sands reflecting alternations between
storm and fairweather conditions characterise this area.   This
transition leads to the shoreface area which is the sub-tidal part
of the beach extending from fairweather wave base to mean low tide
level.   Physically, the shoreface is an area of transition between
offshore or shelf waves and nearshore waves.   The area tends to
be dominated by sands which become coarser towards the wave breaker
zone in the upper shoreface.   These sands may either be silty,
bioturbated sands with hints of wave ripple lamination (low wave
energy), or rather coarser better sorted sands exhibiting wave

ripple lamination, cross-bedding or flat lamination with minimal bioturbation (high wave energy). The foreshore is the inter-tidal part of the beach face which is dominated by the surf zone and, more particularly, swash-backwash processes. Sands domina-ted by flat lamination with occasional sets of landward-oriented cross bedding tend to accumulate in this area. Beyond the fore-shore is the supratidal backshore area and the aeolian dune ridge which is produced by wind reworking of sand emplaced on the upper beach face by storm waves. In wave-dominated barrier islands a narrow, low energy lagoon occurs landward of the aeolian dune ridge. The central parts of the lagoon tend to accumulate a varied suite of facies in shallow, rather low energy waters which often have abnormal salinity values (brackish to hypersaline). During storms, sediment-laden storm waves locally breach the aeolian dunes and pour into the seaward side of the lagoon, de-positing sands derived from erosion of the beach face. The wash-over fans produced by this process commonly coalesce laterally to produce a continuous, though irregular apron of sands projecting into the lagoon (Andrews, 1970; McGowen and Scott, 1975).

Seaward progradation of a beach or wave-dominated barrier island produces an elongate, coast-parallel sand-body which may extend laterally for many 10's of kilometres (Curray et al., 1969; Bernard et al., 1962). The sand-bodies may be lenticular or sheet-like in cross section depending on the extent of prograd-ation. Vertical sequences through these sand-bodies reveal grad-ational coarsening-upwards trends from offshore muds, via inter-laminated muds, silts and sands into nearshore sand member. The sequences are generally 10-30 m thick, and the facies record in-creasing wave energy as the sequence coarsens upwards. Recent comparative studies of modern beach faces of radially differing wave regime have stressed that details of the facies and overall sequence vary with the wave regime. A suite of wave-dominated coarsening-upwards sequences is thus to be expected, rather than a single, idealised sequence (Clifton et al., 1971; Howard and Reineck, 1972; Hunter et al., 1979).

Wave-dominated coarsening-upwards sequences reflecting beach or barrier island progradation have been widely recognised in the geological record (Campbell, 1971; Davies et al., 1971; Clifton, 1981; Figs. 15 and 16). The offshore mudstones at the base of the sequences are often devoid of laminations due to intense bio-turbation, but soon acquire thin lenticular laminations or wave-ripple form sets of well sorted siltstones as storm wave base is encountered. Thin, graded, turbidite-like beds also occur on occasions and are interpreted as density currents generated by ebbing storm waves (Hamblin and Walker, 1979). Inferred shore-face deposits in the central part of these sequences are sand-dominated and often exhibit a repeated small-scale alternation between erosive-based units of flat laminated or hummocky cross

Fig. 15. Two wave-dominated coarsening-upwards sequen-
          ces occupy the right and central parts of
          this photograph; the first sequence termin-
          ates at the slight ridge centre-right, and
          the upper sequence culminates in the major
          ridge-forming sandstone which strikes across
          the photograph; both sequences are attributed
          to beach progradation.   Cretaceous, Aren
          Sandstone, Spanish Pyrenees; see Ghibaudo et
          al., 1974.

stratified sandstone (storm deposits) and intensely bioturbated
and/or wave ripple laminated sandstones (fairweather deposits).
Primary sedimentary structures increase in importance relative
to bioturbation structures upwards through the shoreface deposits
as wave energy increases, and the style of bioturbation also
changes (Howard, 1972).   Sandstones at the top of the sequence
are predominantly flat laminated and reflect deposition in the
foreshore-backshore area.   Aeolian dune facies are rarely represented, due presumably to their low preservation potential.
The beach face sandstones are often mature orthoquartzites reflecting intense abrasion and attrition of the grains in the nearshore wave zone.   Many examples have carbonate cements, though
quartz cements are also common (Heward, 1981).

Fig. 16.   Small-scale (10 m), wave-dominated coarsening-
upwards sequence produced by beach progradation; note the low-angle dipping surfaces extending through the sandstone unit at the top
of the sequence which are interpreted as beach-
face accretion surfaces.   Upper Carboniferous,
Co. Clare, Eire.

Wave- and tide-influenced barrier island/tidal inlet systems.
Barrier islands in mesotidal areas (2-4 m tidal range) display a
number of important differences with the wave-dominated barrier
islands previously discussed.   Numerous tidal inlets aligned
normal to the shoreline dissect the barrier island into short segments.   Flood- and ebb-tidal deltas form at the landward and seaward mouths of the inlets, and large areas of the lagoon assume
the characteristics of tidal flats.   Studies of modern systems
of this type have recently demonstrated that the inlets may be
extremely significant in the overall facies pattern as they tend
to migrate laterally in the direction of longshore drift (Kumar
and Sanders, 1974).   Tidal currents operating in the inlets
arrest alongshore drift sediment transport causing the side of
the inlet which is upstream with respect to the longshore current
to accumulate whilst the opposite side is eroded.   The tidal
inlets therefore migrate laterally in the direction of longshore

drift, reworking previously deposited barrier sands and depositing
a substantial body of sediment on the updrift margin.   Lateral
migration rates of present-day inlets are variable, but can be
extremely high.   For example, Fire Island inlet in Long Island
has migrated 8 km in 115 years at an average rate of 64 m per
year (Kumar and Sanders, 1974).   Tidal inlets are represented by
erosive-based sand-bodies in which a shell gravel lag is overlain
by well sorted sands.   In the Fire Island example the lower part
of the sand-body is dominated by tidal current produced cross-
bedding and ripple lamination, but the upper part exhibits wave-
produced structures deposited as a beach-spit caps the sequence
as migration continues (Kumar and Sanders, 1974).   As the inlets
scour well below sea level their deposits have a high preservation
potential, and if high rates of migration are sustained in a constant direction, inlet deposits may constitute the entire depositional record of a mesotidal barrier island system.

Ancient mesotidal barrier island deposits have been inferred
in the Carboniferous of eastern U.S.A. (Horne and Ferm, 1976;
Hobday and Horne, 1977).   Orthoquartzite sandstone bodies 11-26
m thick, 2-8 km wide and several 10's of kilometres long comprise
a coarsening-upwards sequence produced by beach face progradation
which is often truncated by multistorey channel sandstones interpreted as tidal inlets on the evidence of palaeocurrents from sets
of cross-bedding.

So far in this account, beach and barrier island systems
have been assumed to be prograding seawards under conditions of
stable sea level, moderate subsidence rates and steady sand supply.
However, under contrasted conditions of rising sea level, high
subsidence rates or limited sand supply, these systems can become
transgressive and migrate landwards.   Most major present-day
beaches and barrier islands migrated landwards across the shelf
during the Holocene transgression and studies of deposits which
accumulated during this event have made a substantial contribution
to our understanding of transgressive shoreline deposits (Fischer,
1961; Swift, 1968; 1975; Sanders and Kumar, 1975a, b; Fig. 17).
It appears that the shorelines either swept continuously across
the area leaving a thin remnant sand-sheet of washover-dominated
lagoonal facies abruptly overlain by lower shoreface sands
('shoreface retreat'), or skipped across the area leaving a series
of coast-parallel linear sand-bodies representing former shoreline positions ('in place drowning').   Slow rates of sea-level
rise are felt to favour shoreface retreat whereas more rapid
rates favour in-place drowning, though it is doubtful that these
mechanisms are such distinct end-members as is currently implied.
Where mesotidal barrier islands migrated landward it is considered
that the deposits are dominated by tidal inlet deposits.

In the geological record the most important deposits of

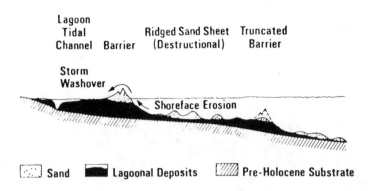

Fig. 17.   Landward migration of barrier islands by
            shoreface retreat during the Holocene trans-
            gression; after Swift, 1975.

transgressive shoreline systems are the laterally extensive bodies
of well sorted, mature sandstones commonly preserved above unconformity surfaces (Fig. 18).   Detailed analysis of sandstone-bodies
of this type in recent years has begun to unravel the complexity
of their facies patterns and has also revealed that the upper
parts of the sandstone bodies are often composed of progradational
shoreline facies associations building out across their earlier
transgressive deposits, presumably following the stabilisation of
sea level (Driese et al., 1981; Richards, 1981).

    Tidal flat/estuarine systems.   In areas of high tidal range
(macrotidal areas   4 m range), tidal flats and estuaries often
form extensive shoreline areas.   Sedimentation is dominated by
tidal currents which are unsteady in a periodic fashion on a
number of short time scales (semi-diurnal or diurnal ebb-flood
cycles, neap-spring cycles and seasonal equinoctial cycles).   Many
of the criteria for recognising ancient tidal deposits depend upon
this repeated unsteadiness: herringbone cross-bedding and bimodal/
bipolar palaeocurrent patterns, repeated evidence of small-scale
fluctuations between bedload transport and deposition from suspension (e.g. mud-draped foresets), and an abundance of reactivation surfaces in cross-bedded sandstones.   In addition, evidence
for periodicity related to neap-spring tidal cycles has recently
been described from ancient tidal facies (Visser, 1980; Allen,
1981).

    Sandstones deposited by tidal processes in tidal flat and
estuarine settings are often compositionally mature in view of

Fig. 18.   Basal Trias Quartzite overlying Permian red
           beds with slight unconformity, Alpes-
           Maritimes, France.   The Quartzite uncon-
           formably overlies cover and basement rocks
           throughout the Alps as a sheet sandstone and
           includes a lower unit of transgressive shore-
           line facies, a middle unit of tide-dominated
           shelf facies and an upper unit of prograding
           tidal inlet-barrier island facies.   Basal
           Quartzite thickness in photograph is 35 m;
           see Richards, 1981.

the intense reworking which characterises tide-dominated settings.
However, the sandstones commonly contain innumerable thin flasers
or more continuous beds of mud which act as permeability barriers
to migrating fluids and can therefore influence diagenesis.

     Tidal flats comprise relatively featureless plains dissected
by a network of tidal channels and creeks.   During flood periods,
tidal waters enter the channels, overtop the channel banks
(particularly during spring tides) and inundate the adjacent flats.
Following a brief still-stand the tidal waters ebb via the channels and the flats are re-exposed.   Tidal flats are conventionally
sub-divided into supratidal, intertidal and sub-tidal areas.   The

supratidal flats are mud-dominated areas located above mean high
tide level.   In temperate and humid-tropical areas they comprise
salt marshes and mangrove swamps respectively and accumulate interlaminated clays and silts pervaded by rootlets (Reineck, 1967).
In arid and semi-arid areas they are desiccated mudflats characterised by evaporitic growth of gypsum and halite crystals
(Thompson, 1968).   Intertidal flats are smooth, seaward-dipping
areas which range from mud-dominated near the high-water mark to
sand-dominated near the low-water mark.   Relatively weak tidal
currents and waves interact on the sand-dominated intertidal flats
to produce extensive areas of asymmetrical and symmetrical ripples,
often with complex interference patterns.   Intertidal flat facies
are dominated by interlaminated clays, silts and sands exhibiting
prolific lenticular, wavy and flaser bedding reflecting constantly
fluctuating but relatively low energy conditions.   Tidal channels
and creeks dissect the supratidal and intertidal flats.   The main
channels are often highly sinuous, with laterally migrating point
bars on the inner depositional banks.   The tidal channels are
floored by relatively coarse sands with shell debris and abundant
mud clasts.   This basal lag is overlain by point bar facies composed of thin, interlaminated clay-silt and sand beds characterised
by lateral accretion bedding dipping gently into the channel.
Depending on channel spacing and the rate and direction of channel
migration, a large part of the intertidal flats may be underlain
by an erosive-based sheet of point bar sediments.   The subtidal
area in sand-dominated systems such as the German Bay and Bay of
Fundy is a complex of channels, bars and shoals.   In the German
Bay the channels are deep, broad, funnel-shaped extensions of
estuaries with sand-wave, megaripple and ripple bedforms.   Sand
shoals or bars exist at the mouth of the channels and between the
channels, creating a sand-dominated depositional province which
extends over several hundred square kilometres (Reineck and Singh,
1973).

       Estuaries can be regarded as the tidally influenced stretches
of the lower courses of certain rivers.   Major sand-dominated
estuaries are preferentially developed in areas of moderate to
high tidal range (Hayes, 1975) where they occur as broad funnel-
shaped channels.   Within the overall channel are a series of
either intertidal sand bars or linear tidal sand ridges which
separate ebb- and flood-dominated sub-channels of varying depth
(e.g. Knight, 1980).   In central areas of the estuary, sands are
deposited via a diverse suite of ripple, megaripple and sandwave
bedforms.   Extensive areas of ripple bedforms tend to be confined to the higher parts of the intertidal bars or ridges which
do not experience maximum tidal current velocities, whereas megaripple and sandwave bedforms dominate on the flanks and floors of
the sub-channels dissecting the bars.   The megaripples and sand-
waves are frequently organised as distinct ebb- or flood-oriented
fields which retain their orientation during the ebb-flood tidal

cycle.   Each field therefore produces unidirectional sets of
cross-bedding which are either ebb- or flood oriented and repeatedly interruped by reactivation surfaces and/or mud drapes.

Estuarine channels of this type deposit erosive-based sequences which commence with a lag of shell debris and mud pebbles
overlain by complex cosets of cross-bedding with evidence of
opposed flow directions.   The upper parts of these channel sequences grade into finer grained deposits of intertidal and supratidal
flats which commonly flank margins of the estuary (see Greer,
1975).

Sequences and sandstone bodies produced by tidal flat progradation have been widely recognised in the geological record
(see Ginsburg, 1975 for review).   Fining-upwards sequences
recording the progradation of tidal flats seem most common.
Extensively cross-bedded sandstones with numerous features sugges-
tive of tidal current processes (e.g. flow reversals, spaced
reactivation surfaces or mud drapes) at the base of the sequences
are interpreted as sub-tidal facies.   These sandstones grade upwards into thinly inter-bedded sandstones and siltstones with
small-scale sets of cross-bedding and current-ripple lamination.
Upper bedding plane surfaces in this facies are often covered by
asymmetrical and symmetrical ripple forms, occasionally with interference patterns.   This facies is interpreted as low- to mid-
intertidal flat facies and passes upwards into finer-grained facies
with evidence of repeated exposure (e.g. desiccation cracks,
evaporite mineral growth, rootlets) interpreted as supratidal flat
facies.   Erosive-based tidal channel sequences often replace a
substantial part of the tidal flat facies.

In contrast, few major estuarine sandstone bodies have been
described, probably because of their relatively low preservation
potential due to later fluvial erosion.   Their preservation is
perhaps greatest in a transgressive succession where they may be
locally preserved at the base of the succession, perhaps in depressions in the unconformity surface.

SHALLOW MARINE ENVIRONMENTS AND SAND-BODIES

Shallow marine seas or shelves occupy the area between the
realms of nearshore and oceanic processes and are well represented
at the present day in view of the high stand of sea level following the Holocene transgression.   Whilst it is clearly an advantage to have extensive shallow marine areas available for study,
the effects of the Holocene transgression are not entirely
advantageous.   One of the prime reasons for the comparatively
slow development of ideas on shallow marine sedimentation is the
problem of recognising the extent to which modern shelves are

adjusted to present-day processes, or merely reflect relict facies patterns related to the Holocene transgression.   The current con- census is that modern shelf sand-bodies originated during the Holocene transgression and are presently being maintained and in some cases reworked by the present shelf regime.   Though the solution to this problem is now fairly well understood in most cases, it still limits the application of modern shelf facies patterns to ancient successions.

Modern shelves range in depth from approximately 10-200 m and comprise marginal continental shelf seas and partially enclos- ed epeiric seas.   Consideration of the main processes operating in these areas permits them to be divided into two main groups: tide-dominated and storm-dominated, the latter involving wind- and wave-induced currents (Johnson, 1978).

ZONE I	**ZONE II**	**ZONE III**	**ZONE IV**	**ZONE V**
BEDROCK AND GRAVEL	**SAND RIBBONS**	**SAND WAVES**	**SMOOTH SAND**	**SAND PATCHES**

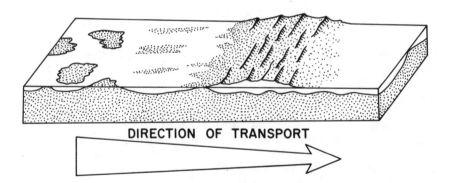

DIRECTION OF TRANSPORT

DECREASING MID-TIDE SURFACE VELOCITY & BOTTOM GRAIN SIZE

Fig. 19.   Facies zones developed along a tidal current transport path on the U.K. continental shelf; after Belderson et al., 1970.

Modern tide-dominated shallow marine seas.   Tidal currents are particularly important in gulfs and semi-enclosed seas which amplify the tidal wave.   The shelf areas of North-West Europe have been intensively studied and have revealed a pattern of

discrete tidal current transport paths characterised by a series
of depositional zones (Kenyon and Stride, 1970; Belderson et al.,
1970; Fig. 19).   The transport paths record decreasing current
velocity and grain size and include fields of large-scale sand
ribbons and sand waves formed by tidal current reworking of pre-
Holocene sands.   Sand ribbons are flow-parallel ridges up to 15
km long and 200 m wide, but only 1 m maximum thickness.   The
areas between ribbons are current swept gravel lags and it is
considered that the ribbons are produced by helical flow cells
within a appreciable tidal current (Kenyon, 1970).   The sand
waves are straight-crested, transverse bedforms between 3-15 m
in height and 150-500 m in wavelength.   Some of these bedforms
are asymmetrical, but others have weak asymmetry or are symmetrical.   In addition, the lee faces of these bedforms are often
only 5   or so and superimposed megaripples often have opposed
directions on different parts of the sandwave.   Unfortunately,
the internal structures of these complex bedforms are poorly
understood.   At present, it is debatable whether they are likely
to exhibit large-scale sets of avalanche cross-bedding, complex
cosets of smaller-scale cross-bedding, or some blend of the two.
One other important type of tidal current bedform, not represented in the transport path, is the tidal sand ridge which is
common in certain tide-dominated shelf seas, estuaries and deltas
(Off, 1963; Caston, 1972).   These features occur in extensive
fields aligned parallel to flow and tend to be a few 10's of
metres high, a few kilometres in width and several 10's of kilometres in length.   Once again, the internal structures of these
bedforms are poorly known,but it seems likely that they are built
of complexly inter-woven sets and cosets of moderate-scale cross-
bedding, possibly interruped by low-angle accretion surfaces or
master bedding planes (Houbolt, 1968).   It is considered that
the majority of tidal current ridges in shallow marine areas
formed during the Holocene transgression by the detachment and
subsequent modification of earlier ridge features.

Modern storm-dominated shallow marine seas
     Sedimentation in many shelf seas is dominated by storm-
generated waves and wind-driven currents, and in several cases
these areas appear to be largely equilibrated to present conditions (e.g. Oregon-Washington shelf; Kulm et al., 1975).   The
supply of sand to these shelves is considered to be controlled
largely by storm surge currents.   Initially the storm surge is
driven landwards towards the nearshore area from which it erodes
substantial amounts of sand.   Then as the surge ebbs offshore
the sand is redeposited as a discrete, sheet-like bed on the
shelf (see Hayes, 1967).   Storm-enhanced rip-currents may also
emplace sand on the shelf in a similar manner (Morton, 1979).
Since these beds are emplaced below fairweather wave base they
are unlikely to be physically reworked and will instead be
draped by finer grained facies during the post-storm and fair-

weather periods.   They may however, be biogenically reworked.

Ancient shallow marine facies and sand-bodies
     The recognition of ancient shallow marine facies is, where
possible, achieved by means of assemblages of body or trace
fossils.   There are however numerous studies which have had to
rely solely on sedimentological arguments (e.g. in early Pre-
Cambrian successions) and it is apparent from these studies that
once a shallow marine interpretation is confidently established,
the sedimentological evidence permits a detailed characterisation
of the setting.   Important sedimentological characteristics of
shallow marine facies include: i) abundant evidence of wave and/
or tidal current processes; ii) laterally extensive, planar erosion surfaces often with winnowed lags; iii) textural and mineralogical maturity of the sandstones; and iv) complex bimodal or
polymodal palaeocurrent patterns (Anderton, 1976; Johnson, 1977;
Levell, 1980).

     A comprehensive facies scheme exists for shallow marine
deposits, involving three broad facies (sandstone, heterolithic
and mud facies) which are further sub-divided according to
internal structures, sand/mud ratio, bed thickness and faunal
evidence (Johnson, 1978).   Using this scheme it is possible to
assess the relative importance of tidal and storm-generated processes in ancient shallow marine successions.

     Ancient tide-dominated sandstone facies are recognised by an
abundance of moderate to large-scale cross-bedding, involving sets
up to 3-4 m high.   The sets are repeatedly interrupted by reactivation surfaces, clay drapes or zones of intense bioturbation reflecting the intermitten nature of bedform migration, and generally
show some evidence of bidirectional currents.   A few instances
of very large-scale, avalanche cross-bedding interpreted as sand-
saves have been recorded in ancient shallow marine successions
(Nio, 1976; Nio and Siegenthaler, 1978).   More commonly, moderate-
scale sets of cross-bedding (0.1-1.0 m) are observed to descend
or traverse gently dipping master bedding planes, implying that
these bedforms were migrating across the lee surface or flank of
a larger scale, ?sandwave (e.g. Levell, 1980).

     Sandstone-bodies produced in tide-dominated shallow marine
settings tend to be either isolated, linear features developed
locally in finer grained marine facies, or laterally extensive,
and often very thick, sheet sandstones.   Numerous examples of
isolated, linear sandstone bodies broadly comparable to tidal
sand ridges have been described from the Jurassic-Cretaceous
epeiric seaway of the Western Interior (Brenner and Davies, 1973;
1974; Berg, 1975; Fig. 20).   Each sandstone body is composed of
highly mature, extensively cross-bedded orthoquartzites and is
characterised internally by a 10-30 m coarsening-upwards sequence

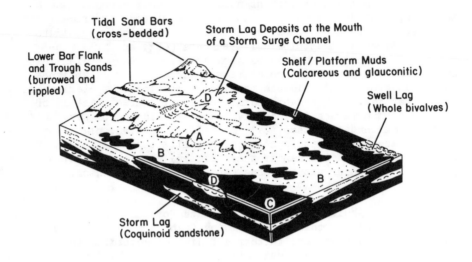

Tidal Sand Bars
(cross-bedded)

Storm Lag Deposits at the Mouth
of a Storm Surge Channel

Lower Bar Flank
and Trough Sands
(burrowed and
rippled)

Shelf / Platform Muds
(Calcareous and glauconitic)

Swell Lag
(Whole bivalves)

Storm Lag
(Coquinoid sandstone)

Facies
A. Cross-bedded well-sorted sands
B. Interbedded glauconitic sands and burrowed ripple sands
C. Calcareous glauconitic
D. Coquinoid sands

Fig. 20. Facies model for linear tidal current sand
          ridges in the Jurassic (Oxfordian) of the
          Western Interior, U.S.A.; after Brenner and
          Davies, 1973.

which records migration of the sand ridge across the surrounding
fine-dominated shelf.   Sedimentary structures reflect increasing
levels of tidal current as the sequence is ascended, coupled with
evidence of intermittent storm action.   These sandstone bodies
range from 2-60 km in width, are up to 160 km long, and often
form important stratigraphic traps because of their encasement in
impermeable, hydrocarbon-bearing marine mudstones.   The best
described sheet sandstone bodies are from later Precambrian and
Lower Palaeozoic successions (Swett and Smit, 1972; Banks, 1973a;
Anderton, 1976; Levell, 1980).   Once again they comprise mature,
cross-bedded orthoquartzites, but these bodies can often be traced
over hundreds of kilometres and attain thicknesses of several
kilometres.   The Precambrian Lower Sandford Formation in north
Norway includes a 1.5 km thick succession of extensively cross-
bedded, mature sandstones interpreted as a tide-dominated shallow
marine deposit (Levell, 1980).   A diverse array of cross-bedding

types with a predominantly unidirectional palaeocurrent trend is
interpreted as the product of bedforms generated by a tidal
current regime which possibly reinforced an independent shelf
current.   The thickness of the body, which far exceeds the
thickness of modern shelf sands, is attributed to the erosion of
tide-dominated deltaic shorelines as abandoned lobes were trans-
gressed, thus libterating sand to the adjacent shelf.

    Ancient storm-dominated sandstone facies in shallow marine
settings typically comprise a regular, turbidite-like alternation of thinly interbedded sandstones and shales.   This alternation is considered to reflect storm and fairweather processes
with the sandstone beds (termed sublittoral sheet sands, Goldring
and Bridges, 1973) representing the deposits of ebbing storm
surges.   The sandstone beds range from 5-30 cm thickness and
have erosive bases which are occasionally ornamented by sole
marks.   Internally the beds are either normally graded reflecting deposition from storm-generated density currents below storm
wave base, or exhibit flat lamination and hummocky cross-strati-
fication deposited by waning storm waves above storm wave base
(Harms, 1975; Hamblin and Walker, 1979).   The upper parts of
the beds are frequently bioturbated and grade into fairweather
siltstones and mudstones.   In some cases these facies occur in
large-scale (30-100 m) coarsening-upwards sequences attributed
to the progradation of a storm-dominated shelf (Banks, 1973b).
In contrast, a 200 m fining-upwards sequence dominated by
hummocky cross-bedding has recently been interpreted as the product of a transgressive, storm-dominated shelf on a tectonically
active continental margin (Bourgeois, 1980).

DEEP BASIN ENVIRONMENTS AND SAND BODIES

    The emplacement of relatively coarse-grained sediment in
deep basin environments (and some of the finer grained 'background' sediment) takes place via a series of mass gravity transport processes of which the most important are sediment gravity
flows (Middleton and Hampton, 1973; Lowe, 1979).   Four 'ideal'
processes have been distinguished in the latter suite by means
of the manner in which the sediment is supported during transport: debris flow (matrix strength), grain flow (grain interaction),
liquefied flow (upward directed intergranular flow) and the
turbidity current (turbulence).   Depositional textures and
structures have been proposed for each of these ideal processes
including, for example, the Bouma sequence and all its variants
for turbidity currents (Middleton and Hampton, 1973).   However,
it was stressed in this work that in reality these processes form
a gradational and inter-changeable series rather than discrete
flow types.   Several sediment gravity flow processes may contribute to the development of a bed at different times.   The

current concensus on the relative importance of the four 'ideal'
flows is that turbidity currents and debris flows are the main
processes responsible for long distance transport. Grain flow
and liquefied flow conditions are considered to be important at
various stages in the flow history such as the initiation of the
flow and during or immediately after deposition (Walker, 1978).
The latter is particularly important since it means that the textures and structures of a bed may reflect the mode of deposition
rather than the mode of transport.

Sandstone-dominated deep basin successions are usually thick,
well bedded successions of regularly alternating sandstones and
shales in which each sandstone bed (unless amalgamated) is the
product of a single sediment gravity flow event. The composition
of the sediment deposited by these processes is entirely dependent
on the source of the sediment and is therefore extremely varied.
Initially turbidites were equated with matrix-rich greywackes
because of the turbulent, 'muddy' nature of the flows. However,
in many cases it is now accepted that much of this matrix is of
secondary rather than primary origin, and besides many low matrix,
relatively mature turbidites have been described.

Deposition in deep basin areas takes place in three main
environments: basin slopes, submarine canyons and fans, and basin
plains, and it is considered that each produces a distinctive
facies association (Mutti and Ricci Lucchi, 1972; Walker and Mutti,
1973; Fig. 21).

Basin slopes. Basin slopes tend to comprise thick successions
of fine-grained facies deposited by hemi-pelagic settling and a
variety of gravity-driven, mud-dominated suspensions generated
either by sediment spilling over the shelf/slope edge, or by slump
events on the upper slope. The latter is an important process
in basin slope sedimentation. It can result in the preservation
of large, slump scars in the upper slope which are subsequently
infilled, and can also produce slide units, folded slump sheets
and debris flows in the lower slope deposits.

On occasions basin slopes are locally cut by submarine canyons which act as preferred pathways for high density, coarse-
grained, sediment gravity flows. The canyons may intercept shelf
sediment transport paths, or may directly tap deltaic and non-
deltaic shorelines, particularly where the shelf is relatively
narrow or non-existent. These canyons and their extensions into
the inner part of the submarine fan (see below) can produce significant sandstone bodies, particularly where the lower courses
of the canyon are deeply incised. Linear, valley-shaped sandstone bodies 10's (or in some cases 100's) of kilometres long,
several kilometres wide and 10's to 100's of metres thick are
produced (e.g. Normark and Piper, 1969; Stanley and Unrug, 1972).

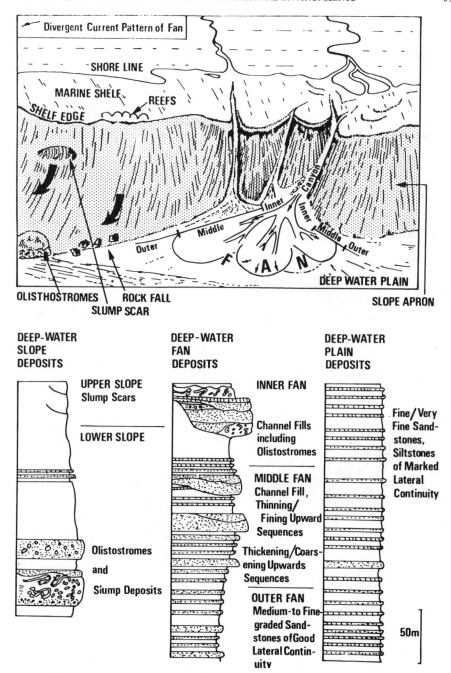

Fig. 21.   Deep basin environments and representative stratigraphic
           successions for slope, fan and basin plain settings; after
           Walker and Mutti, 1973; Kruit et al., 1975.

These bodies are often composed of extremely thick, massive turbidites (AE, commonly) frequently with dish-and-pillar structure
suggesting late-stage liquefaction.   Thick conglomerate units
which are either matrix-supported (debris flows) or clast-
supported are important in some ancient canyons and fan valleys.
Walker (1975) has recently proposed a descriptive model for the
clast-supported conglomerates which separates disorganised conglomerates from three classes of organised, imbricated conglomerates (a-axis imbrication) which exhibit: i) inverse to
normal grading, ii) normal grading, and iii) normal grading overlain by cross-bedding.   It is considered that these facies are
deposited primarily in lower canyon/fan valley settings and each
of the organised facies as listed is inferred to occur progressively further down-channel.   Large slide and slump units are
also common in canyon-fill sequences.

Fig. 22.   Thick, laterally extensive deep basin
           deposits dominated by turbidites; Tertiary
           Gres d'Annot, Alpes-Maritimes, France.

Submarine fans.   The concentration of high-density sediment
gravity flows in canyons, and consequent expansion and deceleration of the flows at the canyon mouth lead to the construction

of submarine fans. Modern submarine fans are extremely varied
in terms of their size, form and sedimentation. The difficulties
of studying these important depositional sites are increased by
veneers or blankets of hemipelagic deposits which have draped
many of the fans since the Holocene rise of sea-level and consequent widening of the shelves. Nevertheless a general, rather
idealised image of these fans has emerged in recent years, though
not without considerable debate (Mutti and Ricci Lucchi, 1972;
Walker and Mutti, 1973; Nilsen, 1980; Normark, 1980; Walker, 1980).
The fans are generally assumed to be fairly symmetrical, lobe-
shaped features, but they may also be skewed if the basin floor
dips laterally away from the canyon mouth, or in some cases they
may be more linear features perhaps confined within structural
depressions on the basin floor. Conventionally, the fans are
sub-divided into inner, mid- and outer fan zones which are
morphologically distinct. The inner fan is dominated by one or
more deeply incised channels or fan valleys which extend from
the canyon mouth and are flanked by prominent levee ridges. High-
density sediment gravity flows are concentrated in the channels
whereas lower density, turbidity currents occur on the channel
flanks and levees when the upper, finer-grained levels of the
flows spill-over the channel banks. In the periods immediately
following and between sediment gravity flow events, slumping of
the levees take place both into and away from the channel. The
mid-fan area comprises more numerous though smaller-scale and
less incised channels separated by inter-channel areas. These
channels are inferred to shift and migrate with time, often leading to instances of channel abandonment. At the mouth of the
channels are distinctive depositional lobes which then blend
into the smoother, more uniform outer fan which is generally
devoid of deeply incised channels. Moderate to low density,
sheet-like turbidity currents characterise the outer fan.

At this point it is worth emphasising that there is relatively little in the current pool of knowledge on modern fans which
is directly helpful to the geologist faced with a pile of turbidites and related deposits. The scale of features mapped on
modern fans is often much greater than outcrop scale, and the
amount of coring is severely limited both in coverage and depth
of penetration. Nevertheless in recent years there have been
numerous attempts at recognising ancient submarine fans and their
constituent parts using something akin to the generalised model
presented above. Criteria for the recognition of submarine fans
in the geological record include: i) the presence of significant
channelised bodies of turbidites on a variety of scales; ii) the
breakdown of 'proximal' and 'distal' relationships with the
result that these facies are finely interbedded both vertically
and laterally; iii) the presence of ordered sequences within
turbidite succesions; and iv) an overall fan-shaped sediment body
with a radial palaeocurrent pattern. Inner fan deposits comprise

Fig. 23. Lenticular, erosive-based sandstone body
          (arrowed) composed of turbidites and inter-
          preted as a submarine fan channel; exposed
          width of sandstone body in photograph is
          approximately 1 km, maximum thickness in
          axial part of channel is ca. 12m; Tertiary,
          Hecha Group, Spanish Pyrenees; see Mutti,
          1977; Mutti et al., 1981.

large, channelised bodies of debris flow deposits, thick 'proximal'
turbidites and slump units which are flanked laterally by thinner,
finer-grained 'distal' turbidites of the levees (Figs. 23 and 24).
The mid-fan area is characterised by vertical sequences of turbidites, the recognition of which by Italian sedimentologists has
been so crucial to the development of research into ancient submarine fan successions (e.g. Mutti and Ricci Lucchi, 1972; Ricci
Lucchi, 1975).   These workers recognised large-scale sequences
involving hundreds or even thousands of metres which reflect long
term changes such as fan progradation or retreat, but more import-
antly they recognise smaller scale thinning/fining-upwards sequences and thickening/coarsening-upwards sequences within the
large-scale sequences.   Thinning upwards sequences characteristically occur above major erosion surfaces and are therefore interpreted as mid-fan channels which were either gradually abandoned
or migrated.   Thickening-upwards sequences involve a gradational

change from relatively 'distal' turbidites at the base of the
sequence upwards into 'proximal' turbidites, though the sequences
are often complex or multiple.   These sequences are considered
to result from the progradation of mid-fan lobes, and they are
sometimes overlain by erosive-based thinning upwards sequences
thus forming a lobe-channel couplet.   The outer fan is composed
of thin, laterally persistent, 'distal' turbidites with no dis-
cernible trends, reflecting the more uniform topography of this
area.

Fig. 24.   Detail of near-axial part of the submarine
           channel of Fig. 23 illustrating the erosive
           base of the channel and the predominance of
           thick, generally structureless, coarse-grained
           turbidites in the channel fill;   note the
           lenticular, cross-cutting nature of many of
           the lower turbidites and the tendency towards
           slightly thinner, more persistent turbidites
           in the upper part.   Tertiary, Hecho Group,
           Spanish Pyrenees; see Mutti, 1977; Mutti et
           al., 1981

The recognition of ordered vertical sequences of turbidites

is central to this research, but recently the methods employed
have been questioned (see Ricci Lucchi, 1975a for an account of
the methods and Hiscott, 1981 and Ghibaudo, 1981 for critical
discussion).   Recognition is difficult because of the absence
of dramatic process or environmental changes as occur in beach
face or delta front sequences for example.   In many turbidite
sequences only the bed thickness and sandstone:shale ratio
change.   Where this occurs in moderately thick successions and
the trend involves a significant number of beds the identification
of a sequence is probably justified.   However, small sequences
involving only a few beds, and sequences where the 'trend' is
only really apparent in one or two beds are perhaps dubious.

     Basin plains.   Basin plains are generally smooth, feature-
less areas over which turbidity currents can travel for considerable distances.   The basement of the basin plain often shows
appreciable topography but the sediments tend to blanket this
topography and eventually establish a smooth basin plain with
minimal gradient.   The main deposits of basin plains are sand-
or silt-dominated turbidites, lower density mud-dominated turbidites, hemi-pelagic muds and in some cases pelagic-facies (Mutti
and Ricci Lucchi, 1972; Fig. 25).   Criteria for distinguishing
whether a succession of turbidites was deposited in a basin
plain rather than a submarine fan setting include: i) basin
plain successions comprise regular alternations of sandstone and
shale which are vertically monotonous over 10's - 100's of metres
and devoid of sequences; ii) individual turbidite beds tend to be
laterally persistent transverse to flow and in the flow direction,
with some beds being traceable for 100's of kilometres; iii) comparison of stratigraphically equivalent sections in a downcurrent
direction indicate that proximal-distal relationships are valid
in these successions.   Increased understanding of basin plain
sedimentation and the geometry or architecture of the deposits
has stemmed from the recognition of key-or mega-beds in certain
turbidite successions (Ricci Lucchi, 1975; Rupke, 1976).   Mega-
beds are the deposits of exceptionally large turbidity currents
which are usually triggered by a tectonic event in the vicinity
of the basin.   On occasions these beds are composed of sediment
different to that of the 'normal' turbidites, and this fact
allied with their large size permits correlation of the beds
over large areas.   In the Miocene Marnoso-arenacea of the
Italian Apennines, basin-plain turbidites include a number of
mega-beds of which the most notorious is the Contessa-bed
(Ricci Lucchi, 1978; Ricci Lucchi and Valmori, 1980; Fig. 26).
Detailed correlation and tracing of turbidite beds across the
basin reveals that many of the average or 'normal' turbidites
(40 cm thick) are basin-wide in extent, and also demonstrate
the presence of subtle lobe-shaped sand-bodies within the basin
plain succession.

Fig. 25. Thin, laterally extensive, turbidites interbedded with thin pelagic limestones and red
marls in a vertically monotonous succession
of basin plain origin; Tertiary, Zumaya
Flysch, San Sebastian, north Spain; see
Crimes, 1973; Kruit et al., 1975.

Fig. 26.   Turbidites mega-bed (the Contessa bed) in a
           succession of basin plain turbidites; note
           the exceptional thickness of the turbidite
           sandstone and overlying marl interval;
           Tertiary, Marnoso-arenacea, Italian
           Apennines; see Ricci Lucchi and Valmori,
           1980.

ACKNOWLEDGEMENTS

    I wish to thank C.D. Atkinson and M.T. Richards for dis-
cussions on various points covered in the text, Mrs. V. Jenkins
for typing the manuscript, and Mr. A.Smith for photographic
assistance.

REFERENCES

ALLEN, J.R.L. 1965a. A review of the origin and characteristics
   of Recent Alluvial Sediments. Sedimentology, 5, 89-191.
ALLEN, J.R.L. 1965b. Late Quaternary Niger delta, and adjacent
   areas: sedimentary environments and lithofacies. Bull. Am.
   Ass. Petrol. Geol., 49, 547-600.

ALLEN, J.R.L. 1970a. A quantitative model of grain size and sedimentary structures in lateral deposits. Geol. J., 7, 129-146.
ALLEN, J.R.L. 1970b. Studies in fluviatile sedimentation: A
  comparison of fining-upwards cyclothems with special reference
  to coarse-member composition and interpretation. J. sedim.
  Petrol., 40, 298-323.
ALLEN, J.R.L. 1974. Studies in fluviatile sedimentation:
  implications of pedogenic carbonate units, Lower Old Red Sandstone, Anglo-Welsh outcrop. Geol. j., 9, 181-208.
ALLEN, J.R.L. 1978. Studies in fluviatile sedimentation: an
  exploratory quantitative model for the architecture of avulsion-
  controlled alluvial suites. Sedim. Geol., 21, 129-147.
ALLEN, J.R.L. 1981. Lower Cretaceous tides revealed by cross
  bedded sets with mud drapes. Nature, 289, 579-581.
ANDERTON, R. 1976. Tidal shelf sedimentation: an example from
  the Scottish Dalradian. Sedimentology, 23, 429-458.
ANDREWS, P.B. 1970. Facies and genesis of a hurrican washover
  fan, St. Joseph Island, Central Texas Coast. Rep. Invest. Bur.
  econ. Geol., 67, 147, Austin, Texas.
BANKS, N. L. 1973a. Tide-dominated offshore sedimentation, Lower
  Cambrian, North Norway. Sedimentology, 20, 213-228.
BANKS, N.L. 1973b. Innerely Member: Late Precambrian marine shelf
  deposit, East Finnmark. Norg. geol. Unders., 288, 7-25.
BELDERSON, R.H., KENYON, N.H. and STRIDE, A.H. 1970. Holocene
  sediments on the continental shelf west of the British Isles.
  Inst. Geol. Sci. Rept. No. 70, 157-170.
BERG, R.R. 1975. Depositional environment of Upper Cretaceous
  Sussex Sandstone, House Creek Field, Wyoming. Bull. Am. Ass.
  Petrol. Geol., 59, 2099-2110.
BERNARD, H.A., LeBLANC, R.J. and MAJOR, C.F. Jr. 1962. Recent
  and Pleistocene geology of southeast Texas. Geol. Gulf Coast
  and Central Texas and guidebook of excursion, 175-225. Houston
  Geol.Soc.
BLISSENBACH, E. 1954. Geology of alluvial fans in semi-arid
  regions. Bull. geol. Soc.Am., 65, 175-189.
BLUCK, B.J. 1964. Sedimentation of an alluvial fan in Southern
  Nevada, J. sedim. Petrol., 34, 395-400.
BLUCK, B. J. 1974. Structure and directional properties of some
  valley sandur deposits in Southern Ireland. Sedimentology, 21,
  533-554.
BOOTHROYD, J.C. and ASHLEY, G.M. 1975. Process, bar morphology
  and sedimentary structures on braided outwash fans, North-
  eastern Gulf of Alaska. In: Glaciofluvial and Glaciolacustrine
  Sedimentation (Ed. A.V.Jopling and B.C. McDonald). Spec.Publ.
  Soc.econ.Paleont.Miner., Tulsa, 23, 193-222.
BOURGEOIS, J. 1980. A transgressive shelf sequence exhibiting
  hummocky cross stratification: The Cape Sebastian sandstone
  (Upper Cretaceous), southwestern Oregon. J. sedim. Petrol.,
  50, 681-702.

BREED, C.S., FRYBERGER, S.C. ANDREWS,S. McCAULEY, C. LENNARTZ, F. GEBEL, D. and HORSTMAN, K. 1979. Regional studies of sand seas using Landsat (ERTS) imagery. In: A study of global sand seas. (Ed. E.D.McKee). Prof.Pap. U.S. geol. Surv., 1052, 305-397.

BRENNER, R.L. and DAVIES, D.K. 1973. Storm-generated coquinoid sandstone: genesis of high energy marine sediments from the Upper Jurassic of Wyoming and Montana. Bull. geol. Soc. Am., 84, 1685-1698.

BRENNER, R.L. and DAVIES, D.K. 1974. Oxfordian sedimentation in Western Interior United States. Bull. Am.Ass. Petrol. Geol., 58, 407-428.

BRIDGE, J.S. and LEEDER, M.R. 1979. A simulation model of alluvial stratigraphy. Sedimentology, 26, 617-644.

BRUCE, C.H. 1973. Pressured shale and related sediment deformation: mechanism for development of regional contemporaneous faults. Bull. Am. Ass. Petrol. Geol., 57, 878-886.

BULL, W.B. 1972. Recognition of alluvial-fan deposits in the stratigraphic record. In: Recognition of Ancient Sedimentary Environments. (Ed. K.J. Rigby and W.K. Hamblin). Spec.Publ. Soc. econ. Paleont. Miner., Tulsa, 16, 68-83.

BULL, W.R. 1977. The alluvial fan environment. Prog. phys.Geogr. 1, 222-270.

CAMPBELL, C.V. 1971. Depositional model-Upper Cretaceous Gallup beach shoreline, Ship Rock area, northwestern New Mexico. J. sedim. Petrol., 41, 395-409.

CANT, D.J. and WALKER, R.G. 1976. Development of a braided-fluvial facies model for the Devonian Battery Point Sandstone, Quebec. Can.J.Earth Sci., 13, 102-119.

CANT, D.J. and WALKER, R.G. 1978. Fluvial processes and facies sequences in the sandy braided South Saskatchewan River, Canada. Sedimentology, 25, 625-648.

CASTON, V.N.D. 1972. Linear sand banks in the southern North Sea. Sedimentology, 18, 63-78.

CLIFTON, H.E., HUNTER, R.E. and PHILLIPS, R.L. 1971. Depositional structures and processes in the non-barred, high energy nearshore. J. sedim. Petrol., 41, 651-670.

COLEMAN, J.M., GAGLIANO, S.M. and WEBB, J.E. 1964. Minor sedimentary structures in a prograding distributary. Mar.Geol., 1, 240-258.

COLEMAN, J.M., SUHAYDA, J.N. WHELAN, T. and WRIGHT, L.D. 1974. Mass movement of Mississippi river delta sediments. Trans. Gulf-Cst. Ass.geol.Socs., 24, 49-68.

COLEMAN, J.M. and WRIGHT, L.D. 1975. Modern river deltas: variability of processes and sand bodies. In: Deltas, Models for Exploration (Ed. M.L. Broussard), Houston Geol.Soc.Houston, 99-149.

COLLINSON, J.D. 1970. Bedforms of the Tana River, Norway. Geogr. Annalr., 52-A, 31-56.

COLLINSON, J.D. 1978a. Alluvial sediments. In: Sedimentary
Environments and Facies (Ed. H.G.Reading), Blackwell Scientific
Publications, Oxford, 15-60.
COLLINSON, J.D. 1978b. Vertical sequence and sand body shape in
alluvial sequences. In: Fluvial Sedimentology (Ed. A.D.Miall),
Mem.Can.Soc.Petrol.Geol., Calgary, 5, 577-586.
COLLINSON, J.D. and BANKS, N.L. 1975. The Haslingden Flags
(Namurian G_1) of south-east Lancashire: bar finger sands in
the Pennine basin. Proc.Yorks.geol.Soc., 40, 431-458.
CRIMES, T.P. 1973. From limestones to distal turbidites: a
facies and trace fossil analysis in the Zumaya flysch (Pal-
aeocene-Eocene, north Spain. Sedimentology, 20, 105-131.
CURRAY, J.R., EMMEL, F.J. and CRAMPTON, P.J.S. 1969. Holocene
history of a strand plain, lagoonal coast, Nayarit, Mexico.
In: Coastal Lagoons - A Symposium (Ed. A.A. Castanares and
F.B. Phleger), Universidad Nacional Autonoma, Mexico, 63-100.
DAVIES, D.K., ETHRIDGE, F.G. and BERG, R.R. 1971. Recognition
of barrier environments. Bull.Am.Ass.Petrol.Geol., 55, 550-565.
DENNY, C.S. 1967. Fans and pediments. Am. J. Sci. 265, 81-105.
DRIESE, S.G., BYERS, C.W. and DOTT, R.H. Jr. 1981. Tidal depo-
sition in the basal Upper Cambrian Mt. Simon Formation in
Winconsin. J. sedim.Petrol., 51, 367-381.
EDWARDS, M.B. 1976. Growth faults in Upper Triassic deltaic
sediments, Svalbard. Bull.Am.Ass.Petrol.Geol., 60, 341-355.
ELLIOTT, T. 1974. Interdistributary bay sequences and their
genesis. Sedimentology, 21, 611-622.
ELLIOTT, T. 1976a. Upper Carboniferous sedimentary cycles pro-
duced by river-dominated, elongate deltas. J.Geol.Soc., 132,
199-208.
ELLIOTT, T. 1976b. The morphology, magnitude and regime of a
Carboniferous fluvial-distributary channel. J.sedim.Petrol.,
46, 70-76.
ELLIOTT, T. 1978. Deltas. In: Sedimentary Environments and
Facies (Ed. H.G.Reading), Blackwell Scientific Publications,
Oxford, 97-142.
EYNON, G. and WALKER, R.G. 1974. Facies relationships in
Pleistocene outwash gravels, Southern Ontario: a model for bar
growth in braided rivers. Sedimentology, 21, 43-70.
FERM, J.C. and CAVAROC, V.V. Jr. 1968. A non-marine sedimentary
model for the Allegheny Rocks of West Virginia. In: Late
Paleozoic and Mesozoic Continental Sedimentation, Northeastern
North America (Ed. G. de V.Klein), Spec.Paper geol.Soc.Am.,
106, 1-19.
FISHER, W.L., BROWN, L.F. SCOTT, A.J. and McGOWEN, J.H. 1969.
Delta systems in the exploration for oil and gas. Bur.econ.
Geol., Univ. Texas, Austin, 78pp.
FISCHER, A.G. 1961. Stratigraphic record of transgressing seas
in the light of sedimentation on Atlantic coast of New Jersey.
Bull.Am.Ass.Petrol.Geol., 45, 1656-1666.

FISK, H.N. 1961. Bar finger sands of the Mississippi delta. In:
  Geometry of Sandstone Bodies - a Symposium (Ed. J.A.Peterson
  and J.C. Osmond), Am.Ass.Petrol.Geol., Tulsa, 29-52.

GAGLIANO, S.M. and Van BEEK, J.L. 1975. An approach to multiuse
  management in the Mississippi delta system. In: Deltas, Models
  for Exploration (Ed. M.L. Broussard), Houston Geological
  Society, Houston, 223-238.

GALLOWAY, W.E. 1975. Process framework for describing the morphologic and stratigraphi evolution of deltaic depositional
  systems. In: Deltas, Models for Exploration (Ed. M.L.Broussard),
  Houston Geological Society, Houston, 87-98.

GALLOWAY, W.E. 1981. Depositional architecture of Cenozoic Gulf
  Coastal Plain fluvial systems. In: Recent and Ancient Nonmarine
  Depositional Environments: Models for Exploration (Ed. F.G.
  Ethridge and R.M. Flores), Spec.Publ.Soc.econ.Paleont.Miner.,
  31, 127-155.

GHIBAUDO, G. 1981. Deep sea fan deposits in the Macigno Formation
  (Middle-Upper Oligocene) of the Gordana Valley, northern
  Apennines, Italy - Reply J.sedim.Petrol., 51, 1021-1026.

GHIBAUDO, G., MUTTI, E. and ROSELL, J. 1974. Le spiagge fossili
  delle Arenarie di Aren (Cretacico superiore) nella valle
  Noguera-Ribagorzana (Pirenei centro-meridionali, Province di
  Lerida e Huesca, Spagna). Mem.Soc.geol.ital., 13, 497-537.

GINSBURG, R.N. 1975. Tidal Deposits: A Casebook of Recent
  Examples and Fossil Counterparts, Springer-Verlag, Berlin,
  428pp.

GLENNIE, K.W. 1970. Desert sedimentary environments. Developments
  in Sedimentology, Elsevier, Amsterdam, 14, 222pp.

GLENNIE, K.W. 1972. Permian Rotliegendes of Northwest Europe
  interpreted in light of modern desert sedimentation studies.
  Bull.Am.Ass.Petrol.Geol., 56, 1048-1071.

GLENNIE, K.W., MUDD, G.C. and NAGTEGAAL, P.J.C. 1978. Depositional
  environment and diagenesis of Permian Rotliegendes sandstones in
  Leman Bank and Sole Pit areas of the U.K. southern North Sea.
  Jl.geol.Soc.Lond., 135, 25-34.

GOLDRING, R. and BRIDGES, P. 1973. Sublittoral sheet sandstones.
  J.sedim.Petrol., 43, 736-747.

GOLE, C.V. and CHITALE, S.V. 1966. Inland delta building activity
  of Kosi River. J.Hydraul.Proc.Am.Soc.civ.Engrs. 92, 111-126.

GRAAFF, F.R. VAN DE. 1972. Fluvial-deltaic facies of the Castle-
  gate Sandstone (Cretaceous), East Central Utah. J.sedim.Petrol.,
  42, 558-571.

GREER, S.A. 1975. Sandbody geometry and sedimentary facies at the
  estuary-marine transition zone, Ossabaw Sound, Georgia: a
  stratigraphic model. Senckenberg.Mar., 7, 105-135.

GUSTAVSON, T.C. 1978. Bed forms and stratification types of
  modern gravel meander lobes, Nueces River, Texas. Sedimentology,
  25, 401-426.

HAMBLIN, A.P. and WALKER, R.G. 1979. Storm-dominated shallow
  marine deposits: the Fernie-Kootenay (Jurassic) transition,

southern Rocky Mountains. Can.J.Earth Sci., 16, 1673-1690.
HARMS, J.C. 1975. Stratification and sequence in prograding
    shoreline deposits. In: Depositional environments as inter-
    preted from primary sedimentary structures and stratification
    sequences. Soc.econ.Paleont. Miner., Short Course, 2, Dallas,
    81-102.
HAYES, J.B. 1979. Sandstone diagenesis - the hole truth. In:
    Aspects of diagenesis (Ed. P.A.Scholle and P.R.Schluger),
    Spec.Publ.Soc.econ.Paleont.Miner., Tulsa, 26, 127-139.
HAYES, M.O. 1967. Hurricanes as geological agents: case studies
    of Hurricanes Carla, 1961, and Cindy, 1963. Rep.Invest.Bur.
    econ.Geol., Austin, Texas, 61, 54pp.
HAYES, M.O. 1975. Morphology of sand accumulation in estuaries:
    an introduction to the symposium. In: Estuarine Research,
    Vol. II Geology and Engineering (Ed. L.E.Cronin), Academic
    Press, London, 3-22.
HEWARD, A.P. 1978. Alluvial fan sequence and megasequence models:
    with examples from Westphalian D-Stephanian B coalfields,
    Northern Spain. In: Fluvial Sedimentology (Ed. A.D.Miall),
    Mem.Can.Soc.Petrol.Geol., Calgary, 5, 669-702.
HISCOTT, R.N. 1981. Deep sea fan deposits in the Macigno Forma-
    tion (Middle-Upper Oligocene) of the Gordana Valley, northern
    Apennines, Italy - Discussion. J.sedim.Petrol., 51, 1015-1021.
HO, C. and COLEMAN, J.M. 1969. Consolidation and cementation of
    Recent sediments in the Atchafalaya basin. Bull.geol.Soc.Am.,
    80, 183-192.
HOBDAY, D.K. and HORNE, J.C. 1977. Tidally influenced barrier
    island and estuarine sedimentation in the Upper Carboniferous
    of southern West Virginia. Sedim.Geol., 18, 97-122.
HORNE, J.C. and FERM, J.C. 1976. Carboniferous depositional
    environments in the Pocahontas Basin, Eastern Kentucky and
    Southern West Virginia; A Field Guide, 129pp. Department of
    Geology, University of South Carolina.
HOUBOLT, J.J.H.C. 1968. Recent sediments in the southern bight
    of the North Sea. Geol.Mijnb, 47, 245-273.
HOWARD, J.D. 1972. Trace fossils as criteria for recognizing
    shorelines in the stratigraphic record. In: Recognition of
    Ancient Sedimentary Environments (Ed. J.K.Rigby and W.K.
    Hamblin), Spec.Publ.Soc.econ.Paleont.Miner., Tulsa, 16, 215-
    225.
HOWARD, J.D. and REINECK, H.E. 1972. Physical and biogenic
    sedimentary structures of the nearshore shelf. Senckenberg.
    Mar., 4, 81-123.
HUBERT, J.F., BUTERA, J.G. and RICE, R.F. 1972. Sedimentology of
    Upper Cretaceous Cody-Parkman delta, southwestern Powder
    River Basin, Wyoming. Bull.geol.Soc.Am., 83, 1649-1670.
HUNTER, R.E., CLIFTON, H.E. and PHILLIPS, R.L. 1979. Depositional
    processes, sedimentary structures and predicted vertical
    sequences in barred nearshore systems, southern Oregon coast.
    J.sedim.Petrol., 49, 711-726.

JACKSON, R.G. II. 1975. Velocity-bed-form-texture patterns of
    meander bends in the Lower Wabash River of Illinois and Indiana.
    Bull.geol.Soc.Am., 86, 1511-1522.
JOHNSON, H.D. 1977. Shallow marine sand bar sequences: an example
    from the late Precambrian of North Norway. Sedimentology, 24,
    245-270.
JOHNSON, H.D. 1978. Shallow siliciclastic seas. In: Sedimentary
    Environments and Facies (Ed. H.G.Reading), Blackwell Scientific
    Publications, Oxford, 207-258.
KENYON, N.H. 1970. Sand ribbons of European tidal seas. Mar.Geol.,
    9, 25-39.
KENYON, N.H. and STRIDE, A.H. 1970. The tide-swept continental
    shelf sediments between the Shetland Isles and France,
    Sedimentology, 14, 159-173.
KNIGHT, R.J. 1980. Linear sand bar development and tidal current
    flow in Cobequid Bay, Bay of Fundy, Nova Scotia. In: The coast-
    line of Canada (Ed. S.B.McCann). Geol.Surv.Canada, Paper 80-10,
    123-152.
KRUIT, C., BROUWER, J. KNOX, G. SCHOLLNBERGER, W. and VLIET VAN.
    A. 1975. Une excursion aux cones d'alluvions en eau profonde
    d'age Tertiaire pres de San Sebastian (province de Guipuzcoa,
    Espagne). 9th Int.Congr. Sedimentol., Nice 1975, excursion 23,
    75pp.
KULM, L.D., ROUSCH, R.G. HARLETT, J.C. NEUDECK, R.H. CHAMBERS,
    D.M. and RUNGE, E.J. 1975. Oregon continental shelf sedimenta-
    tion: inter-relationships of facies distribution and sedimentary
    processes.
KUMAR, N. and SANDERS, J.E. 1974. Inlet sequences: a vertical
    succession of sedimentary structures and textures created by
    the lateral migration of tidal inlets. Sedimentology, 21,
    491-532.
LEEDER, M.R. 1975. Pedogenic carbonate and flood sediment
    accretion rates: a quantitative model for alluvial, arid-zone
    lithofacies. Geol.Mag., 112, 257-270.
LEEDER, M.R. 1978. A quantitative stratigraphic model for
    alluvium, with special reference to channel deposit density
    and interconnectedness. In: Fluvial Sedimentology (Ed. A.D.
    Miall), Mem.Can.Soc.Petrol.Geol., 5, 587-596.
LEGGET, R.F., BROWN, R.J.E. and JOHNSTON, G.H. 1966. Alluvial fan
    formation near Aklavik, Northwest Territories, Canada. Bull.
    geol.Soc.Am., 77, 15-30.
LEVELL, B.K. 1980. A late Precambrian tidal shelf deposit, the
    Lower Sandfjord Formation, Finnmark, North Norway.
    Sedimentology, 27, 539-557.
LOWE, D.R. 1979. Sediment gravity flows: their classification and
    some problems of application to natural flows and deposits.
    In: Geology of Continental Slopes (Ed. L.J.Doyle and O.H.
    Pilkey). Spec.Publ.Soc.econ.Paleont.Miner., Tulsa, 27, 84-115.
MAINGUET, M. and CALLOT, Y. 1974. Air photo study of typology
    and inter-relations between the texture and structure of dune

patterns in the Fachi-Bilma Erg. Sahara. Zeit. f. Geomorph.,
Suppl.Bd. 20, 62-69.
McCABE, P.J. 1977. Deep distributary channels and giant bedforms
in the Upper Carboniferous of the Central Pennines, northern
England. Sedimentology, 24, 271-290.
McGOWEN, J.H. and SCOTT, A.J. 1975. Hurricanes as geologic agents
on the Texas Coast. In: Estuarine Research, Vol.II Geology and
Engineering (Ed. L.E.Cronin), Academic Press, London, 23-46.
McKEE, E.D. (Ed.). 1979. A study of global sand seas. Prof.Pap.
U.S. geol.Surv., 1052, 429, 5.2.
MIALL, A.D. 1977. A review of the braided-river depositional
environment. Earth-Sci.Rev., 13, 1-62.
MIDDLETON, G.V. and HAMPTON, M.A. 1973. Sediment gravity flows:
mechanics of flow and deposition. In: Turbidites and Deep
Water Sedimentation, 1-38. Soc.econ.Paleont.Miner.Pacific
Section, Short Course, Anaheim.
MOODY-STUART, M. 1966. High- and low-sinuosity stream deposits,
with examples from the Devonian of Spitsbergen. J.sedim.Petrol.
36, 1102-1117.
MORTON, R.A. 1979. Reevaluation of storm surge processes as agents
in the formation of graded shelf deposits. I.A.S.Int.Mtg. on
Holocene Marine Sedimentation in the North Sea Basin; Texel,
Netherlands, 1979, Abs. 87.
MUTTI, E. 1977. Distinctive thin-bedded turbidite facies and
related depositional environments in the Eocene Hecho Group
(South-central Pyrenees, Spain). Sedimentology, 24, 107-132.
MUTTI, E., FONNESU, F. RAMPONE, G. and SONNINO, M. 1981. Channel-
fill and associated overbank deposits in the Eocene Hecho Group,
Ainsa-Boltana region (south-central Pyrenees). I.A.S. 2nd Eur.
Mtg. Bologna, 1981, Abs., 113-116.
MUTTI, E. and RICCI-LUCCHI, F. 1972. Le torbiditi dell-Appennino
settentrionale: introduzione all'analisi di facies. Mem.Soc.
geol.Ital., 11, 161-199. (English translation in Int.Geol.
Rev., 20, 125-166).
NAGTEGAAL, P.J.C. 1980. Clastic reservoir rocks - origin, dia-
genesis and quality. In: The Sedimentation of the North Sea
Reservoir Rocks. Norwegian Petroleum Society, Geilo, May
1980, 1-23.
NAMI, M. 1976. An exhumed Jurassic meander belt from Yorkshire,
England. Geol.Mag., 113, 47-52.
NILSEN, T.H. 1980. Modern and ancient submarine fans: discussion
of papers by R.G.Walker and W.R.Normark. Bull.Am.Ass.Petrol.
Geol., 64, 1094-1101.
NIO, S-D. 1976. Marine transgressions as a factor in the forma-
tion of sand wave complexes. Geol.Mijnb., 55, 18-40.
NIO, S-D and SIEGENTHALER, J.C. 1978. A lower Eocene estuarine-
shelf complex in the Isabena Valley. State Univ. Utrecht,
Sedimentology Group, Rept. No. 18, 1-44.
NORMARK, W.R. 1980. Modern and ancient submarine fans: reply.
Bull.Am.Ass.Petrol.Geol., 64, 1108-1112.

NORMARK, W.R. and PIPER, D.J.W. 1969. Deep-sea fan-valleys, past
  and present. Bull.geol.Soc.Am., 80, 1859-1866.
OFF, T. 1963. Rhythmic linear sand bodies caused by tidal currents.
  Bull.Am.Ass.Petrol.Geol., 47, 324-341.
OOMKENS, E. 1967. Depositional sequences and sand distribution in
  a deltaic complex. Geol. Mijnb., 46, 265-278.
OOMKENS, E. 1974. Lithofacies relations in the Late Quaternary
  Niger delta complex. Sedimentology, 21, 195-222.
PUIGDEFABREGAS, C. 1973. Miocene point bar deposits in the Ebro
  Basin, Northern Spain. Sedimentology, 20, 133-144.
PUIGDEFABREGAS, C. and VLIET, A. VAN. 1978. Meandering stream
  deposits from the Tertiary of the Southern Pyrenees. In:
  Fluvial Sedimentology (Ed. A.D.Miall), Mem.Can.Soc.Petrol.
  Geol., Calgary, 5, 469-485.
PUTNAM, P.E. and OLIVER, T.A. 1980. Stratigraphic traps in
  channel sandstones in the Upper Mannville (Albian) of east-
  central Canada. Bull.Can.Pet.Geol., 28, 489-508.
Reading, H.G. (Ed.) 1978. Sedimentary Environments and Facies.
  Blackwell Scientific Publications, Oxford 557pp.
REINECK, H.E. 1967. Layered sediments of tidal flats, beaches and
  shelf bottoms of the North Sea. In: Estuaries (Ed. G.D.Lauff),
  Am.Assoc.Adv.Sci., Washington D.C. 191-206.
REINECK, H.E. and SINGH, I.B. 1973. Depositional Sedimentary
  Environments. Springer-Verlag, New York, 430pp.
RICCI-LUCCHI, F. 1975. Depositional cycles in two turbidite form-
  ations of the northern Apennines (Italy). J.sedim.Petrol., 45,
  3-43.
RICCI-LUCCHI, F. 1978. Turbidite dispersal in a Miocene deep-sea
  plain. Geol. en Mijnbouw, 57, 559-576.
RICCI-LUCCHI, F. and VALMORI, E. 1980. Basin-wide turbidites in a
  Miocene, oversupplied deep-sea plain: a geometrical analysis.
  Sedimentology, 27, 241-270.
RICHARDS, M.T. 1981. Transgressive and progradational shoreline
  sequences developed on a high relief unconformity - The Lower
  Trias, Western Alps. I.A.S. 2nd Eur.Mtg.Bologna, 1981, Abs.,
  164-167.
RIDER, M.H. 1978. Growth faults in the Carboniferous of western
  Ireland. Bull.Am.Ass.Petrol.Geol., 62, 2191-2213.
RUPKE, N.A. 1976. Sedimentology of very thick calcarenite-marl-
  stone beds in a flysch succession, southwestern Pyrenees.
  Sedimentology, 23, 43-65.
SANDERS, J.E. and KUMAR, N. 1975a. Evidence of shoreface retreat
  and in-place 'drowning' during Holocene submergence of barriers,
  shelf off Fire Island, New York. Bull.geol.Soc.Am., 86, 65-76.
SANDERS, J.E. and KUMAR, N. 1975b. Holocene shoestring sand on
  inner continental shelf off Long Island, New York. Bull.Am.Ass.
  Petrol.Geol., 59, 997-1009.
SCHUMM, S.A. 1981. Evolution and response of the fluvial system,
  sedimentological implications. In: Recent and Ancient Nonmarine
  Depositional Environments: Models for Exploration (Ed. F.G.

Ethridge and R.M.Flores). Spec.Publ.Soc.econ.Paleont.Miner., <u>31</u>, 19-29.

SMITH, D.G. and PUTNAM, P.E. 1980. Anastomosed river deposits: modern and ancient examples in Alberta Canada. Can.J.Earth Sci., <u>17</u>, 1396-1406.

SMITH, N.D. 1970. The braided stream depositional environment: Comparison of the Platte River with some Silurian clastic rocks, North-Central Appalachians. Bull.geol.Soc.Am., <u>81</u>, 2993-3014.

SMITH, N.D. 1971. Transverse bars and braiding in the Lower Platte River, Nebraska. Bull.geol.Soc.Am., <u>82</u>, 3407-3420.

STANLEY, D.J. and UNRUG, R. 1972. Submarine channel deposits, fluxoturbidites and other indicators of slope and base-of-slope environments in modern and ancient marine basins. In: Recognition of Ancient Sedimentary Environments (Ed. J.K.Rigby and W.K. Hamblin), Spec.Publ.Soc.econ.Paleont.Miner., <u>16</u>, 287-340.

STEEL, R.J. 1974. New Red Sandstone floodplain and piedmont sedimentation in the Bebridean Province. J.sedim.Petrol., <u>44</u>, 336-357.

STEEL, R.J. 1976. Devonian basins of Western Norway - Sedimentary response to tectonism and to varying tectonic context. Tectonophysics, <u>36</u>, 207-224.

STEEL, R.J., MAEHLE, S. NILSEN, H. RØE, S.L. and SPINNANGR, A. 1977. Coarsening upwards cycles in the alluvium of Hornelen Basin (Devonian), Norway: sedimentary response to tectonic events. Bull.geol.Soc.Am., <u>88</u>, 1124-1134.

SWETT, K. and SMIT, D.E. 1972. Paleogeography and depositional environments of the Cambro-Ordovician shallow marine facies of the North Atlantic. Bull.geol.Soc.Am., <u>83</u>, 3223-3248.

SWIFT, D.J.P. 1968. Coastal erosion and transgressive stratigraphy. J.Geol., <u>76</u>, 444-456.

SWIFT, D.J.P. 1975. Barrier island genesis: evidence from the Middle Atlantic Shelf of North America. Sedim.Geol., <u>14</u>, 1-43.

THOMPSON, R.W. 1968. Tidal flat sedimentation on the Colorado River delta northwestern Gulf of California. Mem.geol.Soc. Am., <u>107</u>, 1-133.

VISSER, M.J. 1980. Neap-spring cycles reflected in Holocene subtidal large-scale bedform deposits: a preliminary note. Geology, <u>8</u>, 543-546.

WALKER, R.G. 1975. Generalized facies models for resedimented conglomerates of turbidite association. Bull.geol.Soc.Am., <u>86</u>, 737-748.

WALKER, R.G. 1978. Deep-water sandstone facies and ancient submarine fans: models for exploration for stratigraphic traps. Bull.Am.Assoc.Petrol.Geol., <u>62</u>, 932-966.

WALKER, R.G. (Ed.). 1979. Facies Models. Geoscience Canada Reprint Series <u>1</u>, Geol.Ass.Canada, 211pp.

WALKER, R.G. 1980. Modern and ancient submarine fans: reply Bull.Am.Ass.Petrol.Geol., <u>64</u>, 1101-1108.

WALKER, R.G. and MUTTI, E. 1973. Turbidite facies and facies associations.   In:Turbidites and Deep Water Sedimentation, Soc. econ.Paleont.Miner., Pacific Section, Short Course, Anaheim. 119-157.

WALKER, T.R. 1967. Formation of red beds in ancient and modern deserts. Bull.geol.Soc.Am., 78, 353-368.

WEBER, K.J. and DAUKORU, E. 1975. Petroleum geology of the Niger delta. Proc. 9th World Petrol.Conf., 209-221.

WILSON, I.G. 1973. Ergs. Sediment.Geol., 10, 77-106.

WOODYER, K.D., TAYLOR, G. and CROOK, K.A.W. 1979. Depositional processes along a very low-gradient suspended-load stream: The Barwon River, New South Wales. Sedim.Geol., 22, 97-120.

WRIGHT, L.D. and COLEMAN, J.M. 1973. Variations in morphology of major river deltas as functions of ocean wave and river discharge regimes. Bull.Am.Ass.Petrol.Geol., 57, 370-398.

COMPACTIONAL DIAGENESIS

George V. Chilingarian
Petroleum Engineering Department
University of Southern California
Los Angeles, California 90007
USA

INTRODUCTION

The importance of compaction as a diagenetic process has resulted in numerous publications on the subject. For example, Rieke and Chilingarian (1974), Chilingarian and Rieke (1976), Singer and Muller (1981) and Brenner et al. (1981) discussed the compaction of argillaceous sediments. An elaborate discussion of deep-sea carbonates was presented by Cook and Egbert (1981), whereas Chilingarian and Wolf (1975, 1976) discussed the compaction of coarse-grained sediments. Kisch (1981) made an excellent contribution on deep-burial diagenesis. By necessity, in writing the present overview the writer used freely the material published by him and his co-authors (e.g., K.H.Wolf, H.H. Rieke, W.H. Fertl, R.P. Brenner and D.R. Allen), with permission of the Elsevier Publishing Co. The reader is referred to the original books for greater details.

Mechanical compaction processes include the rearrangement, plastic deformation and breakage of grains. Chemical compaction includes pressure solutions between grains, between grains and various kinds of cements, and within protective and skeletal structures of various organisms.

Diagenesis is commonly defined as "all processes occurring between deposition and metamorphism". In the opinion of the writer, this definition is confusing, because in nature there are two distinct and different stages of alteration of rocks subsequent to deposition and prior to metamorphism: (1) diagenesis - conversion of sediments into sedimentary rocks, and (2) catagenesis - changes occurring in sedimentary rocks prior to metamorphism. These two

A. Parker and B. W. Sellwood (eds.), Sediment Diagenesis, 57–167.

stages differ not only in the character of physicochemical processes occurring in the materials composing sediments and sedimentary rocks, but also in the sources of energy required for the
processes.   The source of energy for diagenetic processes is the
combination of various mineral components of the sediment and
interstitial fluids.   During the early diagenetic stage, important roles are played by the organic components (organic acids, etc),
microbial activity, solar energy, excess amounts of water present,
etc.   During the catagenetic stage, on the other hand, the major
role is played by temperature and pressure in the alterations
which occur in the already-formed sedimentary rocks.   Briefly
stated, diagenesis of sediments and catagenesis (or epigenesis)
of rocks differ in sources of energy, physiochemical processes,
and resulting products of alteration.   Diagenesis is still a
rather poorly understood field of knowledge and, although the
term "diagenesis" has been in the technical literature for nearly
80 years, it is not even listed in the general indexes of some
encyclopaedias.   The term was first introduced by Von Guembel in
1868.   Walter (1894) defined it as "all those physical and
chemical changes which a rock undergoes after its deposition,
without the introduction of rock pressure or igneous heat".   It
is clearly understood that "rock" was used in a traditional
geological sense, which in this case implies an initially soft,
unconsolidated sediment.

The stages of diagenesis that pertain to soft clay can be
divided into: (1) pre-burial stage, and (2) shallow-burial and
consolidation stage.   In soft clays, the depth to which the sediments are buried under younger deposits (stage 2) is just as
important to the diagenetic evolution as the length of time which
passes after the deposition of the clay (stage 1).   This is
especially true of the physical changes which take place in the
sediment in the course of diagenesis.   The time during which a
sediment is deposited in a certain physicochemical environment is
important as far as the chemical and mineralogical processes are
concerned.

Compaction, which is one of the main processes of diagenesis,
can be defined as the expulsion of pore fluids and pore volume
decreases in a sedimentary column as a result of normal shear-
compressional stresses due to the overburden load.

Compaction of sediments, which is the process of volume reduction, can be expressed as a percentage of the original voids
present.   The degree of compaction depends largely on the ratio
of fine to coarse material and on the character of the sediment
framework (Chilingarian et al., 1967).   Soft clay undergoes the
highest degree of compaction in the first 25 cm (Ginsburg, 1957).
Ginsburg suggested that the mixing by organisms, the gel-like
character of the sediment and the escape of bacterial gases

contribute to rapid packing.   The burial pressure of sediment
accumulation becomes effective somewhat later to produce some sort
of physical cohesion between the particles.

The marine sediments, which are deposited rapidly, are not
normally compacted near the water-sediment interface, because the
overburden stress is largely borne by the interstitial fluid
rather than by the sedimentary particles.   These fluid-saturated
sediments have low compressive and shear strengths and are of
elasticoplastic nature.

As pointed out by Allen and Chilingarian (1975, in:
Chilingarian and Wolf, 1975), the mechanics of compaction have
been approached from two different viewpoints, that of a soils
engineer and that of a geologist-sedimentologist.   Inasmuch as
the terms compaction and consolidation are used by both disciplines, but with different connotations, it is necessary to present
various definitions as used in each field (see Table 1).   Although
the largest body of knowledge on properties affecting sediment
compaction has been accumulated by soils engineers, there is
voluminous geological literature on the subject.   This available
knowledge has been largely neglected by the soil scientists and
vice versa and resulted in duplication of effort in many instances.
Obviously, the existing situation should be immediately corrected.
Geologists and soil scientists will also find many useful references in the petroleum engineering literature.

Possible Loading Conditions During Sedimentation.   Wolf et
al. (1976, p.24) showed possible loading conditions on a hypothetical sediment cube (Fig. 1).   The first condition presented
(Fig. 1-A) is polyaxial loading, in which the three principal
stresses are of different magnitude.   Some investigators prefer
to call this stress condition triaxial loading.   The second
loading condition presented (Fig. 1-B) is hydrostatic, in which
the three principal stresses applied are equal.   The third type
of loading (Fig. 1-C) is triaxial, in which two of the three
principal stresses are equal.   Although some investigators
justifiably refer to it as biaxial stress, the term triaxial is
strongly embedded in the earth sciences literature.   In the
uniaxial loading condition (Fig. 1-D), the applied force acts in
one direction only and is perpendicular to one surface of the
sediment sample.   The four faces of the cube parallel to the
direction of the stress remain stationary.   This arrangement can
be achieved by placing the sample in a thick-walled, cylindrical
chamber, the sides of which are stationary.   The pressure can be
applied with either one or two pistons, and the change in the
volume of the sample is reflected by the change in the length of
the sample.   In the field of soil mechanics, this method is
sometimes referred to as triaxial testing.   It should be mentioned
also that some investigators reserve the term uniaxial for cases

TABLE 1. Definition of compaction and consolidation as used in the field of geology and soil mechanics (after Allen and Chilingarian, 1975, p.43).

Geology-sedimentology usage	Soils engineering usage
Compaction	
A lessening of sedimentary volume owing to overburden loading, grain rearrangement, etc.	An increase in bulk specific weight (or bulk specific gravity) caused by mechanical or hydraulic means such as vibrating, loading or wetting.
Consolidation	
The acquisition of structural competency by sediment owing to reduction in volume induration, cementation, etc.	The lessening volume of a porous material (such as clay), by the expulsion of pore water upon the application of a load.

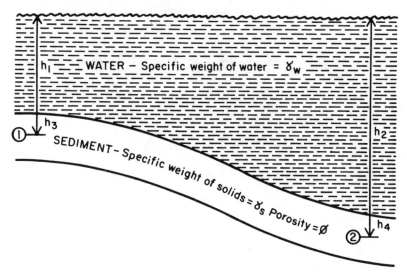

Fig. 1. Effect of submergence on effective (grain - to - grain) stress in a sediment layer. If h_2 and h_1 are equal, the effective stresses at point 1 and 2 are equal. (Wolf et al.1976, fig.1-19, p.25).

when there is a vertical stress with no lateral constraint and, therefore, lateral strain does occur.   In biaxial loading (Fig. 1-E) the two principal stresses are equal, while two faces of the cube are held stationary.   During the initial stages of sedimentation, the loading condition is probably hydrostatic (Fig. 1-B). As sedimentation progresses the loading becomes triaxial (Fig. 1-C). Finally, as the overburden load becomes large, the uniaxial loading condition (Fig. 1-D) is approached.

It seems important to examine the frequently posed question, namely: "Does the height of the overlying water column in a basin affect the compaction of coarse-grained sediments or not?".

It is commonly felt that compactive pressure at a certain depth below the sediment-water interface is higher if the depth of overlying water is greater.   The compactive (or effective) pressure, p_e which actually causes compaction, is equal to the total overburden pressure, p_t, minus the pore pressure, p_p.   The mathematical calculations presented here, however, show that the effective pressure at a certain depth below the sediment-water interface is independent of the depth of overlying basin water. If the depth of water is h, specific weight of water is γ_w, depth of point 1 below the sediment-water interface is h_3, porosity of the sediment is ϕ , and specific weight of solids is γ_s, then the total overburden pressure, p_t, at point 1 is equal to (see Fig. 2):

$$P_{t1} = \underline{h}_1 \, ^\gamma \, _w + \underline{h}_3 \, ^\gamma \, _s \, (1 - \phi) + \underline{h}_3{}^{\phi\gamma} \, _w. \tag{1}$$

Inasmuch as the pore pressure, $\underline{P}_p$, at point 1 is equal to the hydrostatic head,

$$P_{p1} = \underline{h}_1 \, ^\gamma \, _w + \underline{h}_3 \, ^\gamma \, _w. \tag{2}$$

Thus the effective pressure, $\underline{P}_e$, at point 1 is equal to

$$\underline{P}_{e1} = \underline{P}_{t1} - \underline{P}_{p1} = \underline{h}_3 \, ^\gamma \, _s \, (1 - \phi) + \underline{h}_3 \, ^\gamma \, _w \, (\phi - 1) \tag{3}$$

Similarly, at point 2:

$$P_{t2} = \underline{h}_2 \, ^\gamma{}_w = \underline{h}_3 \, ^\gamma \, _s \, (1 - \phi) + \underline{h}_3 \, ^{\phi\gamma} \, _w \tag{4}$$

$$P_{p2} = \underline{h}_2 \, ^\gamma \, _w + \underline{h}_3 \, ^\gamma \, _w \tag{5}$$

and

$$\underline{P}_{e2} = \underline{P}_{t2}$$

and

$$\underline{P}_{e2} = \underline{P}_{t2} - \underline{P}_{p2} = \underline{h}_3 \, ^\gamma \, _s \, (1 - \phi) + \underline{h}_3 \, ^\gamma \, _w \, (\phi - 1). \tag{6}$$

Inasmuch as right-hand side of the Eqs. 3 and 6 are identical, effective pressure at point 1 is equal to the effective presssure at point 2. This is true, however, only if the water above the water-sediment interface is connected with the intergranular water (intercommunicating pore space). If there is an impermeable shale layer above the coarse-grained sediment, then the effective (compactive) pressure at point 2 will be greater, <u>if</u> the pore pressure at points 1 and 2 are less than those indicated by Eqs. 2 and 5 respectively.

POROSITY AND DENSITY

The decrease in porosity per unit of the original bulk volume can be expressed as:

$$\sigma\phi = \frac{\sigma e}{1 + e_o} \tag{7}$$

where $\sigma\phi$ = change in porosity related to the original $\underline{V}_b$, $\underline{e}_o$ = orginal void ratio, and σe = change in void ratio.

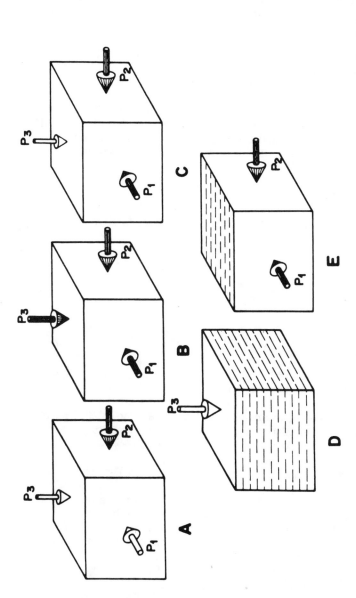

Fig.2. Compaction loading classification (after Sawabini et al., 1974, fig. 1, p.133). (a) Polyaxial loading ($p_1 \neq p_2 \neq p_1$), called triaxial loading by some investigators. Hydrostatic loading ($p_1 = p_2 = p_3$). (c) Triaxial loading ($p_1 = p_2 \neq p_3$), called biaxial loading by some investigators. (d) Uniaxial loading (four sides parallel to the stress p, are kept stationary), referred to as biaxial loading by some investigators. (e) Biaxial loading ($p_1 = p_2$ and sides parallel to these two stresses are kept stationary).

Specific weight, γ, is often used in conjunction with porosity and void ratio. It is defined as the weight per unit volume, whereas density, ρ, is the mass per unit volume, and is equal to γ where g is gravitational acceleration. The term "density", however, is often used to designate specific weight, which often results in erroneous calculations. Mass ρ is attracted by the earth with a force γ (= ρ xg). For example, if the specific weight of water is equal to 62.4 lb/ft^3 and the gravitational acceleration, g is equal to 32.174 ft/sec^2, then the density expressed in terms of slugs/ft^3 is equal to 1.94 (=62.4/32.174).

Bulk specific weight can be either "dry" or "wet", depending upon the nature of the fluid in the pore spaces. The unit weight of a dray sand (only air is present in pore spaces) is equal to:

$$\gamma_{db} = (1 - \phi) \, \gamma_s \qquad\qquad (8)$$

where γ_{db} = dry weight per unit of bulk volume, and γ_s = specific weight of solids (grains).

The unit weight of a wet sand is expressed as:

$$\gamma_{wb} = (1 - \phi) \, \gamma_s + \phi\gamma_f \qquad\qquad (9)$$

or:

$$\gamma_{wb} = \gamma_s - \phi (\gamma_s - \gamma_f) \qquad\qquad (10)$$

where γ_{wb} = weight per unit of wet bulk volume, and γ_f = specific weight of fluid in the pores.

Quartz sands have an average specific gravity of 2.65 with reference to water (specific gravity = specific weight of a material at 60°F: specific weight of water at 60°F)*.

*(°C x 1.8) + 32 = °F; $[(°F + 40)/1.8]$ - 40 = °C.

If the weight of one cu.ft. of freshwater is assumed to be equal to 62.4 lb, then one cu.ft. of solid silica having a specific gravity of 2.65 would weigh 165.4 lb. If one cu.ft. of dry sand weighs 137.3 lb. then its dry bulk specific gravity would be equal to 2.2 (= 137.3/62.4). The porosity of this dry sand can be calculated using Eq. 8: γ_{db} = (1-ϕ) γ_s or ϕ = (γ_s - γ_{db})/γ_s = (2.65 - 2.2)/2.65 = 0.17 or 17%. If the sand were saturated with water, its bulk specific weight would be equal to the weight of the solids (137.3 lb) plus the weight of the water (0.17 x 62.4 lb) or 147.91 lb/ft^3.

It can be easily illustrated on using idealised spheres that porosity is dependent upon the method of packing.   If packed cubically, spheres of equal size would have a maximum possible void space of about 47.6% (Slichter, 1897-1898).   If packed rhombohedrally, the porosity is reduced to a minimum of about 26% (Fig. 3).   It is obvious that sphere size does not change

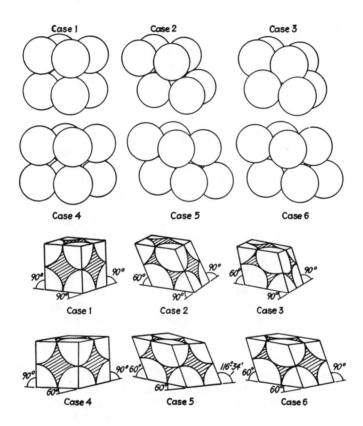

Fig. 3.   Several different ways of packing spheres
(after Graton and Frazer, 1935).

porosity when unit volumes with sides at least 2 radii in length are examined.   In nature, owing to variation in size of grains and their angularity, usually the porosity of a sand or sand- stone will be less than the values specified for spherical grains. It has also been demonstrated that in a mixture of spherical particles of different sizes, stacking arrangement does affect porosity. The introduction of a second set of spheres, small enough to fit in the pore space between the larger set, can reduce porosity to about 13%. (See Wolf and Chilingarian, 1976, Vol. II, Chapter 3, for more details).

Usually, finer-grained sediments exhibit greater porosity when deposited than coarse-grained ones. A well-sorted, well-rounded, loosely compacted medium- to coarse-grained sand may have a porosity of about 37%, whereas a poorly-sorted fine-grained sand with irregularly-shaped grains may have a porosity in excess of 50%. An admixture of irregular-shaped, tabular and bladed particles usually gives rise to a higher porosity because of particle bridging. The wide variability of porosity owing to depositional environment is best illustrated by the greywackes, which may have either a high or a very low porosity, depending on the amount of fine-grained material filling the pores. Clays and silts may have porosities as high as 50-80% when freshly deposited.

The terms "effective porosity" and "total porosity" are often used in petroleum geology and petroleum reservoir engineering studies. These terms differentiate between the interconnected pores through which fluids can move and the total pore space, regardless of its ability to transmit fluid.

In the opinion of the writer, however, the existing definition of the effective porosity should be changed. One should define the effective porosity as the porosity of all inter-communicating pores minus the irreducible fluid saturation (whether of oil or water). The minute crevices, cracks and pores occupied by the irreducible fluid do not influence the flow of fluids through porous media. Consequently, there is a much better correlation between the permeability and the effective porosity as defined by the writer than the effective porosity as usually defined.

According to Singer and Muller (1981), argillaceous muds, having an initial porosity of 70-90%, are compacted during shallow burial (burial depth down to 500 m) diagenesis to mud-stone (or shale if fissile) with a porosity of 30%. Argillite having a porosity of < 10% forms on further compaction. The decrease in porosity (and water content) is rapid down to about 150-300 m burial depth and show below those depths. The writer strongly recommends that compressibility values for various sedi-ments be used to quantify compaction. As pointed out by Singer and Muller (1981), degree of compaction is not only a function of burial depth (overburden) and diagenesis, but also of (1) time, (2) geothermal gradient, (3) tectonic stresses, (4) lithology, (5) depositional environment, and (6) overpressured fluid zones.

Cook and Egbert (1981) pointed out that at shallow-burial

phase (1 to < 250 m) early dewatering of deep-sea carbonates as a result of gravitational compaction in the major process for reducing porosity (from ≈ 80% to ≈60%). There is a corresponding change in bulk density from 1.35 to 1.64 (Schlanger and Douglas, 1974). According to Schlanger and Douglas (1974), porosity decreases from about 60% down to 35-40% upon burial from 200 m to 1,200 m. Gravitational compaction appears to play a minor role during this deep-burial phase, with precipitation of low-magnesium calcite being predominant.

BUOYANCY

When a body is completely or partly immersed in a static fluid, there is an upward vertical buoyant force on this body equal in magnitude to the weight of displaced fluid. This force is a resultant of all forces acting on the body by the fluid. The pressure is greater on the parts of the fluid more deeply immersed. The pressures at different points on the immersed body are independent of the body material. For example, if the same fluid is substituted for the immersed body, this fluid will remain at rest. This means that the buoyant, upward force on the substituted fluid is equal to its weight.

If the immersed body is in static equilibrium, the buoyant force and the weight of the body are equal in magnitude and opposite in direction, passing through the centre of gravity of the body. For a comprehensive treatment of fluid statics, the reader is referred to an excellent book on fluid mechanics by Binder (1962).

The buoyancy principle is illustrated in Fig. 4. The weight of a solid body having a volume of 1 cu ft and a specific weight of 162.24 lb/cu ft (sp.gr.=2.6) is registered on the scale as 162.24 lb (Fig. 4A). If the container is filled with water, the buoyant force acting upwards will be equal to the weight of displaced fluid (=62.4 lb) and consequently the weight registered on the scale will be equal to 99.84 lb (=162.24 - 62.4 = 99.84 lb) as shown in Fig. 4B.

If a porous medium, having a porosity of 20% is saturated with water and the specific gravity of the solid grains is equal to 2.6, the weight registered on the scale is equal to the sum of the weights of the solid grains and of the fluid, i.e., 2.6 x 62.4 (1-0.20) + 62.4 (0.20) = 142.27 lb (Fig. 4C). The weight registered with water inside the container is equal to 142.27 - 62.4 = 79.87 lb, i.e., the weight of the saturated porous medium (142.27 lb) acting downwards minus the buoyant force (62.4 lb), which acts upwards and is equal to the weight of the displaced fluid.

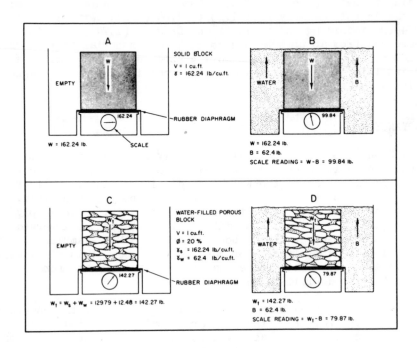

Fig. 4.   Illustration of the effect of buoyancy on the
          weight of solid and water-filled porous materials.
          W = weight in lb; B = buoyant force in lb; V =
          volume in cu.ft.; γ = specific weight in lb/
          cu ft; φ = porosity. Subscripts: t = total,
          s = solids, and w = water.   Inside container,
          solid blocks are placed on top of the rubber
          diaphragm, which, in turn, rests over the scale.
          (After Rieke and Chilingarian, 1974, Fig. 2).

     The latter experiment by Rieke and Chilingarian (1974) also
demonstrates that Archimedes principle can be applied to the porous
rocks underground.   The force of uplift owing to pore pressure is
not proportional to surface porosity, and the fluid pressure P_p
is effective over the entire area of any surface passed through
the system, irrespective of the fraction of the area lying within
the solid-filled part of the space (see stimulating discussions
by Laubscher, 1960 and Hubbert and Rubey, 1960).   Unfortunately,

many investigators are not familiar with the above described
experiment.

INTRODUCTION TO MARINE CLAYS

Marine pelagic clay sediments account for approximately 15%
of the total sediments laid down in geologic time. Clays and
colloidal particles are mechanically transported to the area of
deposition primarily by water (in suspension or adsorbed onto
organic gels) and sometimes by wind. When the sediment is transported in particulate form, the eventual character of the deposited
minerals is dictated by the physicochemical and sometimes biological environment of the source region, then by the transportation
and depositional environments and, finally, by diagenesis. Upon
burial in a predominantly clayey marine facies, there is a natural
tendency toward the creation of a relatively impervious shield,
covering the deposit, so that further alteration of the mineral is
restricted by the supply of ions originally present in the transporting water originally surrounding the clay mineral. During
this stage, an almost closed system is generated, and for this
reason one finds that the marine clay minerals commonly show
little difference in columns of 1,000 m or more of clays as shown
by Milne and Early (1958). Later tectonic movements, changes in
sea levels, or any process that could result in the displacement
of interstitial water with water of a different ionic concentration
or having different types of ions would likely result in an
alteration of the volume of sorbed fluid on the clay particles
and hence affect the engineering properties of the clay deposit.

As soft clay sediments accumulate on the sea floor, the
underlying layers are compacted. If interstitial water can
escape, it mingles with the waters in overlying sediments and
eventually escapes into the waters above. Clay sediments with
an initial porosity of 80%, for example, undergo a marked physical
change during compaction. There is a general decrease in
porosity with increasing depth of burial. This occurs very
rapidly to a depth of 500 m and much slower below that depth. In
those instances where the interstitial water cannot escape, the
porosity remains higher and the pore pressure increases. The
important factors which affect the porosity or void-volume reduct-
ions of a clay deposit are as follows: (1) lithology, (2) initial
thickness of sediment, (3) maximum effective stress, (4) rate of
loading, (5) decomposition of organic material, (6) recrystallization, (7) particle size and distribution, (8) type of clay minerals
deposited, (9) adsorbed cations on clays, (10) interstitial
electrolyte solutions, and (11) pH of transporting agent, depositional medium, and interstitial fluids, and (12) diagenetic and
catagenetic processes.

Due to the intimacy of the clay-water relationship, the reduction in pore volume of clays with increasing overburden pressure
can best be analysed in terms of water removal.   The effects of
particle size and distribution types of clay minerals, types of
adsorbed cations, and chemistry of interstitial electrolyte solutions are discussed in this paper (see Fig. 5).   The overall
compaction of clays is accomplished chiefly by elimination of
pore space and the resulting expulsion of the interstitial fluids.
Decomposition of organic matter and the transformation and recrystallization of clays also can play an important role.

The organic matter content in recent marine and fresh-water
clay sediments ranges from about 2 to 5%.   The organic matter
consists of less than 10% amino-acids and 60-80% of undefined
"organic residue" (Degens, 1967).   The important point to keep
in mind is the effect of organic matter on clay hydration.

DIFFERENCES BETWEEN FRESH-WATER AND MARINE CLAYS

There are several differences between marine and fresh-water
clays.   These are due to the chemical-mineralogical and, partly
the physical changes in a sediment during diagenesis.   Diagenetic
processes in large measure depend on the chemistry of the interstitial water in contact with the clay-mineral particles.   The
initial porosity of clays and, thus, their water content, is very
much higher than that of sands.   Clay muds from recent sea and
lake bottoms as a rule have a porosity of 70-90% corresponding
to a water content of 50-80% (expressed as a percentage of the
wet weight).   From a purely geometrical viewpoint, clay minerals
could compact to a sediment with a very low porosity.   The fact
that this is not the case with soft clay sediments shows that
their porosity cannot be understood only on the basis of geometrical models as is quantitatively possible with sands (Muller,
1967, p.136).   The large surface areas and associated surface
charges of the clay particles must be considered.   Sediments
of different sedimentary basins with the same amount of clay-size
fraction have distinctly different water contents (see Fig. 6).
The differences expressed in Fig. 6 are likely to be due to the
following important factors: (1) influence of clay-mineral
composition, (2) influence of interstitial electrolyte solutions,
(3) influence of exchangeable cations, (4) rate of sedimentation
(Meade, 1964).   In addition, pH and associated organic material
play an important role.

In natural and artificially produced soft clay sediments
there is a relation between the water content or porosity of a
freshly deposited sediment and the electrolyte content of the
(1) depositional medium (Muller, 1967, p.137) and (2) the interstitial fluids.   For waters of higher electrolyte contents,

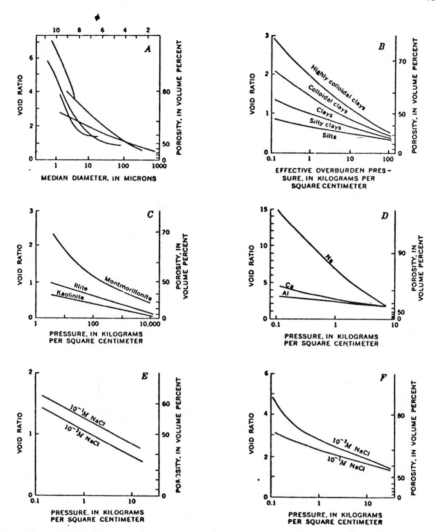

Fig. 5. Influence of different factors on the relationship between void ratio and pressure in clayey materials. A = Relationship between void ratio and median particle diameter at overburden presssures less than 1 kg/cm² (after Meade, 1964, p. B6). B = Generalized influence of particle size (modified from Skempton, 1953, p.55). C = Influence of clay-mineral species (modified from Chilingar and Knight, 1960, p. 104). D = Influence of cations adsorbed by montmorillonite (modified from Samuels, 1950). E = Influence of NaCl concentrations in unfractional illite, about 60% of which was coarser than 2 m in size (modified from Mitchell, 1960, Fig. M3). F = Influence of NaCl concentration in illite finer than 0.2 m (modified from Bolt, 1956, p. 92. (After Meade, 1968, fig. 1, p.D4).

such as those found in marine environment, the porosity decreases.
Fresh-water clays, therefore, as a rule have a higher water content and porosity than similar marine clay sediments.   According
to Muller (1967, p. 137), this phenomenon can be explained by van
Waals' attractive forces between the clay plates and repulsive
forces due to the presence of diffuse electrical double layers on
charged clay plates.

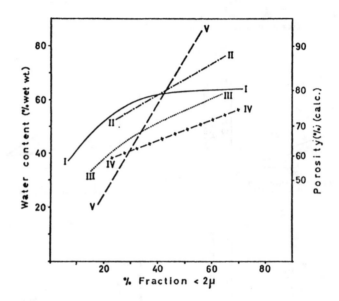

Fig. 6.   Correlation between water content and per-
          centage of clay fraction (< 2 μm) in different
          environments: I = Lake Constance (after G.
          Muller, unpublished data). II = Zuider Zee
          (after Wiggers and Smits, cf. Seibold, 1956).
          III = Mississippi delta (after Shepard and
          Moore, 1955).   IV = Rockport area, Texas (after
          Shepard and Moore, 1955). V = California
          Basin (after Emery and Rittenberg, 1952).
          Environments: I = fresh-water. II - IV =
          brackish. V = marine (after Muller, 1976,
          p. 136).

     High electrolyte contents of depositional medium result in
the coagulation of clays, with coagulation effect of the ions
increasing rapidly with increasing valency (Schulze-Hardy valency
rule).   The concentrations of uni-, di-, and trivalent cations,

which are necessary for flocculation, are in the ratio 500/10/1. The sequence of the flocculation intensity is as follows:

Al > Ca > Mg > K > Na (Muller, 1967, p.137).

EFFECT OF SEDIMENTATION RATE ON DIAGENESIS

Studies on the influence of sedimentation rate on geological processes in contemporary sedimentary basin indicate that degree of diagenetic alterations of sedimentary mineral components and associated organic matter in sedimentary deposits is a direct function of sedimentation rate and its duration (see Nazarkin, 1979).

Sedimentation rate controls the degree of compaction of sediments and consequently determines direction, dynamics, and degree of organic matter alteration during diagenesis, which influence the intensity of hydrocarbon generation. As the sediments compact, dehydration occurs as a result of release of interlayer and adsorbed water. This dehydration intensifies hydrocarbon generation and migration. Associated liberation of heat supplements the petroleum generation potential. Thus, in the case of low sedimentation rate one cannot expect presence of large oil and gas accumulations. Other factors being favourable, petroleum-generating potential of source rocks increases with increasing rate of sedimentation and its duration. Porosity and permeability of reservoir rocks would also be greater. Nazarkin (1979, p.286) proposed the following from categories of petroleum-generating potential of palaeobasins:

(1)   High petroleum-generating potential - average sedimentation rate range from 300 to 900 tons/km^2/year.

(2)   Average petroleum-generating potential - average sedimentation rate ranges from 160 to 300 tons/km^2/year.

(3)   Low petroleum-generating potential - average sedimentation rate ranges from 60 to 160 tons/km^2/year.

The critical rate of sedimentation necessary for generation of sufficient quantities of hydrocarbons for accumulation of economical deposits is 60-100 tons/km^2/year. In the case of majority of source rocks in petroliferous economic basins, sediments were accumulating at rates ranging from 140 to 650 tons/km^2/year. Rate of sedimentation determines many diagenetic processes. In order to determine these relationships, however, they should be thoroughly studied.

SOME PHYSICAL PROPERTIES OF CLAYS

As pointed out by Meade (1966), the water content of
saturated clays is a complex function of the physiochemical influences on the sorption of water on surfaces of clay minerals.
Water content is related to the concentration and types of interstitial electrolytes in clays.   It seems to be directly related
to the exchangeable-cation valence in kaolinite, but inversely
related in montmorillonite.

Particles of clay minerals seem to develop preferred orientation at very early stages of compaction at pressures near
1 kg/cm^{-2}, which is enhanced by greater initial water contents
and greater amounts of organic matter.   Degree of preferred
orientation also depends on the concentration and types of interstitial porewater electrolytes.   According to Meade (1966), a
tendency to be compressed into domains or submicroscopic oriented
aggregates may be typical of fine-grained clays mixed with concentrated Na solutions or solutions of other cations.

Sedimentologists have shown that rapidly-deposited marine
sediments are not normally compacted near the water-sediment
interface, because the overburden stress is largely borne by the
interstitial fluids rather than by the sedimentary particles
(Richards and Hamilton, 1967).   These fluid-saturated sediments
have low compressive and shear strengths and higher porosities.
The elastic properties of various recent marine sediments have
been measured in the laboratory using dynamic tests (Table 2).
Hamilton's (1971, p. 580) study is based upon the application of
the Hookean elasticity equation to results from dynamic tests.

The elastic resistance of a sediment to shearing forces is
called rigidity.   In argillaceous sediments, dynamic rigidity
is related to cohesion between fine clay-size particles.   Cohesion
is the resistance to shear stress and primarily caused by physicochemical forces between particles.   Cohen (1968) demonstrated
an increase in dynamic rigidity in flocculated structures formed
by bentonite and kaolinite with time.   Hamilton (1971, p. 595)
pointed out that slow deposition rates and longer time may lead
to an increase in sediment cohesion owing to compaction and
cementation of the particles.   Humphries and Wahls (1968) showed
that the dynamic modulus of rigidity increases with increasing
effective stress and decreasing void ratios in clays.   Kaolinite
shear-modulus values were on the order of 6-10 times higher than
those of bentonite.   The dynamic rigidity of clays increases
with increasing effective stress.

Poisson's ratio is defined as the ratio of the unit transverse strain to the unit longitudinal strain.   The Poisson's
ratio of a liquid (having zero rigidity) is 0.50.   Judd (1959)

stated that for rough computations, Poisson's ratio for rocks is generally regarded as 0.25. Hamilton (1971, p.559) has reported average values of 0.478 for clayey silts from continental terrace, 0.484 for silty-clay marine turbidites (abyssal plain), and 0.487 for pelagic silt clays. Poisson's ratio varies with porosity and density in the same manner as does dynamic rigidity (Rieke and Chilingarian (1974, p. 167).

TABLE 2. Measured and computed clastic constants for North Pacific sediments. Laboratory measurements at a temperature of 23°C and a pressure of 1 atm[1] (After Hamilton, 1971a, table 1, p.587. Courtesy of Journal of Geophysical Research).

Sediment type	Measured			Computed			
	ϕ (%)	ρ (g/cm³)	v_p (m/sec)	κ (dynes/cm·10)	v	G (dynes/cm·10)	v_s (m/sec)
Continental terrace (shelf and slope)							
Sand							
Coarse	38.6	2.03	1836	6.6859	0.491	0.1289	250
Fine	43.9	1.98	1742	5.6877	0.469	0.3212	382
Very fine	47.4	1.91	1711	5.1182	0.453	0.5035	503
Silty sand	52.8	1.83	1677	4.6812	0.457	0.3926	457
Sandy silt	68.3	1.56	1552	3.4152	0.461	0.2809	379
Sand-silt-clay	67.5	1.58	1578	3.5781	0.463	0.2731	409
Clayey silt	75.0	1.43	1535	3.1720	0.478	0.1427	364
Silty clay	76.0	1.42	1519	3.1476	0.480	0.1323	287
Abyssal plain (turbidite)							
Clayey silt	78.6	1.38	1535	3.0561	0.477	0.1435	312
Silty clay	85.8	1.24	1521	2.7772	0.486	0.0773	240
Clay	85.8	1.26	1505	2.7805	0.491	0.0483	196
Abyssal hill (pelagic)							
Clayey silt	76.4	1.41	1531	3.1213	0.478	0.1498	312
Silty clay	79.4	1.37	1507	3.0316	0.487	0.0795	232
Clay	77.5	1.42	1491	3.0781	0.491	0.0514	195

[1] ϕ= porosity; ρ = bulk saturated density; v_p = compressional-wave velocity; κ = modulus; v= Poisson's ratio; G = rigidity (shear modulus); and v_s = shear-wave velocity.

There is considerable confusion in the literature concerning the term "strength" of a sediment. Ultimate strength of a sedimentary rock is defined as the stress at the time of sudden failure under an applied load. In unconsolidated sediments such as soft clays, the strength of bearing capacity of the material is defined in terms of the amount of load per unit area that causes the sediment to deform, i.e. to become plastic under stress. Hamilton (1971) has noted that the shear strength of silts and clays, in the absence of normal stress, may be defined by the

cohesion alone. For most clay sediments, the shear strength and
dynamic rigidity increase with increasing effective pressure. For
a detailed treatment of the subject, the reader is referred to
Brenner et al. (1981).

Shibakova and Shalimova (1967) investigated the optimal
loading pressure for kaolinite, hydromica, and montmorillonite
clays with an initial moisture content of the samples corresponding to the lower plasticity limit. The strength of each clay
type was determined by its texture and increased with increasing
degree of orientation of individual clay particles. Sedimentation-
compression curves for normally consolidated sediments are presented in Fig. 7. (For details see Rieke and Chilingarian,
1974, p.172).

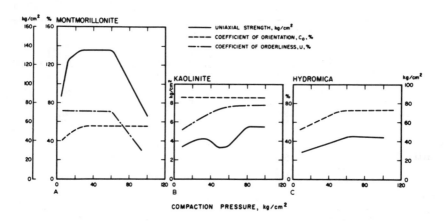

Fig. 7. Variation in strength and textural coefficients
of clays on compaction. A = Montmorillonite
clay from the Askanskiy deposit of Georgia
(95% clay). B = Kaolinite clay of
Novoselitskiy deposit of Ukraine (70% of clay
particles are less than 0.001 mm in size).
C = Cambrian clay (hydromica) of Leningrad
(40% of clay particles less than 0.001 mm and
20-30% are 0.01-0.05 mm in size). (After
Shibakova and Shalimova, 1967, fig. 1. p.91).

MICROSTRUCTURE OF CLAYS UNDERGOING COMPACTION

Overburden pressure appears to cause preferred orientation
of mineral grains. Sections cut in clay samples normal to the
pressure direction commonly show a higher concentration of elong-

ated mineral axes than cut parallel to the pressure direction
(Rieke and Chilingarian, 1974, p.127).   The importance of rearrangement of clay particles is that it changes the permeability,
porosity, and other properties of a clay deposit.

Casagrande (1932), Mitchell (1956), Tan (1957), Lambe (1958),
Rosenqvist (1958), Quigley (1961), Martin (1962), Von Engelhardt
and Gaida (1963), Pusch (1966, 1970), Meade (1968) and others have
investigated the particle arrangement of clays in recent argillaceous sediments and presented various idealized pictures of the
observed clay microstructures.   O'Brien's (1963, p. 20-40)
experiments on illite and kaolinite pastes and by Martin (1965)
on kaolinite slurries indicate that in the case of sufficient
initial water content, complete orientation of the clay-mineral
grains occurs at pressures less than 100 kg cm^{-2}.   Martin noted
that, in a clay slurry a preferred grain orientation could first
be detected at pressures of 1 kg cm^{-2}.   This did not change upon
further compaction up to 32 kg cm^{-2}.   Rosenqvist (1958) further
observed that there was a better particle orientation in fresh-
water clays than in marine clays.

Clay sensitivity may be defined as

$$S_t = \frac{\text{Unconfined compressive strength of undisturbed sample}}{\text{Unconfined compressive strength of disturbed sample}} \quad (11)$$

In general, clays with secondary structural features, such as
cracks, joints, etc. have a sensitivity of less than 1.   Most
soft clay deposits show a sensitivity in the range of 1 to 8.
Sensitive soft clay deposits may be defined as those deposits
having a sensitivity range of 4-8;   extra-sensitive, with a
range of 8-15;   and quick clays, with sensitivities over 15.

Pusch (1966) investigated the microstructure of Swedish
quick clays.   Quick clay is an ultra-sensitive clay deposited
and consolidated in a flocculated state, with subsequent increase
in the double-layer potential either by leaching of electrolytes
or by addition of certain ions (Rosenqvist, 1966, p.445).   Pusch
(1966) observed that the microstructure of the leached marine
quick clay was characterized by a linkage of groups or chains of
small particles in and between denser flocs or aggregates (or
between larger particles, see Fig. 8).   He did not note any
preferential orientation either of the large or small clay
particles in the quick marine clays studied.   The arrangement of
the large clay flocs linked together in chains by small particles
is almost identical to the hypothetical floc structure proposed
by Pusch in 1962 (pp. 53, 55).   Rosenqvist (1958) showed that
virgin marine clay-particles arrangements were similar to those
proposed by Tan (1957), i.e. a card-house structure (see Fig. 9).

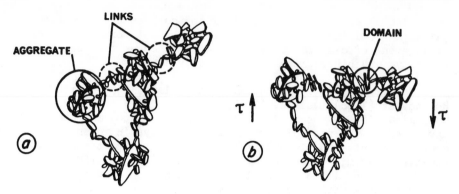

Fig. 8.   Schematic diagram of microstructure in quick
          clay and the failure process.   a = Natural
          microstructural pattern.   b = Breakdown of
          particle links resulting in domain formation.
          (After Pusch, 1970, p. 7, fig. 8.   Courtesy of
          Canadian Geotechnical Journal).

        Pusch (1966) observed that the relative pore areas in marine
clays are considerably larger than those in fresh-water clays with
similar bulk densities and water contents.   This is in close
agreement with the results of laboratory studies at the University
of Southern California by the writer showing that the permeabilities of filter mud cakes, deposited from flocculated systems are
considerably higher than those deposited from a deflocculated
clay-plus-water system (mud).

        Pusch (1970) studied the microstructural changes in a soft
marine quick clay from Molndal, Sweden, undergoing unconfined
compression tests in the laboratory.   Analysis of clay samples
that had failed under compression loading had a microstructural
pattern characterized by a network of small aggregates connected
by links of particles.   Links between particles broken down
successively on increasing shear deformation and formed domain-
like (oriented) groups of particles.   The shear mechanism seems
to involve movement of the aggregates as units in conjunction with
deformation of their links.   Pusch (1970, p.6) concluded that
at high shear stresses, the motion of the aggregates caused large
deformation of the connecting links.   This resulted in a parallel
orientation of the linking particles.   The residual strength
appears to correspond to the state where the links are broken due
to the large deformation.   A reduction in strength (by 30-50%
of the peak shear-strength value) may be caused by dilatancy when
the rigid aggregates are approached.   Remoulding, which causes
successive breakdown of the aggregates, results in further
decrease in shear strength.   The rigidity of the aggregates seems
to govern the shear process.   High viscosity of the pore water

Fig. 9. Schematic diagram of Tan's card-house structure
in clays.   (After Tan, 1957, fig. 3, p. 87;
also Tan, 1966, fig. 1, p. 257. Courtesy of
Springer Verlag).

at low water contents of 15-30% (Rosenqvist, 1959) together with
the strong bonding of the closely located individual particles
in the aggregates, gives the clay mass considerable rigidity
during compaction.   (See Rieke and Chilingarian, 1974, for
detailed discussion).

The problems of particle orientation in clay deposits, is a
complex one.   There are many variables which have to be considered
including type of mineral(s), grain size and shape, initial fluid
content, electrolyte content of the interstitial fluid, the type
of clay-plus-water system from which clay is initially deposited
(deflocculated, flocculated, etc.), and the magnitude of overburden pressure (Rieke and Chilingarian, p.131).

The importance of microstructural studies lie in an understanding of the physical properties, such as compressive strengths,
porosities, and permeabilities of the clay deposits.

FACTORS INFLUENCING COMPACTION OF CLAYS

Meade (1963, 1964) found in some cases an anomalous increase
in pore space with increasing depth when one would expect a decrease in porosity.   Grain-size variation could not explain this

anomaly; therefore, Meade searched for additional factors through
a literature survey and examination of cores. He concluded that
the following variables, in addition to overburden pressure, have
had an influence on pore volume in fine sediments: particle size,
clay minerals and their adsorbed cations, concentration of interstitial electrolyte solutions and their acidity, clay-particle
orientation, and presence or absence of microfossils (i.e. diatoms
in this case). Figure 10 (see also Fig. 5) summarizes some of
the factors of interest here in relationship to compaction and
other diagenetic processes. Figures 10a and 10b demonstrate the
influence of particle size on pore volume: in both unconsolidated
marine sediments and sediments under overburden pressure, the pore
volume increases with decreasing particle size. Figure 10c shows
that in the pressure range of 1-100 kg/100 cm^2, the sequence of
increasing pore volume is kaolinite $\longrightarrow$ illite $\longrightarrow$ montmorillonite
(see also Chilingar and Knight, 1960). This is probably the
result of particle size, because the specific surface measurements show a decreasing order of particle size in the same sequence.

     Figures 10d, e and f are of particular interest to illustrate
a possible relationship between chemical diagenesis and overburden
pressure (i.e. compaction). In sandstones with clay minerals,
either as a matrix or as lenses or beds, the exchangeable cations
adsorbed by clays are also influencing pore volume under low-to-
moderate burial pressure as shown in Fig. 10d. Meade (1963)
mentioned that the smaller the valence and the larger the hydration radius of the adsorbed cation, the greater the pore volume
of the clay sediment. This does not seem to be applicable under
large overburden loads (30-3200 kg cm^{-2}, as demonstrated by
experiments performed by von Engelhardt and Gaida (1963). Figures
10e and 10f illustrate the interrelationships among overburden
pressure, degree of compaction, particle size, and electrolyte
concentration. Further research work, however, is definitely
needed in this area.

MOISTURE CONTENT VERSUS TIME RELATIONSHIP

     The moisture content-versus-pressure curves of clays with
high swelling indices have steeper slopes than those of the clays
which swell to a lesser degree (Fig. 11); however, a longer
time is needed to establish equilibrium for the former clays.
Figures 12 and 13 show the relationship between the moisture content and the time of squeezing for various clays and silicic
acid. As pointed out by Rieke and Chilingarian (1974) most of
the water is squeezed out during a relatively short period of
time (1-4 days) and equilibrium is reached after 7-60 days.
Montmorillonite clay hydrated in distilled water requires more
time to reach equilibrium than montmorillonite clay hydrated in
sea water. Possibly, flocculated montmorillonite clays close-
pack better and faster than do deflocculated ones. Illite, in

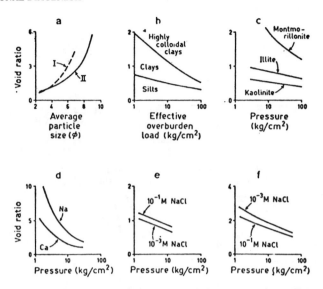

Fig. 10.   Relations of void ratio to other factors, observed in
            natural sediments and in laboratory experiments. Void
            ratio is ordinate in all graphs; note different void-
            ratio scales. (a) Relation to average particle size
            observed in unconsolidated sea-bottom sediments.
            Curve I modified after von Engelhardt (1960, p. 15);
            curve II modified after Shumway (1960, p. 663). (b)
            Generalized relation to effective overburden load and
            particle size in sediments.   Modified after Skempton
            (1953, p.55).   (c) Experimentally determined
            relation to pressure and clay-mineral species.
            Modified after Chilingar and Knights (1960, p.104), to
            show their results to 100 kg/cm^2.   (d) Experimentally
            determined relation to pressure and adsorbed cations
            in < 0.2 μm fraction of montmorillonite.   Modified
            after Bolt (1956, p.91).   (e) Experimentally deter-
            mined relation to pressure and electrolyte concentra-
            tion in unfractionated Fithian illite (about 60% by
            weight coarser than 2 μm).   Modified after Mitchell
            (1960, fig. M3).   (f) Experimentally determined
            relation to pressure and electrolyte concentration
            in < 0.2 μ m fraction of Fithian illite.   Modified
            after Bolt (1956, p.92).   (After Meade, 1963, fig. 2,
            p.237, courtesy Sedimentology).

turn, requires a longer interval than kaolinite for attainment of
equilibrium at a given pressure.   (See Rieke and Chilingarian,
1974, and Chilingar and Knight, 1960).

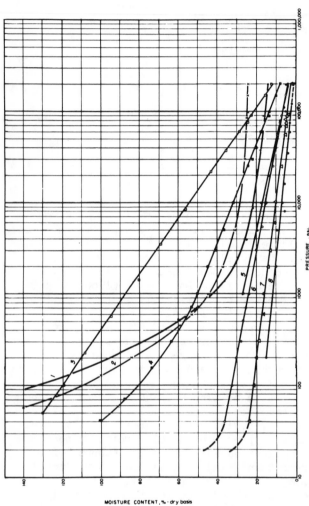

MOISTURE CONTENT, % - dry basis

PRESSURE, psi.

Fig. 11. Relationship between moisture content M (% of dry weight) and pressure p (in psi) for various clays, gum ghatti, and gum tragacanth. 1 = gum ghatti (natural organic colloid); 2 = gum tragacanth (natural organic colloid); 3 = silicic acid. M = 186 - 33 long; 4 - montmorillonite no.25. Upton, Wyoming, John C. Lane Tract bentonite) - M = 104 - 17.06 log p; (straight-line portion of the curve); 5 - montmorillonite no.25. M = 58 - 10.2 log p (hydrated in sea water); 6 = illite no.35, Fithian, Illinois. M - 50 - 8.7 log p; 7 = kaolinite no.4, Macon, Georgia, Oneal Pit. M = 33.9 - 5.96 log p; and 8 = dickite no.15, San Juanito, Chihuahua, Mexico. M = 26.7 - 5.04 log p (after Chilingarian and Rieke, fig.1, p.812).

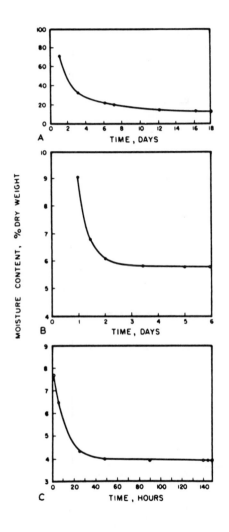

Fig. 12. Moisture content versus time. A. Variation in moisture content (% dry weight of montmorillonite clay (API no.25) with time (in days) at 88,500 psi. B. Variation in moisture content (% dry weight of illite clay (API no.35) with time (days) at 91,000 psi. C. Variation in moisture content (% dry weight) with time (in hours) for kaolinite clay (API no.4) at 88,500 psi. (After Chilingar and Knight, 1960a, figs. 4-6. Courtesy of American Association of Petroleum Geologists).

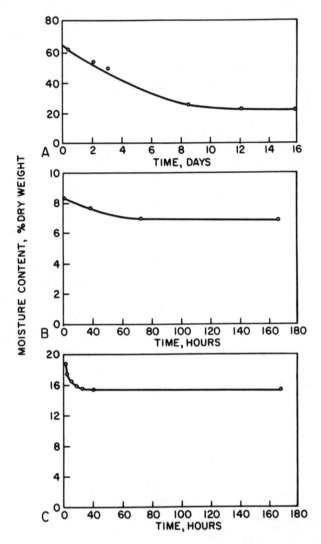

Fig. 13. Moisture content versus time. A = Variation in moist-
          ure content (% dry weight) of silicic acid with time
          (in days) at 88,500 psi. (After Chilingar and Knight,
          1960b). B = Variation in moisture content (% dry
          weight) of montmorillonite (API no.25) hydrated in
          sea water with time (in hours) at 15,000 psi. (After
          Rieke, 1970, p.173). C = Variation in moisture content
          (% dry weight) of soil from weathered limestone terrain
          (Louisville, Kentucky) with time (in hours) at 10,000
          psi.   (After Rieke, 1970, p.173).

   Shishkina (1968) reported that on compacting diatomaceous
marine mud having 70% moisture content at 405 kg cm^2 (5,760 psi),
85% of all interstitial water is squeezed out after 4 hours. After
10 hours, 90.0% of water was squeezed out at this pressure. Subsequent squeezing did not achieve any additional removal of fluids.
On squeezing marine mud having initial moisture content of 49.5%,
after 4 hours 87% of all interstitial fluids was squeezed out at
1,355 kg cm^2.   On increasing the pressure to 3,100 kg cm^2, 93.5%
of water was squeezed out after 7 hours and 96.4% after 10 hours.
Subsequent squeezing for 6 hours did not produce any additional
water.

   At pressures of 400 kg cm^2, depending on type of sediment
and initial moisture content, up to 85% of all water is squeezed
out after 4-6 hours.   At 202 kg cm^2 pressure, 80-85% of all water
is squeezed out after 10 hours; remaining moisture content was
around 16-23% (Shishkina, 1968).

   E.C. Robertson (personal communication, 1967 and 1968) in
analyzing the rate of compaction of aragonite found that the
practical minimum time required to approach maximum compaction
was 3 hours (Fig. 14), as shown by the flattening of the curves.
Compaction experiments up to 9 years in duration lead the writer
to believe that laboratory experiments definitely reproduce
natural conditions.

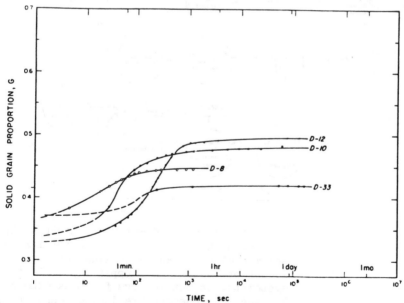

Fig. 14.   Relationship between solid grain proportion
           and duration of compaction for aragonite sedi-
           ments.(After Robertson, 1967, fig.2, p.125).

GENERAL FACTORS CONTROLLING COMPACTION OF SANDSTONES
(after Wolf and Chilingarian, 1976).

Any discussion on the origin of sedimentary rocks should include references to compaction, lithification, and other diagenetic processes as shown in the diagrams of Figs. 15 and 16.
These diagrams demonstrate that compaction is one process that
determines the final property of a rock.   Examination of Fig. 15,
for example, shows that any primary sedimentological factor that
controls the original characteristics of the sedimentary deposit
will also determine the compaction history of the accumulation.
A possible sequence of interdependent, large-scale geologic
parameters, which influence compaction directly or indirectly,
has been given in Fig. 17.   Purely theoretical considerations or,
as in the sequence in Fig. 17, mere "common sense" may be sufficient to think of other similar mega-relationships.

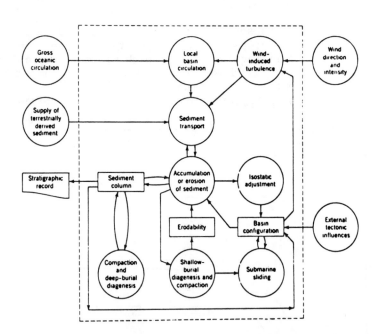

Fig. 15.   Theoretical system diagram showing principal
           processes that affect shallow-water marine
           sedimentation.   Dashed outline separate exo-
           genous processes that supply imputs to the
           system but do not receive feedback from it.
           Endogenous processes are inside dashed outline.
           (After Harbaugh and Merriam, 1968; in
           Harbaugh and Bonham-Carter, 1970, fig. 7-1,
           p.265; courtesy Wiley-Interscience, New York).

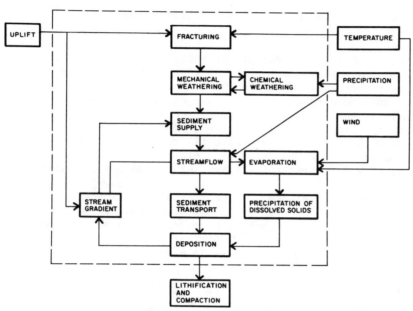

Fig. 16. Diagram of fluvial system.   Uplift and
          climatic factors of temperature, precipitation,
          and wind are outside the system, feeding into
          the system, but not receiving feedback from
          the system.   Lithification and compaction
          are also outside the system, but receive out-
          put from the system instead of feeding into
          the system.   (After Harbaugh and Bonham-
          Carter, 1970, Fig. 1-14, p.22; courtesy
          Wiley-Interscience, New York).

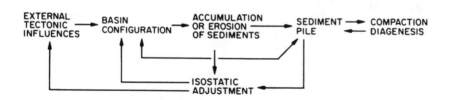

Fig. 17. Flow diagram showing the relationships between
          various large-scale geologic variables during
          the development of a basin and sedimentation
          that influence compactional diagenesis.   These
          in turn, have a reciprocal control on the
          large-scale variables.   (After Wolf and
          Chilingarian, 1976, fig. 3-3, p.70).

In the future, however, more subtle and quantitative inter-
dependencies must be determined to make progress in solving complex problems in sedimentology, as has been attempted by Wolf
(1973a, b), among others.   To do this, it seems best to start
not with the regional but with the microscopic and mesoscopic
processes and factors involved in compaction.   These concepts
should then be applied on a regional scale.   Review of the
literature shows that a large amount of data on compaction are
already at hand, especially on the more local scale.   The information on the vertical and horizontal variations of the effects
of compactional diagenesis, on the other hand, is somewhat meagre.
Several examples, however, are available from the literature.

In compaction studies of sandstones many factors are to be
considered (Table 3).   It is also obvious that certain factors
of importance in the compaction of pyroclastics either need not
be considered in epiclastic (= terrigenous) sediments or are of
relatively minor significance (Table 4).

Allen (1969, 1970) has pointed out in several papers that
theory and experiments based on the sphere as the ideal particle,
fail to account satisfactorily for the concentration of solids
observed in unconsolidated natural sands.   Allen (1969, p.309)
stated that "when the prolate spheroid of moderate axial ratio
is substituted for the sphere, a theory which is satisfied by
observation becomes possible, for the reason that equal spheroids
can be regularly packed in both close and open ways".   Allen
pointed out that the freshly deposited natural sands have a
relatively low particle concentration, even through theory based
on the sphere as the ideal sedimentary particle predicts for them
a relatively high concentration of solid.   Natural sands are
chiefly deposited either as laminae on horizontal or nearly horizontal surfaces or as steeply-inclined laminae on slip faces or
ripples and/or dunes (see results of numerous experiments by
Jopling, 1963, 1965a,b, 1966, 1967, for example).   The concentration of grains, $\underline{C}$ which is equal to (space occupied by solids/
(total space), ranges between 0.50 and 0.65 in the above two
cases.   The variable $\underline{C}$ depends on the average grain size, degree
of sorting, medium of deposition and intensity of deposition;
porosity is equal to (1-$\underline{C}$).   According to Allen (1969, pp. 309-
310), the range of possible particle concentration $\underline{C}$, obtained
experimentally by using haphazardly packed equal solid spheres
is roughly between 0.60 and 0.64.   Theory based on the sphere
as the ideal sedimentary particle, as mentioned in most text and
reference books, predicts $\underline{C}$ to increase with increasing range of
sizes of different sphere populations present.   Consequently, $\underline{C}$
of natural sands should be distinctly larger than the possible
range of $\underline{C}$ for equal spheres, because natural sands are composed
of grains having an appreciable range of sizes.   Observations
do not agree with this, because the lower limit of $\underline{C}$ for natural

Table 3. Factors and processes controlling compaction of sandstones (after Wolf and Chilingarian, 1976, table 3-3, p.71).

Group 1
(Inherited factors[1]: individual properties of grains and fluids)

grain size
grain shape
grain orientation
grain surface features
grain rounding
grain sphericity
grain electrostatic properties
grain composition
impurities on grain's surface
fluid composition

Group II
(Inherited factors[1]: mass properties)

grain/matrix/cement ratios
shale/siltstone/sandstone ratios in stratigraphic section
absorbability
adsorption capacity
degree of cohesion
size sorting
compositional sorting
thixotropic properties
porosity and permeability
grain stacking patterns
packing
packing heterogeneity
grains' frictional properties
grains' total surface area
shear strength
surface area configuration of deposit
topography (e.g. slope) of unit
sedimentary structures

Group III
(Dynamic factors)

rate of fluid movement
replenishing of fluid
removal of fluid
fluid pressure
overburden pressure
subsurface temperature
rate of sedimentation
(- rate of loading)
earthquakes
tectonics
time

Group IV
(Inhibitory factors[1]: reducing compaction)

diagenetic changes
   neomorphism
   cementation
   recrystallization
epigenesis (catagenesis)
metamorphism

[1]This term was adopted from Coogan and Manus (1975).

Table 4. Factors and processes significant in controlling compaction of pyroclastics[1] (after Wolf and Chilingarian, 1976, table 3-2, p.72).

Factors	Remarks
(1) Heat retained in the deposit	may cause welding of the particles, thus increasing the strengths of the deposit
(2) Rate of settling and accumulation of the pyroclastic debris	controls the rate of increase in overburden
(3) Distance from source and thus differential sorting of pyroclastic debris	controls thickness of deposit and, therefore, overburden pressure
(4) Degree of reworking in the depositional environment	controls textures, fabrics, and structures of deposit
(5) Amount of rainfall during volcanism (water precipitation is often reported during volcanicity)	controls initial compaction
(6) Amount of ground water	influences diagenesis, e.g., controls cementation by zeolitization, which increases strength of deposit
(7) Composition (mineralogic and bulk chemical) (a) acidic versus intermediate versus basic volcanic particles (b) ratio of phenocrysts/fine debris (c) bubbles versus groundmass in shards	controls degree of reactibility (or chemical stability) during diagenesis, as well as physical strength prior to and subsequent to compaction
(8) Shape, internal textures, and fabrics of vitric shards	determine degree of resistance to mechanical compaction

[1]Most or all of the factors in Table 3 are applicable here also, but have not been duplicated. Only those of specific importance in pyroclastics have been listed here.

sands is about 0.5, which is below the lower possible limit of C
for haphazardly packed equal spheres by 0.1.   The upper limit of
C for natural sands deviates little from the upper limit for
haphazardly accumulated spheres.   Allen attributed this to the
use of the sphere as an ideal sedimentary particle.   As shown by
numerous investigations, the natural sand particle is approximated
by a triaxial ellipsoid with a long axis about 1.5 times longer
than the intermediate axis and about 2.0 times longer than the
short axis.   Allen (1969, p.310) continued to explain that equal
ellipsoids can be regularly packed in more different ways than
equal solid spheres, because these particles can be differently
oriented in space, which need not be the same for all the ellipsoids (Figs. 18, 19, 20).   Some of their regular packing is extremely open, and it seems that the comparatively low C values of
natural sands can be explained by the presence of only a small
proportion of the grains in an open-packing pattern.   When
spheres are used, one can take into account only sorting or size
variation and not orientation together with size, as spheres have
no dimensional orientation.   The closer regular packing of equal
ellipsoids is highly anisotropic, whereas the open packing is
perfectly isotropic.   Inasmuch as natural sands possess a fabric
which is neither perfectly anisotropic nor perfectly isotropic,
a combination of both close- and open-grain packings may be expected to be present on a microscopic or a local scale.   For
these reasons, Allen (1969) explored the different regular packings of ellipsoids using not the triaxial but the prolate spheroids
(= ellipsoid of revolution about the long axis).

Figures 18 and 19 illustrate the nine different packing
systems.   Figure 20 shows that (1) C for example, decreases
increasingly rapidly as the axial ratio falls below unity (see
V);   (2) C declines increasingly rapidly (more so than in case
V) as the axial ratio decreases below unit (see VI);   (3) C
decreases rapidly with the reduction of the axial ratio below
unit (case VII);   (4) C at first increases with falling axial
ratio, but thereafter decreases rapidly (case VIII); and (5) C
at first increases slightly before starting to decrease rapidly
as the axial ratio is reduced below unity (case IX).   Aside
from the regular, well-defined packings presented here, it is
not clear how they are to be used to form a model if the con-
concentration of real sedimentary particles arranged in a partially disordered manner.   On the basis of theoretical considerations provided by Allen (1969) it seems that on combining the
anisotropic rhombohedral arrangement of packing IV with the two-
dimensional isotropic packing VI a field that partly represents
natural sands is obtained, as shown in Fig. 21.   As pointed out
by Allen 1969, pp.317-318), only a comparatively small proportion
of spheroids in an open, isotropic packing is necessary to give
concentrations in the observed ranged of natural sands.
Theoretically, the particles of natural sand should show a

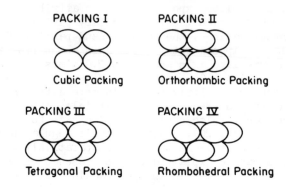

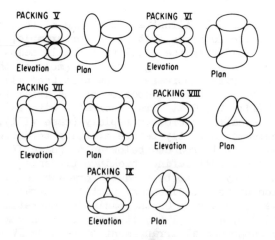

Fig. 18.   The four close packings of equal prolate
            spheroids when the major axes lie in the
            planes of the layers. (After Allen, 1969,
            fig. 1, p.312; courtesy of Geol. Mag.).

Fig. 19.   The five open packings of equal prolate
            spheroids. (After Allen, 1969, fig. 2,
            p. 313; courtesy of Geol. Mag).

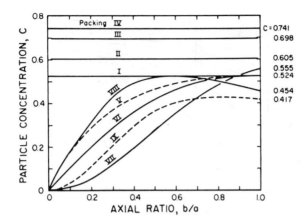

Fig. 20. Concentration as a function of axial ratio
for the spheroid packings of Figs. 3-59 and 3.60.
(After Allen, 1969, fig. 3, p.314; courtesy
of Geol.Mag.).

substantial degree of dimensional ordering, though not a perfect
one.   The degree of ordering should increase with increasing
grain concentration, $\underline{C}$, for a constant axial ratio.

     After the theoretical studies, an attempt was made to compare
the results with natural sands, e.g. the grain relationships were
investigated.   In order to exclude the possibility of secondary
changes of the original depositional fabric by compaction, for
example, the samples collected had relatively high contents of
secondarily introduced cement and lacked: (1) empty pore spaces;
(2) evidence of recementation and corrosion of the detrital
grains; and (3) evidence of physical distortion or breakage of
particles.   It was found that the natural sandstones had local
patches of grains packed according to the V through IX patterns
(Fig. 21) which although uncommon were not rare.   In the future,
maybe statistical analyses can be performed on the occurrence and
abundance of these various packing patterns.

     The results shown in Figure 22 are as follows: a, b = long
axes virtually at right angles, packing V;   c-e = three elongated
grains in a manner as in packings VI and VII; f-i = trios of
grains, similar to packings VIII and IX; j, k = two sets of four
grains each arranged roughly along the edges of a square or
rectangle, similar to packing VI; and l-p = clusters of a large
number of grains arranged in a manner similar to packing VIII.
From these results, Allen (1969, p.320) concluded that the
"concentration of a partially disordered assemblage of spheroids

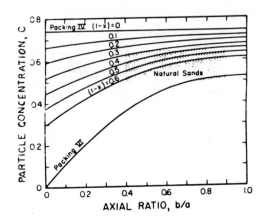

Fig. 21.   Concentration of a spheroid assemblage of
           mixed packings (IV and VI) as a function of
           axial ratio and proportion of packing VI
           given as (1-x).   (After Allen, 1969, fig. 4,
           p.317; courtesy of Geol.Mag.).

can be represented by a model in which one open and one close
regular packing are combined in a proportion to yield observed
concentration in natural sands and sandstones.   In terms of such
a model, the concentration of a natural sand could be achieved
if a comparatively small percentage of an open regular packing
were combined with a comparatively large percentage of a close
regular packing.   Moreover, the percentage would be such that
the sand would still display a substantial degree of dimensional
ordering of the particles.

THE EFFECTS OF COMPACTION AND DEPTH OF BURIAL
(after Wolf and Chilingarian, 1976).

     Fuchtbauer (1967a) presented the results of an investigation
in which he directly considered effects of compaction and depth
of burial.   In the section on quartz diagenesis and mechanical
compaction, he showed (Fig.23A) that the porosity of the "Dogger
beta" quartz sandstone decreases with increasing depth.   The
curve nearly coincides with the dashed curve presented by
Proshlyakov (1960, in Maxwell, 1964).   Examination of the grain
surfaces revealed the cause for this porosity decrease, i.e. the
percentage of quartz grains, which show secondary overgrowths on
crystal faces, increases with depth (Fig.23B).   As the silica is
dissolved as a result of pressure solution at points of contact,

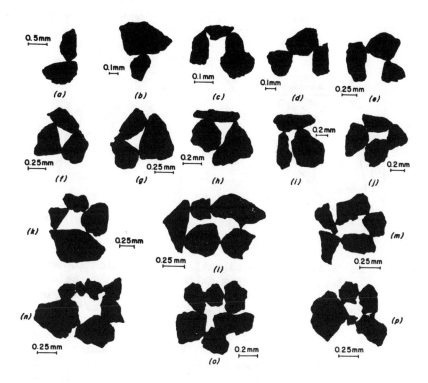

Fig. 22.   Cluster of real grains simulating the packings
           of Fig. 19, as observed in sandstones of Old
           Red Sandstone age. (After Allen, 1969, fig.
           5, p.319; courtesy of Geol.Mag.).

the grains slip into denser packing (Fig. 24A).   The latter
figure also shows that the amount of pressure solution required
is small.   Fuchtbauer (p.355) pointed out that the rock volume
to be dissolved is in general smaller with increasing steepness
of the contact faces of the grains.   In Figure 24B, changes in
volumes are based on the most unfavourable case of horizontal
contact planes.   Only 1.5% of the material must be dissolved
to result in compaction (up to 50%), as graphically shown by curve
A in Figure 24.   The amount of quartz which has to be dissolved
evidently increases with compaction, because contacts become
longer and the number of horizontal contacts increases.   Consequently, as pointed out by Fuchtbauer (p.355) the "mechanical
slipping (rearrangement) of grains dominates during early compaction, whereas chemical compaction (pressure solution is predominant during later stages".   In determining the dissolution
of quartz with depth, as shown in Figure 23B, Fuchtbauer employed

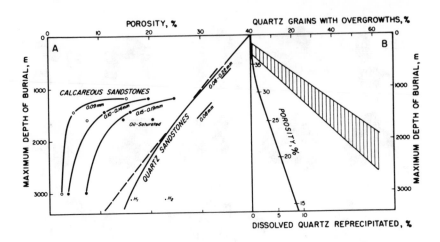

DISSOLVED QUARTZ REPRECIPITATED, %

Fig. 23.   Porosity and SiO_2 migrations in relation to depth of
burial.   The maximum depth of burial is the greatest
depth to which a sandstone was ever buried.   (After
Fuchtbauer, 1967a, fig. 1, p.354; courtesy 7th World
Pet.Congr.).
A = The values for calcareous sandstones were taken from
many measurements of the "Bausteinschichten" (= Chattian
Molasse; Fuchtbauer, 1964, fig.22).   The curve for the
quartz sandstones is for the "Dogger beta" (after
Cuchtbauer and Reineck, 1963). H_1 = average value for
silicified sandstones; H_2 = sandstones with diagenetic
chamosite seams that hindered quartz diagenesis (after
Horn, 1965).   (Dashed curve is after Proshlyakov, 1960,
in Maxwell, 1964).
B =   Lower left = maximum amount of quartz dissolved
by pressure solution (see text).   Upper right = quartz
diagenesis, i.e., percentage of quartz grains with
secondary overgrowth (after Philipp et al., 1963).
Fuchtbauer (1961) defined it as: $1000(0.5 b + c)/(a + b + c)$, where a = number of grains with very few
crystal faces, b = grains with some crystal faces, c =
grains with crystal faces covering the larger part of
the surface.   The counting was done with microscope
using grains of 0.12 to 0.15 mm fraction on a dry glass
slide.   These values, however, do not depend on the
grain size.

curves A and B in Figure 25 and the porosity curve of quartz
sandstone in Figure 23A.   The general shape of the curves is

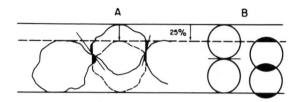

Fig. 24.   Mechanical and chemical compaction.   A.
           Inclined contact planes - mechanical compact-
           ion is predominant.   The rock volume to be
           dissolved (black) is even smaller if rotational
           movement occurs.   B.   Horizontal contact plane
           - exclusively chemical compaction.   Much more
           material must be dissolved in this case in
           order to achieve the same amount of compaction.
           Its amount is independent of grain size as
           deduced from geometric consideration.   (After
           Fuchtbauer, 1967a, fig. 2, p. 355; courtesy
           7th World Pet. Congr.).

probably correct, but the maximum amounts of dissolved quartz
may be unrealistic.   The dissolution of silica is minor in the
upper 1000 m of sediments, but increases roughly linearly with
depth.   This, in turn, correlates with the linear increase of
silica overgrowth below a depth of several hundred metres (Fig.
23B).   The upper 1000-1500 m of sediments are, therefore, the
domain of mechanical compaction.   It seems that both sorting
and roundness of grains influenced compaction, roundness being
particularly important.   The finer-grained sands are more porous
than the coarser-grained ones having a comparable clay content,
which does not agree with the relationship postulated by Weyl
(1959) for the effective range of pressure solution.   The
appreciable difference in porosity is indicated by the short
line marked 0.08 mm in Figure 23A, for some "Dogger beta" rocks.
Fine-grained "Bentheimer Sandstein" has a porosity of 27% at a
depth of 1100 m, whereas the coarse-grained sandstones have a
porosity of 22% (see von Engelhardt, 1960, figs. 11 and 49).
The differences in sorting may be significant, but do not seem to
explain completely the higher porosity of the finer-grained sandstones.   As von Engelhardt (p.21) stated, the higher porosity
values of finer sandstones can be explained by the larger number
of grain contacts/unit volume of sediment resulting in a higher
resistance against compaction.   As this has not been found to
apply to all petrographically similar fine-grained sediments,
other factors must also be active.   A more important factor may

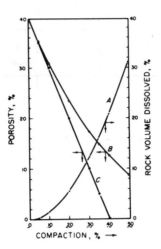

Fig. 25.   Theoretical relationship among degree of
           compaction, porosity (curves B and C), and
           the amount of rock volume dissolved (curve A)
           for the model presented in Fig. 23B. Curve
           A = dissolved quartz (%), ordinate to the
           right; curve B = porosity (quartz removed),
           ordinate to the left; curve C = porosity
           (quartz reprecipitated), ordinate to the left.
           (After Fuchtbauer, 1967a, fig. 3, p.356;
           courtesy 7th World Pet. Congr.).

be the roundness, as it is commonly better developed in coarser
sandstones (excluding "textural inversion of Folk, 1968), and,
therefore, the coarser grains "slip" better.   From Figure 26,
two general relationships become clear: (a) roundness increases
with grain size, as has been reported from numerous other studies;
however, it reaches a maximum in the coarse sand range (curve A);
(b) identical grain fractions are better rounded in the fine-
grained sandstones than in the coarse-grained sediments.   This
can be explained by the differences in transportation mechanisms,
namely, rolling versus suspension.   The median rounding (= rounding of the median-sized grains, shown by circles in Fig. 26) can,
therefore, be larger in coarser sandstones (curves A and C) or
equal to that of fine sandstones (curve B).   It follows then,
that should rounding affect porosity, the curves A and C should
show a relationship between grain size and porosity.   But curve
B (centre of Fig. 26) does not show such an interdependency, as
demonstrated in Table 5.   According to Fuchtbauer (p. 357),
there is an important influence of the degree of rounding on

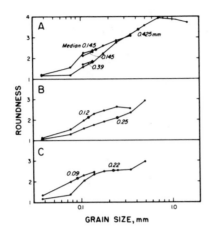

GRAIN SIZE, mm

Fig. 26.   Rounding (Russell-Taylor) in relation to
           grain size.   Ordinate (roundness): 1 =
           angular; 2 = subangular; 3 = subrounded, 4 =
           rounded.   This number was multiplied by the
           number of grains falling into the respective
           roundness class.   Then the sum of these
           products was divided by the number of all
           grains.   Each point corresponds to one grain-
           size fraction.   The median diameter is shown
           by an open circle for each sample.   A =
           "Bentheimer" sandstone from Scheerhorn (Hecht
           et al., 1962), oil-saturated, 4 samples;   B =
           "Dogger beta" sandstone from Hankenbuttel,
           oil-saturated, 2 samples;   C = "Dogger Beta"
           sandstone from Luben-West, water-filled, 2
           samples.   (After Fuchtbauer, 1967a, fig. 4,
           p.356; courtesy of 7th World Pet. Congr.).

compaction during the mechanical stage of diagenesis resulting
in a reduction of porosity.   Inasmuch as the specimens contain
on the average only 11% quartz grains with overgrowths, the amount
of pressure solution must have been comparatively small.

   The second stage of compaction is the chemical one,
characterized by increased pressure solution combined with precipitation of silica on quartz grains.   This can lead to complete silicification of the sandstone as long as no other cementing mineral occurs.   Figure 27 which shows the source of silica,
consists of three parts: (a) top: a section through a water-
filled rocks; (b) centre: a section through an oil-filled rock;
and (c) bottom: a section through a formerly oil-filled, but now

Table 5.   Relationships among grain size, rounding and porosity
           (%) (after Fuchtbauer, 1967a, Table 1, p.356).

Profile A (Fig. 26)

	median	rounding	porosity
Coarse-grained	0.42	3.35	22
Fine-grained	0.12	2.4	27

Profile B (Fig. 26)

	median	rounding	porosity
Coarse-grained	0.23	2.08	27
Fine-grained	0.12	2.10	26

Profile C (Fig. 26)

	median	rounding	porosity
Coarse-grained	0.22	2.53	26
Fine-grained	0.08	2.17	30

water-filled sandstone.   Each one of these cases is considered
separately.

(a) The amount of secondary quartz overgrowths is high in
the water-filled section, except for portions with carbonate
cements.   The latter preserved an early stage of diagenesis,
which is evidence by the presence of a number of unstable heavy
minerals (e.g. garnet and staurolite).

(b) Less quartz precipitation occurred in oil-impregnated
sandstones.   The minor amounts of quartz present apparently have
been preserved by the oil.   Using the upper right-hand side
(hatched area) of Figure 23, one can deduce that the depth at
the time of oil migration and impregnation was about 1000 m.
The importance of this quartz diagenesis to the structural history of one sedimentary trough and the reconstruction of ths history of oil migration have been discussed by Philipp et al. (1963a,
b).   The sections that are close to the sand-shale interface,
have much higher amounts of quartz overgrowths (Fig. 27).   This
increase in quartz content on approaching the sand-shale boundary
is present in all sections of oil-filled sandstones studied by
Fuchtbauer and can be explained only by assuming continuation of
quartz precipitation after oil impregnation, possibly at a reduced rate.   It seems that silica migrated into the connate
water of the oil-saturated sandstones from adjacent shales and
siltstones.   Inasmuch as the existence of large volumes of
connate water is improbable in this case, diffusion must have
played a role.   The latter process always should be considered

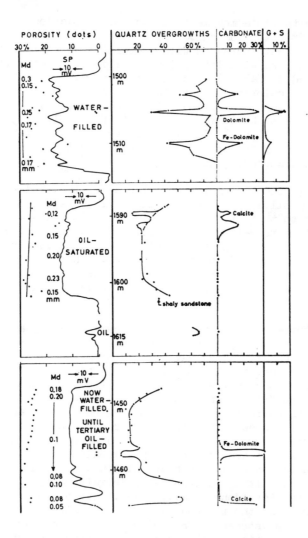

Fig. 27. Relationship between the degree of quartz
diagenesis (center), pore filling (left),
carbonate content (right), and distance from
the sandstone-shale contact for three sandstones of the "Dogger beta" Formation. (After
Fuchtbauer, 1967a, fig.5, p.357; courtesy of
7th World Pet. Congr.). Top = Wesendorf-South
2 (after Fuchtbauer, 1961), center = Hankensbuttel; bottom = Bodenteich; left = median
diameter in mm, porosity, and SP curve; right
= G + S = garnet + staurolite in the heavy
mineral fraction (after Drong, 1965).

together with fluid movements caused by compaction.   That silica
migration from the shales actually occurred is demonstrable.
Chlorite content increases with increasing depth at the expense
of kaolinite and with liberation of silica.   A mineral balance
is presented in Table 6 (Fuchtbauer, 1967a).   Transformation of
kaolinite to chlorite upon burial from 1000 to 3000 m gives rise
to 5.5% increase in SiO_2 content, which can be partly explained
by the change of kaolinite to chlorite (4.3 - 1.1 = 3.2%) (see
Table 6).   Additional silica might have been available from
kaolinization of feldspar and from pressure solution of quartz
at clay mineral contacts.

(c) In the bottom part of Figure 27, the sandstone is not
filled with oil at the present time.   The quartz diagenesis is
so similar to that of the oil-filled rock (b), however, that a
former oil impregnation is assumed.   Fuchtbauer (p.358) stated
that one may find surprisingly high porosities in very deeply
buried coarse-grained quartz sandstones.   Here, pressure solution
is more pronounced in the fine-grained than in the fine-grained
than in the coarse-grained sandstones, which confirms Weyl's
(1959) postulation.

Fuchtbauer found that carbonate cementation (calcite or
dolomitic sandstones) is not uncommon.   The "minus-cement
porosities" (Heald, 1956) have been determined by adding the volume of the cement to the pore volume.   This gives the total
porosity if not cement were present at the particular stage of
burial.   The "minus-cement porosities" generally range from 30
to 35% which correlates with a depth of 600-1100 m according to
Figure 23A.   Those carbonate-cemented sandstones with 40% minus-
cement porosity must have been cemented soon after deposition and
prior to distinct compaction.   This is also confirmed by the
good preservation of labile heavy minerals in such layers (Drong,
1965), as a result of an absence of intraformational corrosion
and leaching (Pettijohn et al., 1972).   Such cemented layers are
most common near the top and bottom of sandstone bodies adjacent
to shales or mudstones.   The anions are retained in the sandstones by filtration at the sand-shale contact (Fothergill, 1955).
The degree of carbonate cementation in the centre of sandstones
are generally related to grain size.   The carbonate-cemented
sandstones are most common in the fine-grained layers that are
overlain by, or intercalated with, coarse-grained sandstones
(Fuchtbauer, 1967a, p.358).

Other authigenic minerals were also mentioned by Fuchtbauer.
Authigenic kaolinite is often confined to the purest sandstone
(e.g. left part of Fig. 28) where its growth was not hindered or
masked by detrital clay minerals, whereas early diagenetic dolomite or calcite cementation may have hindered kaolinite precipitation (right part of Fig. 28).   The pore fluids may also

Table 6. Silica balance during the late diagenesis of Jurassic shales (after Füchtbauer, 1967a, table II, p.347)

Depth (m)	Quartz (%)	Feldspar (%) (estimated)	Kaolinite (%) (ca.45% SiO_2)	Chlorite (%) (ca.28% SiO_2)
1000	14.5	5	63	17.5
3000	20.0	5	53.6	21.4
SiO_2 content % increase	+5.5	0	-4.3 {=(63-53.6) x 0.45}	+1.1 {=(21.4-17.5 x 0.28}

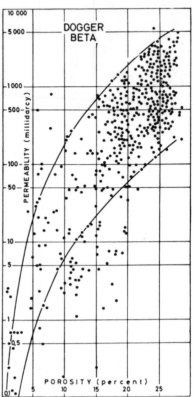

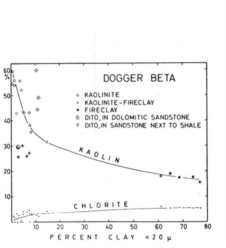

Fig. 28. Relationship between the kaolinite and chlorite content (abscissa, in percent of the clay minerals in the fraction 0.02 mm) and the total clay content (ordinate) in sandstones (left portion of graph) and shales (right) of the "Dogger beta" Formation from Hankensbuttel and Luben-West, Germany. The remainder of clay is illite. (After Fuchtbauer, 1967a, fig.6, p.358; courtesy of 7th World Pet. Congr).

Fig. 29. Relationship between the porosity and permeability of the "Dogger beta" Formation from Gross-Hamburg, Meckelfeld, Vorhop, and Wittingen-South, Germany. These measurements are taken from W. Tunn as are those of Figs.3-229, 3-230A, and 3-230B. (After Fuchtbauer, 1967a, fig.7, p.359; courtesy 7th World Pet. Congr.).

have a controlling influence, e.g. oil impregnation may hinder kaolinite crystallization.   As shown in Figure 28, the kaolinite is poorly developed in the shales, is of primary detrital origin, and can be easily distinguished from the well-formed authigenic kaolinite of the cleaner sandstones.   Similarly the authigenic quartz, the kaolinite in these sandstones is concentrated close to the sand-shale boundaries, whereas diagenetic chlorite appears to be more common in the shales associated with greywackes.

Von Englehardt (1960, p.83) has offered and discussed a formula in his book on the relationship between porosity and permeability for sands with a restricted size range.   Fuchtbauer (p.359) stated that the shape of the "point cluster" in Figure 29 can be explained as follows.   The vertical variation results from grain-size differences, and according to Fuchtbauer (p.359) the "fact that decrease in porosity and permeability does follow the drawn theoretical curves indicates that the specific surface is not higher in the denser samples than in the more porous ones. This is possible only because the clay content of the sandstones of the "Dogger beta" is generally low.   Porosity decreases only through denser packing because of greater depth of burial as well as by carbonate, silicate, anhydrite or pyrite cementation.   Both processes, however, do not appreciably change the specific surface". The points above the curve are indicative of samples cemented by pyrite, whereas those below the lower curve represent argillaceous samples.   The data presented in Figure 30 are different in that the samples are from a narrow depth range and uncemented.   Porosities, therefore, deviate slightly to low values at the same permeability.   There is a decrease in porosity with increasing permeability, resulting from increasing grain size and roundness, which enhances mechanical compaction:

$$\underline{k} = \phi^{3} / \{5(1 - \phi)^{2} \underline{S}_{0}^{2}\} \tag{12}$$

where $\underline{k}$ is the permeability; ϕ is the porosity; and $\underline{S}_{0}$ is the specific surface area which depends on grain-size distribution and shape of the grains.   In Fig. 29 the two solid line curves, which envelope most of the points, were calculated, and plotted for two constant $\underline{S}_{0}$ values.   Accurate formulae relating porosity, permeability, and specific surface area have been presented by Langnes et al. (1972).

According to Fuchtbauer, "the clay content (less than 20 μm in size is responsible for the position of the fine-grained samples (median size is less than 0.12 mm) in Figure 30.   It increases with decreasing porosity from 5 to 30%, thus increasing the internal surface area and, therefore, decreasing the permeability. The open circles in the lower part of the diagram represent relatively coarse-grained sandstone with clay seams from the uppermost part of the sandstone body".   The dashed line

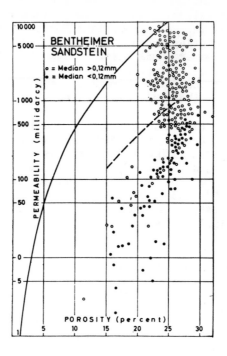

Fig. 30. Relationship between the porosity and
          permeability of the "Bentheimer" Sandstone,
          Scheerhorn, Germany. (After Fuchtbauer,
          1967a, fig.8, p.360; courtesy 7th World Pet.
          Congr.).

corresponds to the median grain size of 0.12 mm and the highest
median found (0.42 mm) is used for calculating the data for the
upper curve. The break between samples with median values higher and lower than 0.12 mm evidently lies in the fact that the
clay content is different.

    As shown on the lefthand side of Figure 23 there is a decrease
in porosity with increasing depth for three groups of sandstones
having different medians.   These arenites are rich in dolomite
and calcite grains and have calcite and clay matrices.   Porosity
decreases with increasing calcite cement, but is independent of
the dolomite content.   The decrease in porosity with depth is
much more distinct than in many other sandstones, and the grain-
size effect is different from that in the quartz arenites in that
the coarsest ones are also the most porous.   The carbonates
(average content of 15%) in coarser calcareous sandstones occur

mainly as detrital carbonate grains (Fig. 23A) and in the form
of recrystallized calcilutite matrix which reduces the porosity
in the finer sandstones.   Figure 23 also illustrates that the
average porosity of oil-saturated sandstones is higher than that
of oil-free sandstones.

Figures 31 and 32 show a clear relationship between porosity
and permeability.   If this relationship were due to an increased
cementation, the points should lie on the smooth curve of equal
specific surface area or even beyond its lower left part.   The
downward-increasing deviation of the data points from the curve,
on the other hand, indicates that the specific surface area increases with decreasing porosity as a result of the sediments
becoming finer-grained and more argillaceous.   The curves in
Figure 33 resemble the curve (Bausteinschichten Sandstone) on the
lefthand side of Figure 23.   There is a different reason, however,
for the dependency on the grain size.   In Figure 23A, the
Bausteinschichten sediments show an increase in calcareous matrix
content with decreasing grain size, so that the net result is a
lowering of the porosity, whereas in the case of sandstones presented in Figure 33 is the particularly high clay content that
lowers porosity.   As shown in Figure 34, the clay content is
lower than 10% only in coarse-grained sandstones (Md larger than
0.5 mm).   In sandstones with less than 30% clay, more than half
of the clay is authigenic kaolinite (Fig. 34; see also Table 7).

In studying compaction, one has to distinguish four types
of clay occurrences described by Fuchtbauer, i.e. (a) layers,
completely separated from the sand; (b) seams, relatively well
separated from the sand; (c) uniformly distributed clays, authigenic (late diagenetic) in origin; (d) uniformly distributed,
detrital matrix and early diagenetic clay minerals; and (e) combinations of the above.   The influences of diagenetic processes,
e.g., compaction and migration of fluids, will be different in
each case, as discussed by Fuchtbauer.

In cases where organic material is present, mainly H_2O (with
dissolved humic acids) and CO_2 are released during the first
stage of coalification, causing an acid environment in which
feldspars are dissolved and replaced by kaolinite.   This occurs
particularly in coarser sandstones that have little detrital
clay and are more permeable to fluids prior to kaolinite authi-
genesis.   The numbers in Figure 34 indicate that the feldspar
content is lowest in the sandstones containing more kaolinite.
Some kaolinite, however, may have formed independently of the
feldspar breakdown.   According to Fuchtbauer (p.363), the
chlorite in shales is largely of secondary origin, whereas in
sandstones it is mostly detrital.   In the latter case, some
chlorite may be the product of alteration of a detrital matrix.
During the second stage of coalification, coal seams only release
chemicallally ineffective methane.   This causes an increase in

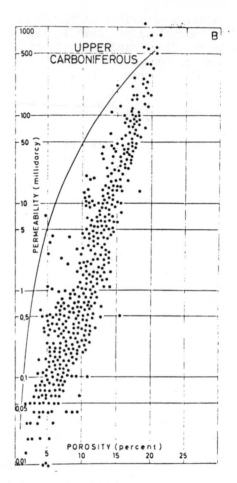

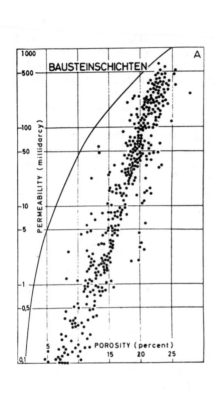

Fig. 31(A). Relationship between porosity and permeability of the "Bausteinschichten" Sandstone
(= calcareous - calcite + dolomite -, clayey
arenites of the Tertiary Molasse). Solid-
line curve is for constant specific surface
area. (After Fuchtbauer, 1967a, fig. 9.
p.360; courtesy 7th World Pet. Congr.).

Fig. 32(B). Relationship between porosity and permeability of the Upper Carboniferous sandstones
from the Ermsland region, Germany. (After
Fuchtbauer, 1967a, fig. 10, p.361; courtesy
7th World Pet. Congr.).

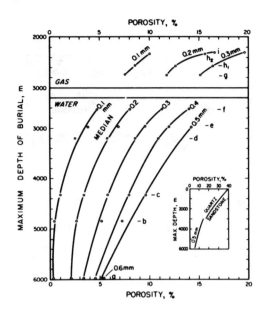

Fig. 33.   Relationship among porosity, maximum depth of
           burial, and median grain size in water- and
           gas-saturated sandstones of the Upper Carbon-
           iferous sandstones in western Lower Saxony.
           (After Fuchtbauer, 1967a, fig. 11, p.362;
           courtesy 7th World Pet. Congr.).   Each one
           of the horizontal lines (a to i) corresponds
           to one drilling well; they were obtained
           from curves showing relationship between
           porosity and median grain size for each bore-
           hole.   Maximum burial probably corresponds
           to the present depth in sections a, b, and g,
           however, greater depth was formerly assumed
           for geologic reasons.   In the lower right,
           the 0.5 mm curve is compared with the
           "Dogger beta" quartz sandstones from Fig.23.

pH, probably leading to a slightly alkaline environment.   During
this stage, mica (sericite) instead of kaolinite replaces the
feldspar.

    As shown by Fuchtbauer, many of the diagenetic alterations
are directly related to mechanical and chemical compaction.   Some
of the chemical constitutents of the cements are of local origin,
whereas others appear to have been derived from more remote sedimentary rocks undergoing compaction.

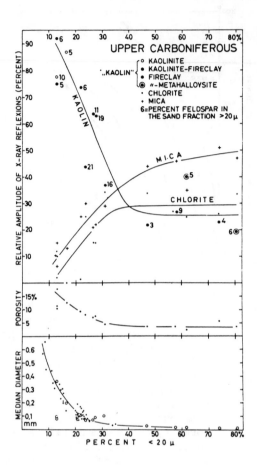

Fig. 34. Relationship among clay (<20 μ) content, median
          diameter, porosity, and the relative amounts of kao-
          linite, mica, and chlorite in clay fraction of Upper
          Carboniferous sandstones.   (After Fuchtbauer, 1867a,
          fig. 12, p.364; courtesy of 7th World Pet. Congr.).
          The numbers inside the diagram represent feldspar
          content in percent.   The ordinate shows the peak
          heights of the 001 mica line, the 002 kaolinite and
          the 004 cannot be read from the graph.   All values
          are for one well from about 2,500 m depth, except
          for the points in the lower graph.

Table 7. Mean clay contents (< 20 µ) in relation to the median
diameter (after Füchtbauer, 1867a, table III, p.361).

Sandstone types	Median diameter (mm)			
	0.1	0.13	0.16	0.2
	Mean clay content (%)			
"Dogger beta"	2.5	1.8	1.3	1.0
Valendis	12.5	2.5	2.5	1.6
Molasse				
Tertiary	12.5	10	8	6.5
Upper Carboniferous	22.	20.5	19	17

Blanche (1973) observed that particularly in fluvial and
sabkha facies, the clay matrix is composed predominantly of illite
together with kaolinite, chlorite and traces of montmorillonite.
These minerals are derived from the decomposition of feldspars
and micas. The permeability in these rocks decreases with decreasing grain size, whereas the porosity remains unchanged,
sorting being its most important controlling factor. Following
the reasoning by Fuchtbauer (1967a) and Taylor (1950), Blanche
expected the textural changes with increasing compaction as outlined by these two researchers. The maximum values of porosity
and permeability, however, appear to be little affected by depth
of burial down to 4000-4250 m. Below this depth a rapid decrease
was obvserved, i.e., the porosity gradient was 1.3% for each 3055 m
of burial, which is in agreement with values published for other
sandstone formations. Blanche also found that the depositional
nature of the sands set physical limits on the reservoir potential:
optimum values tended to occur in eolian dune facies, whereas the
lowest porosities and permeabilities were present in the fluvial
and sabkha facies.

Powers (1967) treated the release mechanisms of fluids from
clay-rich sedimentary units (see also Burst, 1969 for another
example). The stages at which certain amounts of fluids are released from clayey deposits determine how, when and in what quantities these fluids are made available to the sandstones during
compaction. The fluids from the finer sediments influence diagenesis in general, including mechanical and chemical compaction,

as well as cementation, decementation, and oil and ore-fluid
migration.   Together with studies such as done by Hitchon (1968)
on the total volume of sandstones, shales, and limestones within
certain sedimentary basins, the data on the release mechanisms
of fluids from the clayey and other sediments and the information
on the compaction history of the sedimentary pile will enable the
investigator to calculate the amounts of water, hydrocarbons, and
possible, ore fluids that may be available per m^3 of rock.   At
least the total volume of fluids could be estimated.   Some of
these studies have already proved of practical value in estimating
the total amounts of petroleum and ores present in certain regions.

Powers (1967) offered a compaction history (Fig.35), which
is based on laboratory and field data and clay mineralogy.   His
conclusions can be summarized as follows:

(1)   When montmorillonite clays are buried to a depth of
about 3000 ft, most of the water is expelled, except for the last
few bound layers between the basal clay surfaces.   This bound
water may comprise nearly 50% of the volume of the clayey rock,
and, apparently, cannot be squeezed out by further increase of
burial pressure.

(2)   At a depth of about 3000 ft, the effective porosity and
permeability are essentially zero for the mudstone.   The clays
and fixed water are "wrapped-around" the sand and silt particles.

(3)   A change from montmorillinonite to illite begins at a
depth of about 6000 ft. and continues at an increasing rate to
a depth of about 9000-10,000 ft. where no montmorillonite is left.
This alteration mechanism is accompanied by desorption of the last
layers of boundwater from the clays and their release as "free
water".

(4)   The fluids are released suddenly from the montmorillonite in the deep subsurface, maybe during the clay-mineral trans-
formation(s) and origin of mudstone fissility.

(5)   This sudden release of fluids would be absent from
illite and kaolinite deposits because of the absence of a mineralogical change.   Normal fluid movements by compaction would
have occurred earlier.

(6)   During the montmorillonite-to-illite change, desorption
of interlayer hydrocarbon layers and, possibly, trace elements
takes place.

(7)   Below the "no-montmorillonite level" (Fig.35) small
amounts of water continue to be lost from the shale as the montmorillonite fraction of the mixed-layer clays collapses to form

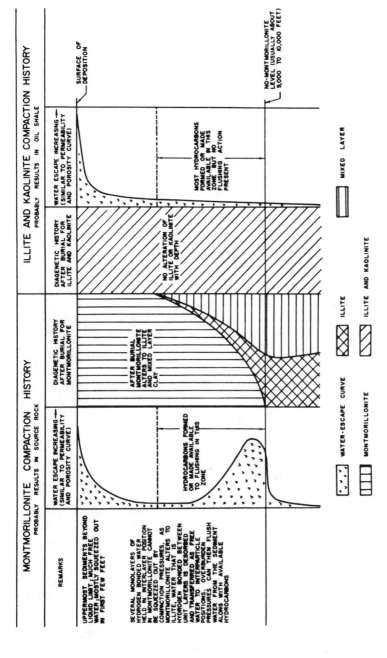

Fig. 35. Compaction history of different clay minerals when deposited in marine environment and
its probable relation to release of hydrocarbons from mud rocks. (After Powers, 1967,
fig.3, p.1245; courtesy Am. Assoc. Pet. Geologists).

illite.   This occurs at about 14,000 ft. in the Gulf Coast area.

Magara (1973) studied (1) relationship between differential
shale compaction and depth;   (2)   relationships among porosity,
permeability, density, and depth; (3) relationship between clay
composition and depth; (4) relationship between the rate of sedimentation and compaction; (5) amount of fluid expelled during
compaction; and (6) porosity distribution in shales as an indicator of permeability of associated sandstones and coarse limestones.   According to Magara, "shale porosity distribution in
incompletely compacted shale zones also may be affected by the
permeability and the extent of adjacent sandstone or carbonate
rock bodies.   A sharp decrease of porosity in shales close to
such rock bodies would suggest that relatively large volumes of
fluids have been expelled from the shales into the adjacent sandstones or carbonates.   If this expelled fluid volume is large,
the possibility of hydrocarbon accumulation in such sandstone or
carbonate rocks is considered to be favourable".   This has been
confirmed by Magara's studies of Mesozoic oil and gas pools in
Western Canada.

Mitra and Beard (1980) analyzed the reduction of porosity
due to pressure solution for theoretical models consisting of
aggregares of spherical grains of uniform size.   All six of the
stable regular packing arrangements suggested by Graton and Frazer
(1935) were considered for two limiting cases (Fig. 36, Table 8):
(1) dissolved material is removed from the system and (2) all of
the dissolved material is locally precipitated.   Mitra and Beard
in their valuable paper stated that as far as the amount of porosity reduction is concerned, solution is more important in the
early stages, whereas cementation becomes more effective during
the later stages.   Even for the most compact packing arrangement
and complete local precipitation, there will be a vertical short=
ening for total porosity reduction in excess of 26%.   Calculations
of Mitra and Beard were checked by the writer.

This study presents a possibility of estimating the role of
pressure solution and the minimum volume of cement that has
entered the system by comparing the "minus cement porosity" with
the volume of cement predicted by a theoretical model.   This
study should be extended to sands of ovoid and other shapes.

As far as the initial porosity necessary for the calculations
is concerned, Manus and Coogan (1974) and Sibly and Blatt (1976)
have suggested that orthorhombic packing arrangement (case 2) is
the most practical because the initial porosity is close to 40%.

Relationship between the loss due to solution ($\Delta\phi_s$) and the
additional porosity loss due to the precipitation of cement
($\Delta\phi_c$) for various packing arrangements are presented in Figure
37.

Table 8. Various properties of different types of regular packing arrangements of spheres (after Mitra and Beard, 1980, table 1, p.1349; courtesy of the Journal of Sedimentary Petrology).

Packing	Shape of Polyhedron	Initial porosity (ϕ_o) in percent	Number of grain contacts		Stability
			Total	Solution	
Case 1 Cubic	Cube	47.64	6	2	Low
Case 2 Orthorhombic (I)	Hexagonal prism	39.54	8	4	Medium
Case 3 Rhombohedral (I)	Dodecahedron	29.95	12	8 when h_f >1 * 10 when h_f <1	High
Case 4 Orthorhombic (II)	Hexagonal prism	39.54	8	2	Low
Case 5 Tetragonal-sphenoidal	Fourteen-sided poly-hedron	30.18	14	4	Medium
Case 6 Rhombohedral (II)	Dodecahedron	25.95	12	6	High

*h_f = final vertical distance between adjacent layers in the compacted state.

Table 9. Type, location and composition of samples analysed by Ebhardt (1968, p.59)

Sample	Provenance	Type of sediment	Content % Carbo-nate	Sand	Silt	Clay	Median diameter Md (μm)	Sorting coefficient $\underline{S}_o = (Q_3/Q_1)$
AdIV	Adria	Sand	86	96	3	1	450	1.6
Rm	Red Sea	Sandy silt	75	20	80	-	20	1.3
P	Persian Gulf	Clayey silt	5	17	56	27	6	3.7
FII	Florida Bay	Clayey silt	86	8	61	31	7	3.8
AdI	Adria	Clayey silt	41	4	78	18	11	3.3
Hz	Persian Gulf	Silty clay	36	-	48	62	2	-
$CaCO_3$	-	Silt	100	-	100	-	22	1.5
Bd	Lake Constance	Sandy silt	91	24	70	6	17	3.5

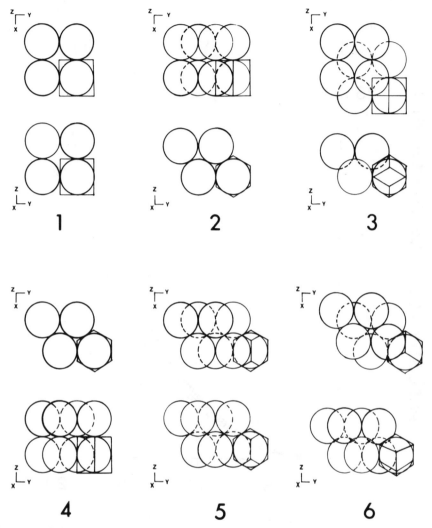

Fig. 36. Orthographic projection of aggregates of
spheres for six stable packing arrangements.
(After Mitra and Beard, 1980, fig.1, p.1348;
courtesy of the Journal of Sedimentary
Petrology). The top diagrams show the projections perpendicular to the Z axis, whereas
the bottom diagrams (in each case) show the
projections perpendicular to the X axis. The
unit polyhedron is shown in each case. 1 -
cubic; 2 - orthorhombic, I; 3 - rhombohedral;
4 - orthorhombic, II: 5 - tetragonal -
sphenoidal; and 6 - rhombohedral, II.

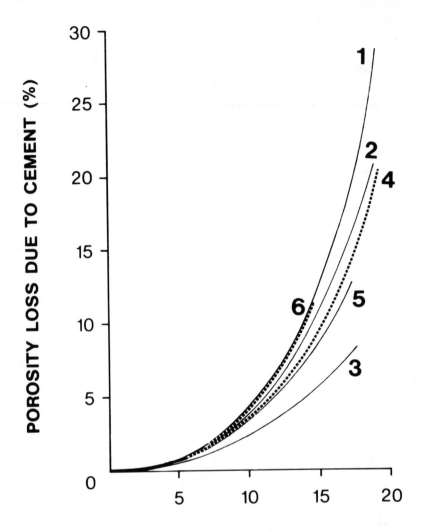

Fig. 37. Relationship between the porosity loss due
          to solution ($\Delta\phi_s$) and porosity loss due to
          precipitation of cement ($\Delta\phi_c$) for the six
          regular packing arrangements. (After Mitra
          and Beard, 1980, fig. 5, p.1355; courtesy of
          the Journal of Sedimentary Petrology).

     Mitra and Beard (1980) also estimated the volume of cement
that can be produced by stylolitization for various original

porosities.

COMPARISON BETWEEN COMPACTION OF CLAYS AND SANDS

Meade (1966) observed that variations in the porosity of
sands and in their water content are not predictably related to
depth and an increase in the overburden load.   Figure 38 shows
this for a selected group of sediments that were exposed to
their approximate maximum overburden loads at the time of sampling.
All these sediments are Cenozoic in age and most are inorganic
and terrigenous clastics, except for sample III containing about
10% siliceous skeletal material, sample II which is clayey, and
sample IV with 10-2% $CaCO_3$.   The curves are drawn in such a way
that the area above the lines gives the volume of solid particles,
whereas the area below represents the interstitial volume.   As
to the influence of particle size on water content and porosity
(the two are numerically equal because both are expressed as
percentage of bulk volume), Meade stated that the particle size
may be as important a control as overburden load on porosity,
especially during the early stages of compaction.   A comparison
of Figures 38A and 39A shows that order-of-magnitude difference
in diameter between 0.001 and 1 mm may cause about the same
amount of porosity variation as an order-of-magnitude difference
in depth burial.   With an increase in pressure, the finer sediments are compacted more rapidly than the coarser ones, and the
relationship between grain size and porosity becomes less
pronounced (Fig. 39B).   A significant influence on grain size
on porosity is still apparent at overburden loads approaching
100 kg cm^{-2}.   For the sediments presented in Figure 39B, the
magnitude of change in porosity with changing grain size is about
equal to that related to depth of burial (cf. the curves in
Figure 39B with curve VII in Figures 38A and 38B).   Another
group of sediments in California exposed to overburden presssure
of 5-60 kg cm^{-2} (Meade, 1963) demonstrated that their downward
decrease in particle size (together with changes in other factors)
reversed the expected effect of increasing load and caused a
systematic downward increase in porosity.   With a further increase in pressure as a result of overburden, an inverse relation
between the grain size and porosity changed to a more direct
correlation.   According to Meade (1963, p.1087), "the more rapid
decrease in the water content of clays eventually overtakes the
slower compaction of sands, perhaps at some depth of burial near
1 km".   This is supported by the observation that deeply buried
coarser sediments and their equivalent rocks are commonly more
porous than adjoining claystones and shales.

In his section on compaction of sands, Meade (1963, p.1096)
stated that most sands are only slowly compacted during the early
stages of compaction;   however, relatively few unequivocal data

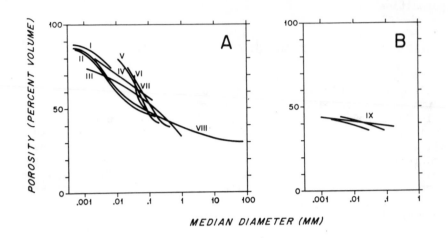

MEDIAN DIAMETER (MM)

Fig. 38.   Relation between porosity and depth of burial in metres
           in selected clays and claystones (A) and selected sands
           and sandstones (B). I = Recent, Lake Mead on Colorado
           River (Gould, 1960, p.176, lower 3 graphs); II = Recent,
           Santa Barbara Basin off Southern California Emery and
           Rittenberg, 1952, p.755); III = Recent and older(?),
           western Bering Sea (Lisitsyn, 1956, fig. 16); IV =
           Recent and older(?), eastern Black Sea (data from
           Ostroumov and Volkov, 1964, pp.94-95); V = Recent and
           older(?), continental slope off Nova Scotia (Richards
           and Keller, 1962); VI = Recent, Orinoco River delta
           (Kidwell and Hunt, 1958, p.808); VII - Pliocene and
           Pleistocene, central California (Meade, 1963a; curves
           adjusted for artesian pressure); VIII = Pliocene to
           Recent, Baku Archipelago (Koperina and Dvoretskaya,
           1965, fig. 1; plus data from Korobanova, Kovaleva,
           Kopylova, and Safokhina, 1965, pp. 128-130); IX =
           Tertiary, Venezuela (Hedberg, 1963, p. 256); X =
           Miocene and Pliocene, Po Valley (Storer, 1959, p.523);
           XI = Miocene, southern Louisiana and southwest Texas
           (Maxwell, 1964, p.704). Where bulk density or water
           content (by weight) were reported, porosity was com-
           puted assuming a particle density of 2.60 g/cm^3.
           (After Meade, 1966, fig. 1, p.1086; courtesy of
           J.Sed.Petrol.).

are available to demonstrate this, because of the difficulty in
collecting unconsolidated sands without disturbing them. Curve
VII in Figure 38B represents silty sands, and it seems that the
silt gave the sand enough cohesion to withstand coring and sampling

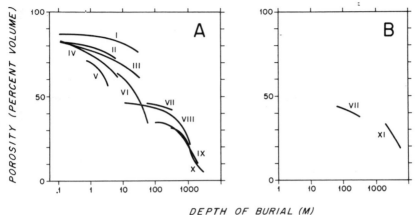

Fig. 39. Relationships between porosity and median
particle diameter in selected Recent surface
or near-surface sediments under overburden
loads less than 1 kg cm^2 (A), and in sediments
under overburden loads between 7 and 70 kg cm^2
(B).   I = Lake Mead on Colorado River
(Sherman, 1953, p.339); II = Lake Maracaibo,
Venezuela (Sarmiento and Kirby, 1962, p.719);
III = reservoirs in western United States
(Hembree, Colby, Swenson, and Davis, 1952,
p.39); IV = Gulf of Paria (van Andel and
Postma, 1954, p.108); V = North Sea
(Fuchtbauer and Reineck, cited by von
Engelhardt, 1960, p.15); VI = Continental
shelf off southern California (Hamilton and
Menard, 1956, p.757); VII = San Diego Bay
and adjacent continental shelf(data from
Shumway, Pleistocene alluvium, central
California (Meade).   (After Meade, 1966,
fig. 2, p.1087; courtesy Am. Assoc. Pet.
Geologists).

without too much distortion.   Curve XI in Figure 38B represents
quartzose sandstones in approximate equilibrium with their present overburden pressure.   During the early stages of compaction,
the parameters that control porosity are mainly textural, i.e.
grain size, sorting, roundness, shape and flexibility of certain
types of grains (von Engelhardt, 1960, pp.3-16; Fraser, 1935;
Gaither, 1953; Hamilton and Menard, 1956).   The results of experiments on the influence of sorting and roundness on compaction
of pure quartz sands are given in Figures 40A and B.   In the
pressure range up to 100 kg cm^{-2}, compaction occurs as a result

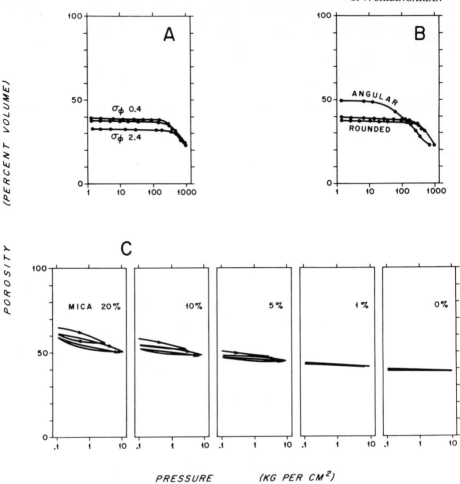

Fig. 40.   Influence of different factors on the relations between
           porosity and pressure in sands, as determined in labor-
           atory experiments. A = Influence of sorting in well-
           rounded quartz sands (Roberts and de Souza, 1958); sort-
           ing index (σ_ϕ) defined by Inman (1952, pp. 135-136;
           median diameter of two better sorted sands = 0.60 mm;
           median diameter of sand with poorer sorting = 0.48 mm.
           B = Influence of rounding of quartz sands, 0.42-0.84 mm
           in size (Roberts and de Souza, 1958). C = Influence
           of mica particles mixed in different proportions with
           rounded quartz sands (Gilboy, 1928, p.560); particles
           of both constituents are 0.42-0.59 mm in size. (After
           Meade, 1966, fig. 10, p.1097; courtesy Am. Assoc.Pet.
           Geologists).

of rearrangement of sand grains into a more dense packing system
and, to a minor degree, by elastic compression of the individual
grains.   At pressures above 100 kg cm^{-2}, compaction increased as
a result of cracking and shattering of the grains in the compression apparatus.   In agreement with other investigators, however,
Meaded doubted whether this cracking and shattering would occur to
such a degree under natural conditions of compaction of sands.

   The slow rate of pressure increase in nature, extending over
thousands of millions of years, permits other processes to be
operative, such as plastic flow, pressure solution, and reprecipitation.   Sands that are composed of minerals other than quartz
may respond differently to pressure, with the softer grains being
more readily deformed.   That the greater porosity of well-sorted
sands in contrast to more poorly-sorted ones, persists during
early compaction is indicated by the results presented in Figure
40A.   The influence of particle roundness is demonstrated in
Figure 40B, where angular sands show greater initial porosities.
This reflects the instability of the intial packing of the angular
grains.   The angular particles are more compactible than rounded
ones of the same size.   The pronounced influence of platy and
flexible mica particles on the behaviour of sands in compaction
experiments is demonstrated in the five graphs of Fig.40C.   The
porosity, compressibility, and elasticity of sand increase with
increasing mica content.   Experiments by McCarthy and Leonard
(1963) support these results and suggest further that the finer
the mica flakes among the sand grains, the greater the increase in
porosity per unit increase in mica content.   It should be noted,
however, that a permanent, rather than elastic, deformation of
mica plates may also take place during the early stages of compaction of micaceous sands.   At pressures between 0 and 100 kg
cm^{-2}, grain size is the most important parameter influencing
compaction.   Not only is it inversely related to water content
and porosity of the sands, but the size of the particles influences most other factors that control mechanical and chemical compaction.   The effect of grain size may be so strong that the
expected decrease in porosity with depth of burial may be obscured.
The presence of mica may have a greater influence on the mass
properties of sands than textural variations, such as those of
rounding and sorting, of the non-mica grains.   It is important
to stress here, that more complex combinations of sand, silt, and
clay should be experimentally examined in order to determine the
influence of their relative proportions on the rate and degree of
compaction.

RELATIONSHIP BETWEEN PRESSURE AND POROSITY OF CARBONATE SEDIMENTS
(after Rieke and Chilingarian, 1974).

   In recent years increased attention has been paid to field

studies and experimental compaction studies of carbonate sediments
(Terzaghi, 1940; Laughton, 1957; McCrossan, 1961; Fruth et al.,
1966; Robertson, 1967; Ebhardt, 1968 and Coogan, 1970).   Early
compaction experiments by Terzaghi (1940) and Laughton (1957) were
carried out under a limited pressure range in piston-sleeve type
apparatuses.   Laughton (1957) studied the velocity of propagation
of compressional and transverse elastic waves in Globigerina oozes
at different compacting pressures.   E.C. Robertson (personal communication) conducted extensive experiments on aragonitic sediments
from the Bahama Banks (Fig. 41).   He found that the consolidation

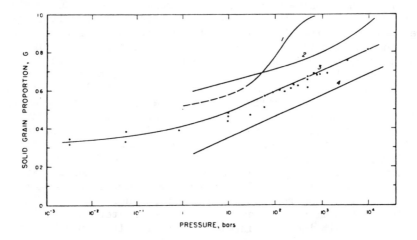

Fig. 41.   Relationship between solid grain proportion
           and pressure. 1 = shale (Athy, 1930); 2 =
           kaolinite (Chilingar and Knight, 1960a); 3 =
           fine-grained aragonitic sediment; 4 = mont-
           morillonite (Chilingar and Knight, 1960a).
           (After Robertson, 1967, fig. 4, p.126).

curve of aragonite lies between the montmorillonite and kaolinite
curves obtained by Chilingar and Knight (1960a) and is nearly
parallel to them.   This could be interpreted as meaning that the
compaction mechanism is similar despite the difference in grain
shapes (plates for the clays and needles for aragonite).   Figure
42 shows than an initial difference in the porosity of Robertson's
aragonite samples (Fig. 42 curves 1, 4 and 6) is maintained through
a large part of the pressure range.   The inference is that the
initial water content does not appreciably affect the change in
compaction with increased pressure (porosity/pressure slope).
Upon exposing a limestone core sample with 10% initial moisture
content to a pressure of 100,000 psi, 59.2% of the original

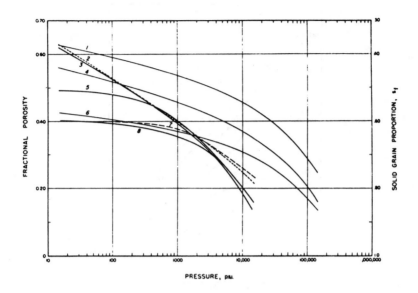

Fig. 42.   Relationship between fractional porosity and
           pressure for various carbonate sediments.
           1 = aragonite sediment from the Bahamas with
           an initial water content of 80% (Robertson,
           1967); 2 = clay-size aragonite grains (mud
           facies of modern Bahamian carbonate sediments)
           (Fruth et al., 1966); 3 = skeletal debris com-
           posed of foraminifer and mollusc shells
           (Fruth et al., 1966); 4 = aragonite sediment
           with an initial water content of 60%
           (Robertson, 1967); 5 = grapestone facies
           (Fruth et al., 1966); 6 = dry aragonite pow-
           der (Robertson, 1967); 7 = oolite facies com-
           posed of polished subspherical ooids and well-
           rounded grains (Fruth et al., 1966); and 8 =
           oolitic facies composed of well-rounded
           oolitically coated grains (Fruth et al.,
           1966).

moisture was squeezed out (Chilingarian and Rieke, 1968).

     Ebhardt (1968 compacted recent carbonate sediments of diff-
erent grain size and with different carbonate content up to
pressures of 650 atm.   The sediment samples tested by Ebhardt
and their properties are presented in Table 9, whereas experimental results are shown in Figures 43 and 44.   The compaction
curves for samples P, Ad, I and Hz, which are characterized by an
increasing clay content, are similar to those of pure clays.

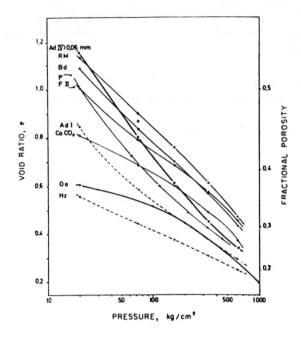

Fig. 43. Relationship between void ratio and confining
          pressure for sediments having different carbon-
          ate content. (After Ebhardt, 1968, p.60).
          Thick solid lines = sand-size sediments,
          dashed lines = clayey sediments, 0$_0$ = oolitic
          sediments. (After Fruth et al., 1966).

Elevated temperature (90°C) cause more intense compaction
and the void ratio is about 10% lower than for runs at room temperature (Fig.45).

Fruth et al. (1966) designed a triaxial cell for systematic
evaluation of pressure, temperature, and strain-rate effects on
calcareous sediments. The amount of compaction caused by increasing confining pressure on aragonite sediments from the Bahamas is
shown in Figure 42. All compaction curves in this figure are
consistent with the expected behaviour of the sediment types:
(1) negligible initial compaction of the well-sorted, hard oolite
grains; (2) high compaction of the fine-grained mud; and (3)
moderate compaction of the grapestone facies (a mixture of friable
and hard grains and aggregates). Greater compaction of the
skeletal type of carbonate sediment reflects initial voids and a
more open structure created by the presence of shells, skeletal
fragments and large grains. With an increase in pressure, the
fragments supporting the open structure yield and the void volume

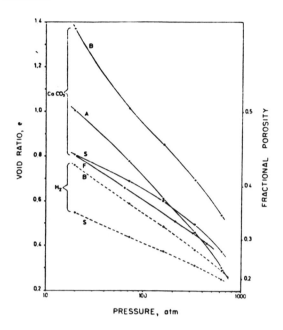

Fig. 44. Relationship between void ratio and pressure
          for sample H$_z$ and CaCO$_3$ samples, saturated in
          benzene (B), air (A), sea water (S), and
          fresh water (F). (After Ebhardt, 1968, fig.4,
          p.61).

is reduced (Fruth et al., 1966, p.750).   It seems that all five
carbonate-sediment types presented in Figure 42 have nearly the
same porosity values (≃ 35%) at a pressure of around 2,000 psi.
According to Fruth et al. (1966), the somewhat poorer sorting and
initial compaction (in the field) of the oolitic facies may account for its lower porosity values than the oolite.   The finer
material fills the void between the larger grains.   The final
porosity of the skeletal facies is lower than that of the aragonite mud, indicating that the skeletal facies compact better.
Finer matrix grains tend to cushion the effects of compaction
in oolitic facies.   In the case of well-sorted oolite, with small
amounts of matrix and hard grains predominating, there is a high
degree of grain fracturing.   There is a negligible initial compaction in the presence of small amounts of matrix and the grains
are in contact with each other during compaction.   Fractures tend
to radiate from points of initial contact;   however, in some
grains a single fracture extends completely across the grain
(Fruth et al., 1966, p.751).

    As pointed out by Rieke and Chilingarian (1974, p.70) it is

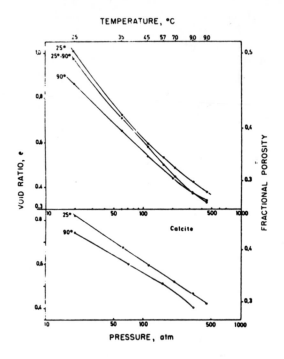

Fig. 45.   Relationship between void ratio and pressure
at room temperature and at 90°C for sample P
and CaCO₃ sample.   In the case of sample
represented by middle curve in the upper
graph, the temperature was raised stepwise
(25-90°C; see upper scale). (After Ebhardt,
1968, fig. 6, p.62).

claimed by many geologists that field observations on carbonate
rocks often indicate that very little compaction has occurred.
Often, the absence of the crushing of delicate fossils is brought
forward as supporting evidence.   Brown (1969, p.492) compacted
pelecypod valves and whole shells in a carbonate-mud matrix up to
a pressure of 15,000 psi.   Only rarely did some shells break,
indicating that the presence of unbroken fossils surrounded by
carbonate mud is not evidence by itself for the lack of compaction.

Battacharyya and Friedman (1979) studied compaction of
various mixtures of ooids and lime mud under continuous and intermittent increase in pressure up to 7,636 psi (555 kg cm⁻²). They
showed that ooids may remain undeformed, and there is a direct
relationship between the increasing proportion of lime mud and
proportion of undeformed ooids.

Experimental compaction also gives rise to buckled, faulted, split, spalled, diagonally fractured, and crushed ooids.   Continuous increase in confining pressure yielded both concave-convex and longitudinal contacts.   In uncompacted samples, the existing grain contacts were point contacts.   (For definitions see Fig.46).

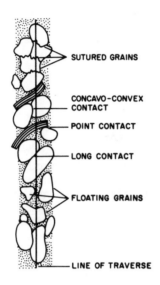

Fig. 46.   Definition sketch of fabric terminology:
quartz (white), mica (lined), and matrix
(stippled).   After Pettijohn et al., 1972,
fig. 3-10; courtesy of Springer, New York.

The lime-mud content determines the way sediments respond to compaction.   Battacharyya and Friedman finally stated that the occurrence of deformation of allochems is less likely in mudstones and wackestones than in grainstones.   The relative condensation of allochems may provide an index of compaction in lithifield rocks.

In an excellent study of compaction in the Mississippian skeletal limestones, Southwestern New Mexico, Meyers (1980) showed that compaction significantly reduced intergranular porosity in over 60% of the packstones and grainstones throughout the 30,000 km^2 of study area.   Both mechanical and chemical compactions were involved, reducing intergranular volumes by up to 50% and, more rarely, by up to 75%.   Bryozoan grains were more susceptible to compaction (both mechanical and chemical) than echinoderms,

which commonly underwent only intergranular pressure solution.
The schematic diagram of sequences of chemical compaction features
resulting from increasing compaction is presented in Figure 47.
According to Meyers (1980), compaction began under a few tens to
a few hundreds of feet of overburden and was completed under less
than 6500 ft (2,000 m) of overburden.   Cementation occurred
during the compaction.   Virtually all non-stylolitic compaction
took place under an overburden of less than 6500 ft.   Although
syntaxial calcite cements inhibited compaction, they did not stop
it.   The fine, microcrystalline, fibrous shell structures of
bryozoans were highly susceptible to mechanical and chemical compaction.

LEACHING AND CRUST FORMATION

        During the past 20,000 years, the sea level has risen approx-
imately 400 ft due to the melting of glaciers.   This has had a
profound effect upon many parts of the world, causing shorelines
to move inland with the resultant deposition of great bodies of
marine soft clays, not only along sea margins, but also far inland in drowned river valleys.   The deposition of these sediments
have been primarily influenced by several factors: (1) eustatic
(world-wide) changes in sea level; (2) local tectonics (movement
of earth crust); and (3) localized loading of the earth's crust
(glaciation).   Change is sea level can result in the change of
type and/or concentration of ions in the interstitial water
present in pores of sediments in contact with clay minerals.
The changes in electrolyte content of the interstitial water can
change the engineering properties of a soft clay deposit by either
shrinking or expanding the volume of "bound-water" surrounding
the clay mineral.   In the case of many Norwegian clay deposits,
where a former marine soft clay deposit has been uplifted and
fresher water replaced a more saline interstitial water, there is
a reduction in electrolyte concentration that results in a diffusion of ions from the bound-water surrounding the clay particles
to the interstitial water.   As the bound-water becomes "fresher",
the volume of bound-water will increase.   The clay-water aggre-
gation increases its volume disrupting the soil structure.   This
results in a reduction of plastic index and shear strength of the
soft clay deposit.   The decrease in electrolyte concentration,
and sometimes the type of ion, often referred to as "leaching" of
the deposit.   The opposite situation, the increase of electrolyte
concentration can occur by sea water displacing a fresher interstitial water.   This can also disrupt the engineering properties
of a soil by shrinking the clay-water aggregates.   The displacement of interstitial water is not a homogeneous process.   The
displacement front will tend to be irregular, resulting in the
localized concentration of low shear strength zones.   As a
result, careful field sampling is required in the case of any

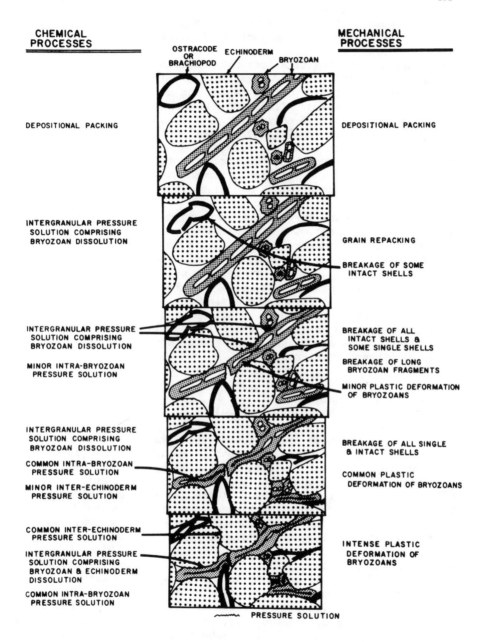

Fig. 47. Schematic diagram of processes that occurred during progressively increasing compaction in calcarenites of the Lake Valley Formation. (After Meyers, 1980, fig.5, p.465; courtesy of the Journal of Sedimentary Petrology).

proposed construction site where soil engineering data is
critical to the design of structural foundation.

Whereas leaching of a soft clay deposit can reduce the electrolyte concentration, near the upper surface of the soft clay
deposit the electrolyte concentration can be increased by such
factors as evaporation.   This process can result in the formation
of a crust generally of only a few feet in thickness over a soft
clay deposit with a significantly lower shear strength and plasticity index.   The depth and formation of a crust are controlled
by the local environment.   The strength of this crust will be
greatly reduced by the addition of fresh water which may reduce
the electrolyte concentration and dissolve cementing material.

If there is any movement of compaction fluids upwards in the
argillaceous sediment, the concentration of salts in the interstitial solution increases as the fluids move upwards.   Finally,
saturation is reached at the upper portions of the sediments and
some salts (e.g. $CaCO_3$) can precipitate out forming a crust
(Rieke and Chilingarian, 1974, p.248).   Formation of the crust,
however, may prevent further squeezing out of the water from
compacting clays.   It should be pointed out that according to
some investigators (Rieke and Chilingarian, 1974, p.274) the
salinity of squeezed-out solutions progressively decreases with
increasing overburden pressure.

POROSITY - DEPTH RELATIONSHIPS
(after Rieke and Chilingarian, 1974)

Many authors have been concerned with variations of porosity
in the shales as a function of depth.   Papers by Athy (1930a);
Hedberg (1936); Hammer (1950); Dallmus (1958); Weller, (1959); von
Engelhardt (1960); McCrossan (1961); Woolard (1962) and McColloh
(1965) are the most noteworthy.   There has been a considerable
amount of laboratory investigation of porosity-pressure relations,
mostly on unconsolidated sediments (Kryukov and Komarova, 1954;
Laughton, 1957; Chilingar and Knight, 1960a; von Engelhardt and
Gaida, 1963; Robertson (1967) and Chilingarian and Rieke,(1968).
Change in porosity is mainly a function of maximum overburden
stress and of time, although it is also affected by lithology,
depositional environment, overpressured fluid zones, diagenesis,
and tectonic stress.   The interrelation between these factors
is quite complex and, consequently, pronounced variations in
porosity-depth curves occur from place to place in porous sedimentary rocks.

Porosity variations with depth.   Porosity decreases with increasing depth and, as shown in Figure 48, there is a marked
decrease in porosity at shallow depths.   The greatest porosity

loss in sediments occur in the first few hundreds of feet of
burial.   Mechanical compaction of muds may occur in geologically
short periods of time if fluid expulsion occurs as porosity
decreases (Magara, 1968).

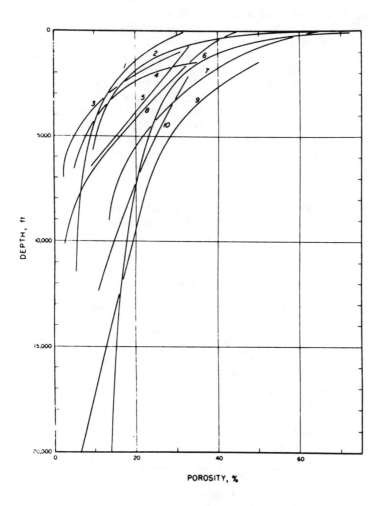

Fig. 48. Relationship between porosity and depth of
          burial for shales and argillaceous sediments.
          1 = Proshlyakov's (1960); 2 = Meade (1966);
          3 = Athy (1930a); 4 = Hosoi (1963); 5 =
          Hedberg (1936); 6 = Dickinson (1953); 7 =
          Magara (1968); 8 = Weller (1969); 9 =
          Ham (1966); 10= Foster and Whalen (1966).

In Figure 48 a comparison can be made of the porosity-depth relations from several different areas. The figure illustrates (1) Proshlyakov's (1960) published porosity data from the Cisaucasus, USSR; (2) Meade's (1966) data as adapted by Brown (1969); (3) Athy's (1930a) curve based on bulk densities of 200 samples of Pennsylvanian and Permian shales in northern Oklahoma; (4) Hosoi's (1963) curve based on density measurements of Tertiary mudstones in Akita and Yamagata Prefectures, Japan; (5) Hedberg's (1936) curve based on bulk densities of samples of Tertiary shales in Venezuela; (6) Dickinson's (1953) average bulk density curve for Tertiary Gulf Coast shales; (7) Magara's (1968) curve from the Shiunji gas field, Nagaoka Plain, Japan; (8) Weller's (1959) combined curve; (9) a curve derived from Ham's (1966) data, and (10) Foster and Whalen's (1966) data. Difference in the position of the curves in Figure 48 is due to the factors such as (1) the amount of induration in the sediment, (2) geologic age, (3) compositional and textural variations, and (4) tectonic stress history.

Weller (1959) constructed the porosity versus depth curve (Fig. 48, curve 8; and Fig. 49) which consisted of a combination

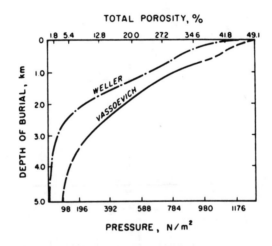

Fig. 49.   Relationship between porosity, depth of
            burial and overburden pressure. N = unit
            of force (Newton) = 102 g-force = 10^5
            dynes. (After Weller and Vassoevich, in:
            Kartsev et al., 1969).

of results from three earlier studies by Terzaghi (1925b), Athy (1930a) and Hedberg (1936). Terzaghi's data was matched with

Hedberg's data by adjusting the latter along the horizontal depth
scale to the point where a smooth junction could be made.   This
shift of 500 ft to the right suggested to Weller that this amount
of overburden may have been removed by erosion.   Athy's curve
required a 3,000 ft shift to the right in order to obtain a smooth
fit.   Athy's (1930a) data, however, can be questioned on the
basis of the geologic age and structural deformation.   It seems
that Weller's density curve is only of limited value, because it
was derived by combining the different data sets which may not
be compatible with each other.   Dickinson's data (Fig. 48; curve
6) agree reasonably well with those of Ham (Fig. 48, curve 9) and
Foster and Whalen (Fig. 48, curve 10) in the low-pressure range,
but appear to give higher porosity values at depths greater than,
8,000 ft.

        Relationship between total porosity and depth of burial,
based on data presented by Vassoevich (1960) and Weller (1959) is
presented in Figure 50.

        McCulloh (1967) plotted total-porosity values found in a
large variety of sedimentary rocks in order to obtain a curve of
maximum porosity versus depth (Fig. 50).   As shown by Figure 50,
the maximum ranges of shale porosity differences vary from about
16% at a depth of 10,000 ft, to about 25.5% at 1,000 ft.   The
ranges appear to be the greatest among the youngest rocks and
narrower for the older rocks.   McCulloh (1967, p.A8) pointed
out that his limits will no doubt be revised as more determinations
of total porosity of sedimentary rocks accumulate.

        Lapinskaya and Proshlyakov (1970, p.116) showed relationships
between porosity and depth of burial for (a) sandy-silty rocks
and (b) clayey rocks in the Pre-Caucasus area and Pre-Caspian
Depression (Fig. 51).   In sandy-silty rocks containing 15-20%
cement, the total (pycnometer) porosity is decreasing 6-9% for
every 1,000 m of depth and at a depth of 3,500-4,000 m constitutes
5-10%.   Initially, clayey rocks are compacted rapidly and at a
depth of 2,000 m their porosity decreases to 20-26%.   The subsequent decrease in porosity at greater depth occurs at a lower
rate.   At a depth of 4,000 m, the total porosities of various
terrigenous rocks become very close to each other and constitute
5-10%.   In the Aralsorskiy borehole (USSR) the porosity is
equal to 3.3.5% at a depth of 6,000 m.

        Relationship between porosity and present depth of burial
for Liassic shales in northwestern Germany is presented in
Figure 52.   The porosities of shales obtained from deep boreholes in zones I - V are more or less what is to be expected at
the present depth of burial, whereas those in zones A, C, Ho and
He are too high in relation to the present depth.   The latter
can be explained by the fact that during the uplift of salt plugs,

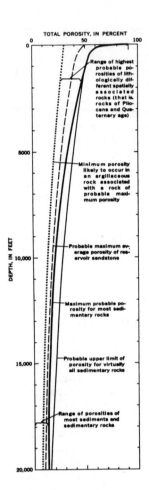

Fig. 50.   Relationship between total porosity and depth
           of burial for sedimentary rocks. Based on
           more than 4,000 laboratory measurements of
           cores from Los Angeles and Ventura basins in
           California, and other scattered localities in
           the United States, and the Po Basin of Italy.
           (After McCulloh, 1967, fig.4, pA9).

the Liassic sediments were raised from a level shown by the arrow
in Figure 52 (Fuchtbauer, in: von Engelhardt, 1960).

    Teodorovich and Chernov (1968) presented two valuable dia-
grams (Figs. 53 and 54) on the relationship between porosity,
depth of burial, and sonic velocity.   They derived the following

Fig. 51. Dependence of
porosity of sandy-silty
and clayey rocks on depth
of burial in Mesozoic and
Upper Paleozoic deposits
of Pre-Caucasus and Pre-
Caspian Depression. 1 =
sandy-silty rocks with
60-80% content of coarse-
grained material (silt +
sand); 2 = clays and
argillites; 3 = epigen-
etically altered sandy-
silty rocks. (After
Lapinskaya and Prosh-
lyakov, 1970, fig. 1,
p.116).

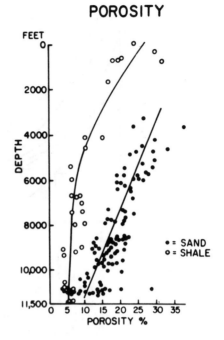

Fig. 52. Porosity and
void ratio of Liassic
shales as related to
present depth. The
arrows indicate the
maximum depth of burial.
(After H.Fuchtbauer,
c.f. von Engelhardt, 1960).

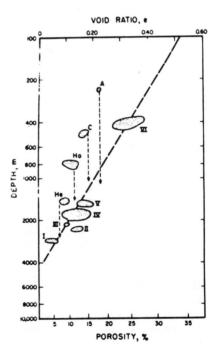

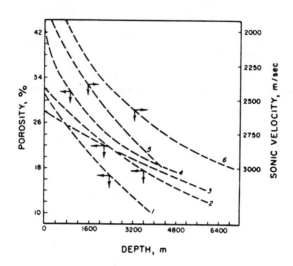

Fig. 53. Relationship between porosity, depth of
burial, and sonic speed. (After Teodorovich
and Chernov, 1968, fig.3, p.86). 1 = clays
(Apsheron Peninsula, Azerbayjan S.S.R.,
USSR); 2 = siltstones (Apsheron Peninsula);
3 = sandstones (Apsheron Peninsula); 4 =
Quaternary and Upper Pliocene clays (Near-
Kurinskaya Depression); 5 = Apsheron
Peninsula. (After R.M. Gadzhiev, 1965); 6 =
Lower Kurinskaya Depression. (After R.M.
Gadzhiev, 1965).

approximate formulas relating porosity   ; permeability, $\underline{k}$; and
depth, $\underline{D}$.

For siltstones:	$\phi = 29.50 - 0.0003\underline{D}$ (0 - 4,500 m)	(13)
or:	$\phi = 31.53 \exp(-0.000152\underline{D}$ (Fig.53, curve 2	(14)
and:	$\log \underline{k} = 2.961 \exp(-0.000147\underline{D})$	(15)
or:	$\log \underline{k} = 2.87 - 0.003\underline{D}$ (500 - 4,500 m)	(16)
For sandstones:	$\phi = 28.21 \exp(1-0.000122\underline{D})$	(17)
or:	$\phi = 28.74 - 0.003\underline{D}$ (Fig.53, curve 3)	(18)
and:	$\log \underline{k} = 2.803 \exp(-0.000074\underline{D})$	(19)
or:	$\log \underline{k} = 2.88 - 0.0002\underline{D}$	(20)

Relationship between porosity and sonic transit time for
Miocene mudstone, Nagaoka Plain, Japan, is presented in Figure
55. Stetyukha (1964) derived the following formula relating
the porosity of shales to depth of burial in the Pre-Caucasus
area:

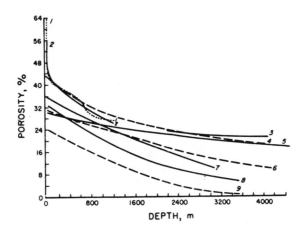

Fig. 54.   Relationship between porosity of various terrigenous
           deposits and depth of burial. (After Teodorovich and
           Chernov, 1968, fig.6, p.89). 1 = clays and 2 - silt-
           stones from Recent to top of Middle Pliocene, Alyaty-
           Sea. (After Koperina and Dvoretskaya, 1965); 3 = fine-
           grained and medium-grained sands with carbonate-clayey
           cement up to 15%, Apsheron Peninsula; 4 = Quaternary and
           Upper Pliocene clays, Near-Kurinskaya Depression. (After
           Prozorovich and Sultanov, 1961); 5 = fine- and medium
           grained sands with carbonate-clayey cementing material
           up to 30%, Apsheron Peninsula; 6 = clays, Apsheron
           Peninsula. (After Prozorovich and Sultanov, 1961); 7 =
           fine-grained quartz sandstones, Devonian, southeastern
           slope of Voronezh Arch. (After Karpov, 1964); 8 =
           quartz siltstones, Devonian, southeastern slope of
           Voronezh Arch. (After Karpov, 1964); 9 = shales,
           Devonian, southeastern slope of Voronezh Arch. (After
           Karpov, 1964).

$$\phi_D = \phi_i e^{-CD} \tag{21}$$

where ϕ_D = porosity at depth $\underline{D}$; ϕ_i = initial porosity, which
is equal to 37.89%; $\underline{D}$ depth; and $\underline{C}$ - coefficient approximately
equal to 10^{-4}.   Linetskiy (1965) believes that a depth of 1,5000
- 2,000 m only oriented (adsorbed?) water is present.   According
to him, at these depths the porosity is of the order of 4-11%.
On the other hand, according to Kartsey et al. (1969), porosity
at 1,500 - 2,000 m depth is around 13.23%.   It is important to
note here that different drying temperatures were used by
different investigators in determining the remaining moisture
content;   this could account for some differences.   In addition,
the types of clays present would affect the results considerably.

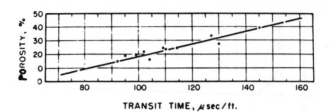

TRANSIT TIME, μ sec / ft.

Fig. 55. Relationship between mudstone porosity, φ
and transit time. Δt$_{log}$. (After Magara,
1968, fig.6, p.2474; courtesy of American
Association of Petroleum Geologists).

Prozorovic (1962) showed that on using the formula $\underline{y} = \underline{c} = \underline{A}e^{-\underline{B}\underline{x}}$ (where $\underline{y}$ is the specific weight in g/cm^3, $\underline{x}$ is depth in m, $\underline{C}$ = 2.8g/cm^3, $\underline{A}$ = 1.23g/cm^3, and $\underline{B}$ = 0.43), the actual specific weight deviates from the calculated one from +0.05 to -0.06 g/cm^3 to a depth of about 3,000 m.

Von Engelhardt (1960) developed equations relating void ratio, $\underline{e}$, and depth of burial, $\underline{D}$, in m for Tertiary deposits of Venezuela (data after Hedberg, 1936), Tertiary sediments for the Po Basin (data after Storer, 1959), and the Liassic shales of northwestern Germany as follows:

$$\underline{e} = \underline{e}_1 - \underline{b} \log \underline{D} \text{ (general equation)} \quad (22)$$
$$\underline{e} = 1.844 - 0.527 \log \underline{D} \text{ (Venezuela)} \quad (23)$$
$$\underline{e} = 1.700 - 0.481 \log \underline{D} \text{ (Po Basin)} \quad (24)$$
$$\underline{e} = 1.160 - 0.317 \log \underline{D} \text{ (Liassic of Germany} \quad (25)$$

where $\underline{e}_1$ is the void ratio at a depth of 1 m and $\underline{b}$ = compressibility of clay. Porosities of these deposits at a depth of 1 m are 65, 64 and 54% respectively.

Mukhin (1968) presented compaction equations for sediments of different geologic ages (Table 10). Relationship between the coefficient of porosity, $\underline{E}$ ($\underline{E}$ = Russian equivalent of void ratio, $\underline{e}$) and pressure is presented for Quaternary deposits of the southern part of the Caspian Sea upon burial under overlying sediments to a depth of 800 m.

The following formula enables one to calculate the volume of squeezed-out solutions during a certain compaction stage (Mukhin, 1968, p.51):

$$\underline{V}_i = \frac{\underline{h}_1}{1 + \underline{E}_1} (\underline{E}_{i\,1}{}^{-\underline{E}}{}_i) \quad (26)$$

Table 10. Compaction equations for sediments of different geologic ages. (After Mukhin, 1968, p.46)

Age	Region	Equation[1]	Initial porosity coefficient, E_o
Lower Cambrian (500 million years)	northwestern area of Russian Platform	$E = 0.67-0.181 \log p$	1.39
Upper Jurassic (140 million years)	central area of Russian Platform	$E = 1.07-0.26 \log p$	2.11
Maykop	Pre-Caucasus	$E = 1.46-0.444 \log p$	3.13
Apsheron	Pre-Caucasus	$E = 1.26-0.45 \log p$	3.06
Tertiary (50-60 million years)	Venezuela	$E = 1.20-0.40 \log p$	2.80
Quarternary (1 million years)	southern part of Caspian Sea	$E = 1.10-0.384 \log p$	2.66

[1] E = porosity coefficient; it is the Russian equivalent of void ratio, e. p = pressure $(Kg \, cm^2)$.

where $\underline{V}_i$ is the volume of interstitial solution squeezed out during a certain compaction stage; $\underline{E}_i$ is the present-day average porosity coefficient of sediment; $\underline{h}_1$ is the present-day thickness of deposit; and $\underline{E}_{i-1}$ and $\underline{E}_i$ are average coefficients of porosity at the beginning and end, respectively, of the compaction stage under investigation (Fig. 56).   As an example, the amounts of squeezed-out interstitial solutions from Quaternary deposits of the Caspian Sea from different depths are presented in Table 21.

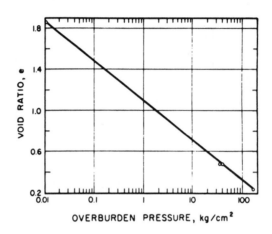

Fig. 56. Relationship between coefficient of porosity,
          E, and pressure for Quaternary deposits of
          southern part of Caspian Sea. E = e (void
          ratio). (After Mukhim, 1968, p.53).

      Dr. Koichi Aoyagi (1980, personal communication) studied
relationship between porosity and depth of the Neogene argillaceous and arenaceous sediments in deep wells drilled in Japan (see Fig. 57).   The relationships for argillaceous rocks were similar both in the Hokkaido and Niigata areas.   The porosity of argillaceous sediments at a burial depth of 1,200 m was 50%; at 1,600 m, 40%; at 2,100 m, 30%; at 2,700 m, 20%; and at 3,700 m, 10%.   On the other hand, relationship between porosity and depth of burial in the arenaceous rocks of these two areas differs considerably, porosity being higher in Niigata area, probably due to the differences in provenance, source materials, mineralogy, interstitial fluids, diagenetic processes, etc.

      In many laboratory investigations on the changes of mass
properties with increasing pressure (simulating burial of sediments in basins), the possible effects of temperature were not

Table 11. Volumes of squeezed-out solutions from various depths of Quaternary deposits of the Caspian Sea. (After Mukhin, 1968, p.54).

Depth of burial (m)	Volume of squeezed-out solution		Depth of burial (m)	Volume of squeezed-out pore-solution	
	ℓ/m³ of sediment	ℓ/m³/m		ℓ/m³ of sediment	ℓ/m³/m
0.003-0.10	280	-	500-800	64	0.21
0.10-10	381	38.2	800-1,000	24.5	0.12
10-60	204	4.1	1,000-1,500	61	0.12
60-100	75	1.9	1,500-2,000	45.5	0.09
100-250	103	0.7	2,000-2,500	37.5	0.075
250-500	90	0.36	2,500-3,500	19.4	0.0385

G. V. CHILINGARIAN

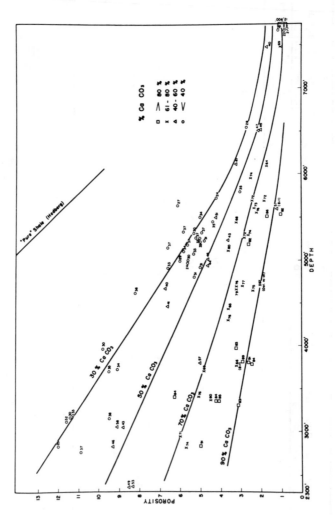

Fig. 57. Relationship between porosity and depth of burial of Ireton rocks. Carbonate content shown in percentage by weight on the axis perpendicular to page. (After McCrossan, 1961, fig. 13, p.461; courtesy of American Association of Petroleum Geologists).

taken into account.   It is, therefore, of particular interest
to note that Somerton and El-Shaarani (1974) found that the compressibility of sandstones increases with increasing temperature.
The effect of temperature was more pronounced at the lower effective pressures.

Sawabin et al. (1974) determined the relationship between
the void ratio (= volume of pores/volume solids) and effective
pressure, p_e ($p_e = p_t - p_p$, where p_t is the total overburden
pressure and p_p is the pore pressure) for unconsolidated sandstones from a depth of 3,000 ft. (Fig. 59).   In the effective
pressure range of 0-3000 psi, the void ratios varied from 0.85 to
0.19.   These authors tested unconsolidated, medium- to fine-
grained, arkosic sand cores obtained from oil-producing formations
of Pliocene and Upper Miocene age in the Los Angeles Basin,
California.   The overburden (external) pressure was held constant
at 3000 psi in a hydrostatic (three principal stresses are equal)
compaction apparatus, at a temperature of 140°F, while producing
the interstitial fluids and thus reducing the pore pressure and
increasing the effective (grain-to-grain) pressure.   It is the
latter stress that causes compaction and, consequently, the one
that should be used in plotting porosity-versus-pressure curves
whenever possible.

Effect of carbonate content on porosity.   McCrossan (1961, p.461)
showed the effect of impurities on degree of Ireton shales, in
which impurities primarily consist of carbonates which are relatively simple to determine.

Figure 57 shows a series of porosity-depth curves for
samples having different carbonate content.   In this graph, three
variables (porosity, depth of burial, and carbonate content) are
considered simultaneously, giving a series of curves which were
drawn in by eye.   These curves represent a surface of best fit
through the points.   As pointed out by Hedberg (1936), even for
purer argillaceous sediments, pressure-porosity relations are
better represented by a band than by a line.   This is also true
for calcareous shales.   Each group on the graph (Fig. 57) is
represented by a band which slightly overlaps the bands on either
side.   The contours are drawn through the middle of the bands
(McCrossand, 1961).

As shown in Figure 58, the more calcareous rocks have a
lower initial porosity and the decrease in porosity of these rocks
with depth is lower than in the case of less calcareous rocks.
Hedberg's (1936) curve for pure shale is also shown for comparison
in the upper part of the figure.   These relationships are probably due to the fact that the carbonate particles reduce the
amount of compactable material.   Even the 30% curve is also
appreciably flatter than Hedberg's, even though the carbonate

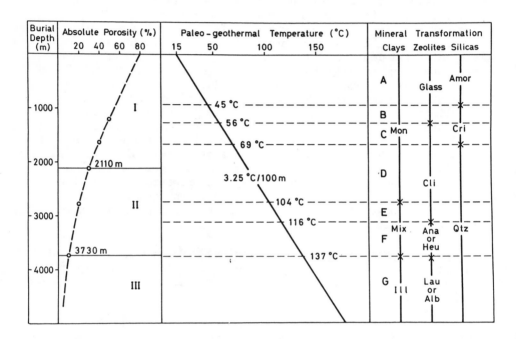

Fig.58. Relationship between porosity and depth of the Neogene
         argillaceous sediments. Paleotemperatures are also
         shown. (After Dr. K. Aoyagi, pers.comm., 1981)

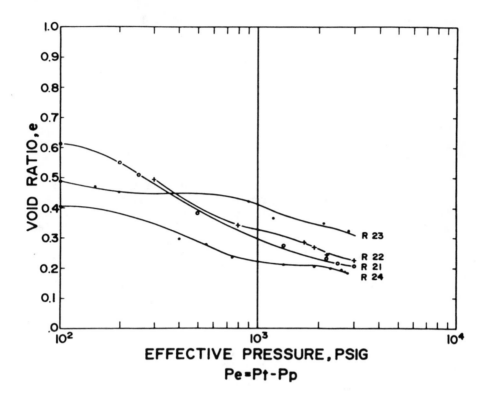

Fig. 59.   Experimental relationship between void ratio
and effective pressure for unconsolidated
sandstones. (After Sawabini et al., 1974,
fig.9, p.136; courtesy Soc. Pet. Eng. AIME).

grains are in the coarse clay size range.   As pointed out by
McCrossan (1961, p.460), comparison with Skempton's (1945) work
is not advisable because the depth ranges of these two studies
overlap only very slightly and extrapolation of either set of
curves is dangerous.

INTERSTITIAL WATER SALINITIES

In a geologically older petroleum-producing basin, the
salinities of normally pressure interstitial fluids in sandstones
may increase with depth of burial to a maximum of about 350,000
ppm (mainly NaCl).   In all cases, the shale waters are fresher
than those in the associate sandstones.   In some Tertiary age
basins, there is an abrupt freshening of the interstitial water

at depth in an abnormally high pressure zone, possibly due to the
presence of large volumes of undercompacted shales containing
fresher water.   This association of high fluid pressures in the
subsurface with low salinities (10,000 to 15,000 ppm) has been
well established for several geologically young petroleum basins
(see Rieke and Chilingarian 1974; Fertl, 1976).

Mekhtiev (1956) showed that in the Azerbayjan SSR oil fields
(USSR) mineralization of waters is decreasing with stratigraphic
depth and calcium-chloride waters are gradually replaced to bio-
carbonate waters.

Chilingarian et al. (1973) concluded that the mineralization
of pore solutions expelled from argillaceous sediments decreases
with increasing overburden pressure.   Thus, as pointed out by
Rieke and Chilingarian (1974), the interstitial water in shales
should be fresher than that in the associated sandstones.   The
concept of relative salinities of interstitial waters in well-
compacted and undercompacted shales and their associated sandstones, as envisioned by Rieke and Chilingarian (1974) are presented in Figure 60.

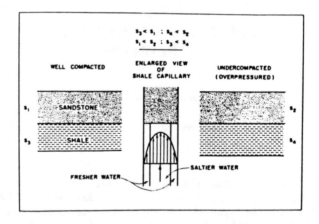

Fig. 60.   Relative salinities of pore fluids in well-
           compacted and undercompacted shales (s =
           salinity). (After Rieke and Chilingarian,
           1974, p.25).

Several reasons have been postulated for the water being
fresher in shales than in the associated sandstones.   One is that
the innerlayer water (adsorbed and oriented) of swelling clay
minerals is fresher than the free water.   According to Po'ster

et al. (1967), when a capillary is filled with an electrolyte solution, the density of the solution next to the capillary walls is
maximum, whereas along the center line of the capillary the density is lowest and approaches a normal one if the radius of the
capillary is large enough (Fig. 61).   Assuming that this density

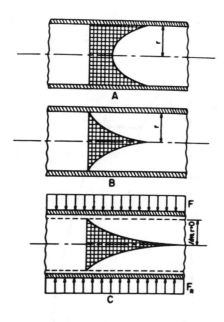

Fig. 61.   Behaviour of electrolyte solution in a cap-
           illary having radius r. A = Distribution of
           interstitial fluid density; maximum density
           occurs at the capillary walls. B = Distribu-
           tion of dissolving capacity of the interstitial
           fluid; maximum dissolving capacity is at
           center. C = Compaction of the capillary by a
           force F.   Illustrating that the part of the
           fluid squeezed out first is the one with the
           highest dissolving capacity (maximum salinity).
           (After Pol'ster el al., 1967, fig.32, p.72).

relationship holds and that the water in the centre of the
capillary is expelled first, then the salinities of interstitial
fluids in shales should be fresher than those in the associated

sandstones if changes in all other variables are identical.
Interstitial water in the undercompacted shales, however, should
be more saline than that in the well-compacted ones, because in
the former case a smaller portion of the more saline fluid present in the centre of capillaries is squeezed out into adjacent
sandstones.   Results of recent laboratory compaction studies
tend to confirm this (Fig. 62).   Chilingarian and Rieke (1976)

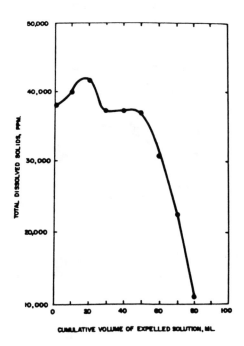

Fig. 62.   Variation in total dissolved solids content
with increasing compaction pressure in subsequent fractions of extruded solutions from
illite plus montmorillonite mixture (50-50)
saturated in sea water. (After Chilingarian
and Rieke, 1975, fig.3, p.677).

compared the chloride content in overpressured undercompacted
shales, well-compacted shales, and associated sandstones (Table
12).   The results give a further indication that water in both
types of shales is fresher than that in associated sandstones.
In addition, water in well-compacted shales is fresher than that
in the associated mineralogically similar (and from approximately

Table 12. Chlorinity of interstitial solutions in some associated undercompacted and well-compacted shales and sandstones. (After Chilingarian and Rieke, 1975, p.676).

Number of samples tested a/b/c	Depth, ft	Chlorinity, mg/l		
		Well-compacted shales (a)	Undercompacted shales (b)	Associated sandstones (c)
3/3/3	2,000-3,000	3,000-4,000	8,000-20,000	70,000-80,000
4/2/2	3,000-4,000	2,000-3,000	10,000-30,000	70,000-90,000
3/3/2	4,000-5,000	1,600-3,500	10,000-40,000	75,000-90,000
2/2/3	5,000-6,000	1,500-3,500	9,000-35,000	60,000-200,000
6/2/3	6,000-7,000	3,000-6,000	8,000-10,000	70,000-130,000
3/3/4	7,000-8,000	4,000-8,000	5,000-9,000	90,000-135,000
3/4	8,000-9,000	10,000-20,000	-	90,000-100,000
4/3/4	10,000-11,000	2,000-3,000	10,000-14,000	15,000-70,000
5/3/2	11,000-12,000	2,000-3,000	8,000-14,000	13,000-17,000
7/3/4	12,000-13,000	1,500-3,000	8,000-14,000	11,000-30,000
2/2/2	13,000-14,000	2,500-4,500	10,000-14,000	11,000-50,000
2/4	14,000-15,000	10,000-14,000	-	90,000-120,000

the same depth) undercompacted shales.

The findings of Fowler (1968) for the Chocolate Bayou Field, Brazoria County, Texas, seem to suggest that the salinity of water in undercompacted shales is higher than in well-compacted ones. He discovered a definite correlation between the high salinity of interstitial fluids and abnormally high pressures.   This is possibly owing to the fact that undercompacted shales did not have a chance to contribute their fresher water to associated sandstones. In addition, he studied the variation in salinity of produced water with time.   The typical pattern is one of decreasing salinity with time, and the freshest water is found in sands receiving most of this water from associated shales.   This is in agreement with the experimental results of several investigators, as discussed earlier, which indicate that salinity of waters in shales is less than that in associated sands.   The pore-water salinities in shales and in associated sandstones are compared in Figure 63.   The examples include field case studies from the

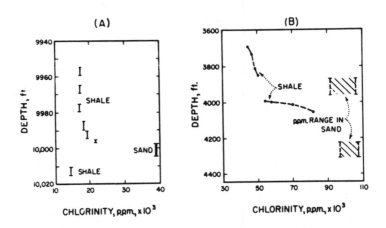

Fig. 63.   Chloride concentration in shales and sands. (In: Fertl and Timko, 1970, fig.4, p.15, based on data by Hedberg, 1967).

Middle East and Texas.   Some data from an offshore well in Louisiana have been given by Fertl and Timko (1970c).   In all cases, the salinities of interstitial fluids in shales are considerably lower than those in associated sandstones.

ABNORMAL FORMATION PRESSURES
(after Chilingarian and Rieke, 1976)

Considerable disagreement exists among engineers and geologists as to the mechanism responsible for the origin of abnormally high formation pressures in the Tertiary basins. A number of causes for the creation and maintenance of abnormal fluid pressures in argillaceous sediments have been proposed: (1) continuous loading and incomplete gravitational compaction of sediments, (2) tectonic compression, (3) faulting, (4) salt and shale diapirism, (5) phase changes in minerals during compaction and diagenesis, (6) geothermal temperature changes creating fluid-volume expansion and/or cracking of organic compounds, (7) osmotic and diffusion pressures, and (8) invasion of water derived from magmatic intrusions (see Rieke and Chilingarian, 1974).

Compaction of sediments under the influence of a vertical monotonic loading has long been a well documented geologic phenomenon (Sorby, 1908; Hedberg, 1926, 1936). This mechanism appears to be a likely candidate for creating an environment conducive to the formation of high fluid potentials in sediments having low permeability. Knowledge of both the vertical and lateral orogenic stress patterns in a depositional basin is of the utmost importance in interpreting abnormal fluid pressure environments and anticipating the location of oil and natural gas reservoirs. The influence of tectonic forces on compaction mechanisms has been discussed by Hubbert and Rubey (1959), Berry (1969), Thompson (1973), Hegert (1973), and Rieke and Chilingarian, (1974).

Overpressured, undercompacted formations in many parts of the world probably owe their origin to rapid sedimentation, which trapped water in compacting clays with practically no permeability. As mentioned previously, laterally-directed stresses of tectonic origin could also give rise to overpressured formations. Dickey (1972), however, stated: "It seems improbable that the comparatively unconsolidated shales could transmit horizontal stress". Anikiev (1964), on the other hand, attributed origin of high pressures in numerous overpressured formations to Recent and Quaternary tectonic movements. He (p.108) stated that gravitational compaction is not a probable cause of overpressures, because compaction occurred in many places where one does not find overpressured formations. Gravitational compaction also could not explain the origin of high pressures present in formations surrounded by halogenous deposits, in fractured limestones and tuffs, and in well-consolidated Palaeozoic rocks.

The forced influx of additional volumes of oil and gas under high pressures into reservoirs having a definite volume also could cause overpressures.

Hedberg (1974) discussed the relation of methane generation
to undercompacted shales, shale diapirs, and mud volcanoes.
The organic matter, which constitutes a substantial part of the
freshly-deposited muds, decomposes during diagenesis as a result
of biochemical and thermochemical processes.   The resulting
methane gas could create, or accentuate, the overpressured undercompacted state of compacting mud sediments, because of (1) building up additional internal pore pressure and (2) by further im-
peding the expulsion of interstitial pore water due to development of a second phase (gas) in the fluid (Hedberg, 1972, p.661).
Presence of gas bubbles dispersed in a liquid phase will reduce
the permeability of the rock to either phase.

The origin of abnormal subsurface pressures and factors
causing these pressures to persist for many millions of years
in sedimentary rocks are discussed in this section.   A mathematical description of the sedimentation and the compaction of
argillaceous sediments is also presented.   A quantitative grasp
of the compaction mechanism is based on the relationship between
overburden stress, effective stress, pore fluid stress, and the
physical properties of the competent and incompetent rocks.
One continuum model, which represents the mechanical behaviour
of compacting sediments, treats the gravitational compaction
process using a deformable one-dimensional porous medium (see
Baghavan, 1974).

In most sedimentary rocks, connate water is the predominant
interstitial fluid.   Degree of expulsion of water from the pore
space by compaction provides a foundation for developing interrelationships between abnormal formation presssures, velocity
of fluid expulsion, and pressure profiles through the sedimentary column.   Previous theories on how free and bound interstitial waters, along with hydrocarbons, are expelled and transported through the porous rocks in the form of an emulsion,
and/or a gaseous phase has been discussed by Chilingar and
Adamson (1964), Smith et al. (1971) and Pandey et al. (1974).

COMPACTION MODEL DEVELOPMENT

Dickinson (1951) performed a very thorough study on the
geologic aspects related to abnormally high fluid pressures in
the Gulf Coast Tertiary basin.   These high-pressure zones
occur frequently in isolated Miocene and Pliocene porous sand
beds surrounded by thick shale sections, which are located below the main deltaic sand series.   Location of the high fluid
potentials is controlled by the regional facies changes in the
basin and appears to be independent of depth and geologic age
of the formation.

In sedimentary environments where sedimentation has been
rapid, the thick accumulation of shales and mudstones of low permeability has retarded the expulsion of water and hydrocarbons.
This excess fluid is trapped as pore fluid, which must bear a
portion of the load that would normally be carried by the grain-
to-grain contacts.

First basic premise in the model development is that the rate
of gravitational compaction of sediments of a depositional basin
is limited by the low permeability of the argillaceous members.
In areas of rapid deposition, the expulsion of water from fine-
grained sediments has been shown to lag behind the loss of water
from sands (see Rieke and Chilingarian, 1974).

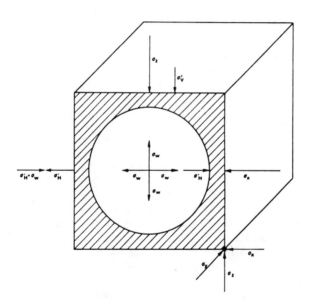

Fig. 64. Diagrammatic sketch of a pore space within a sedimentary
          framework and the corresponding stress state in the
          system, σ'_V is the effective (intergranular) stress in
          the vertical directions, σ'_H is the horizontal effective
          stress, σ_W is the pore water stress, and σ_z is the
          total vertical stress component. The total horizontal
          stress component in the x direction, σ_x, is equal to
          $\sigma'_H + \sigma_W$.

One can visualize the sediments as a two-phase continuum, with sediment grains in mechanical contact with each other and fluid filling every pore space not occupied by solids and wetting everything except the areas of the interparticle contacts. A diagrammatic sketch of stresses in a pore space within a sediment in presented in Figure 64. Each phase, i.e. solid and liquid, is assumed to occupy continuously a portion of the entire space, somewhat analogous to two vapours sharing a space in which they are assumed to exert their own partial pressure. The interstitial fluid is considered to be a slightly compressible homogeneous liquid and the pore pressure may vary from point to point. Pore fluids will flow through the void space under the influence of excess pore pressures. If the structure of the sediment remains rigid during the flow of fluids, a steady state seepage will occur. On the other hand, if the grain arrangement alters to a different packing order, then an unsteady-state flow will arise. The steady-state flow, which requires a rigid stationary framework, is not developed from a mathematical viewpoint in this section. Transient flow implies change in the effective stress which results in deformation of the solid matrix.

Katz and Ibrahim Compaction Model

Katz and Ibrahim (1971) presented a mechanical model for explaining compaction and fluid expulsion from shales (Fig. 65). Their model is based on Terzaghi's simple piston and spring analogy presented in Figure 66. The Katz and Ibrahim model is based on the compaction of an argillaceous layer between two permeable sand layers. As proposed by Terzaghi, the argillaceous sediment is represented by a series of springs and perforated disks. The perforated disks represent low-permeability clays, which restrict the escape of fluids, whereas the springs represent the deformable clay matrix. Sudden loading on the model corresponds to a rapid sedimentation rate. Water contained in the spaces between the perforated disks represents the interstitial fluid. If a stress is applied suddenly to the system the water between the disks initially will support the entire load. After a brief period of time, the water will be forced through the perforations in the disks either in an upward or downward direction, depending on the relative magnitudes of pressure in the compacting system, without lateral flow.

As the top and bottom disks move closer to the internal disks, the springs will begin to carry part of the applied load (Fig.65). Consequently, the fluid pressure between the external disks will decrease. When these disks approach each other, it will become difficult for the pore fluid to escape from inside the system. Katz and Ibrahim (1971) mentioned that the gradual decrease of permeability from the centre toward the top and bottom of the model could be represented either by a decrease in the number of

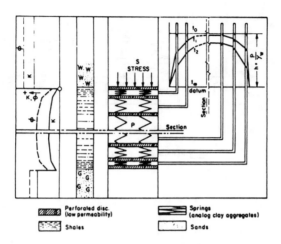

Fig. 65. Schematic representation of clay compaction, porosity
and permeability relationships, and creation of abnormally high formation fluid pressure. k = permeability;
ϕ = porosity; t = time; p = pore pressure; γ_w =
specific weight of water; h = height to which fluid will
rise in the tubes, which represents the pressure head
p; w = water, and G = gas. (After Katz and Ibrahim,
1971, fig.12; courtesy of the Society of Petroleum
Engineers of AIME).

openings in the disks or by an increase of the number of disks
per unit length in the model. Figure 65 also shows the fluid-
pressure distribution as a function of time. Higher fluid potential is shown to exist in the central portion rather than in
the upper or lower portions of the model. This means that it
takes more time for the fluid in the centre of the model to
escape than at the outer boundaries. The permeability and porosity distribution in the model is given on the left-hand side of
Figure 65.

The behaviour of Katz and Ibrahim model is in general agreement with the observed performance of the overpressured formations
in the Gulf Coast area. This model illustrates the reasons for
the higher porosity of undercompacted shales, the extreme drop
in permeability with increasing lithostatic pressure, and the
entrapment of high interstitial fluid pressure in the shales.

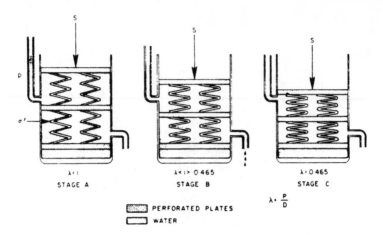

Fig. 66. Simple schematic representation of clay compaction
          (after Terzaghi and Peck, 1948, in: Hottman and
          Johnson, 1965, p.718). σ' = grain-to-grain bearing
          strength, S = axial component of total stress (over-
          burden pressure), p = fluid pressure, and λ = ratio of
          the pore stress to the total stress (e.g. 0.465 psi/ft
          {0.1074 kg cm^{-2} m^{-1}} of depth); σ' = S-p. Stage A:
          overpressured system; water is not allowed to escape.
          Stage B: water is allowed to escape; springs carry part
          of the applied load. Stage C: compaction equilibrium;
          load is supported jointly by the springs and the water
          (water pressure is simply hydrostatic).

REFERENCES

ALLEN, D.R. and CHILINGARIAN, G.V., 1975. Mechanics of sand
   compaction. In: Chilingarian and K.H.Wolf (Eds.). Compaction
   of Coarse-grained Sediments, I. Elsevier, Amsterdam, pp.43-77.
ALLEN, D.R., CHILINGARIAN, G.V. MAYUGA, M.N. and SAWABINI, C.T.
   1971. Studio e previsione della subsidenza. Enciclopedia della
   Scienza e della Tecnica.
ALLEN, D.R. and MAYUGA, M.N., 1969. The mechanics of compaction
   and rebound. Wilmington oil field, Long Beach, California.
   In: Symposium Land Subsidence, IASH/UNESCO, Tokyo, 89 (2):
   410-413.
ALI, S.A. and FRIEDMAN, G.M. (Compilers), 1977. Diagenesis of
   Sandstones. Am.Assoc. Petrol. Geol. Reprint Ser. No. 20:239 pp.
ANIKIEV, K.A., 1964. Anomalously High Formation Pressures in Oil
   and Gas Deposits. Tr. VNIGRI, No. 233, Izd. Nedra, Leningrad,
   167pp.
ATHY, L.F., 1930a. Density, porosity and compaction of sedimentary
   rocks. Bull. Am. Assoc. Pet. Geol., 14:1-24.
ATHY, L.F., 1930b. Compaction and oil migration. Bull. Am. Assoc.
   Pet. Geol., 14, 25-35.
ATHY, L.F., 1934. Compaction and its effect on local structure.
   In: W.E. Wrather et al. (Eds.). Problems of Petroleum Geology -
   a Sequel to Structure of Typical American Oil Fields. Am. Assoc.
   Pet. Geol., Sydney Powers Memorial Volume: 811-823.
AOYAGI, K., 1979. Paleo-temperature analysis by authigenic
   minerals in sedimentary rocks. J. Japanese Assoc. Petrol.
   Technol., 44(6), 1-5.
AOYAGI, K. and KAZAMA, T., 1980. Sedimentary mineralogy of
   argillaceous sediments from deep sea drilling project holes 436,
   438A, and 439, Japan Trench. In: Honza et al. (Eds.), Initial
   Reports of the Deep Sea Drilling Project, Vol. LVI, LVII
   (Washington), 1011-1017.
BHATTACHARYYA, A. and FRIEDMAN, G.M., 1979. Experimental compac-
   tion of ooids and lime mud and its implication for lithification
   during burial. J. sediment. Petrol., 49(4), 1279-1286.
BINDER, R.C., 1962. Fluid Mechanics. Prentice-Hall, Englewood
   Cliffs, N.J., 4th ed. 454 pp.
BLANCHE, J.B., 1973. The Rotliegendes Sandstone Formation of the
   United Kingdom sector of the southern North Sea Basin. Inst.
   Min. Metall., Trans., Sect. B: 85-89.
BOLT, G.H., 1956. Physico-chemical analysis of the compressibility
   of pure clays. Geotechnique, 6, 86-93.
BRENNER, R.P., NUTALAYA, P., CHILINGARIAN, G.V. and ROBERTSON,
   J.O. Jr., 1981. Engineering Geology of Soft Clays, Elsevier.
   (In press).
BROWN, L.F. Jr., 1969. Geometry and distribution of fluvial and
   deltaic sandstones (Pennsylvanian and Permian), north-central
   Texas. Trans. Gulf Coast Assoc. Geol. Soc., 19, 23-47.

BROWN, L.F. Jr., McGOWEN, J.H., SEALS, M.J., WALLER, T.H., and
   RAY, J.R., 1967. Role of compaction in development of geometry
   of superposed elongate sandstone bodies. Bull. Am. Assoc.
   Pet. Geol., 51, 455-456 (abst.).
BURST, J.F., 1969. Diagenesis of Gulf Coast clayey sediments and
   its possible relation to petroleum migration. Bull. Am. Assoc.
   Pet. Geol., 53, 73-93.
CARPENTER, C.B. and SPENCER, G.B., 1940. Measurements of com-
   pressibility of consolidated oil-bearing sandstones. U.S.
   Bur. Mines. R.I. 3540, 20pp.
CASAGRANDE, A., 1932a. The structure of clay and its importance
   in foundation engineering. J. Boston Soc. Civil Engrs., 19,
   168-209.
CASAGRANDE, A., 1932b. Research on the Atterberg limits of
   soils. Public Roads, 13, 121-136.
CHILINGAR, G.V. and ADAMSON, L.G., 1964. Does some migration of
   oil occur in a gaseous form? Proc. 22nd Internat. Geol. Congr.,
   New Delhi, Sec. 1, Part 1, 64-70.
CHILINGAR, G.V. and KNIGHT, L., 1960. Relationship between
   pressure and moisture content of kaolinite, illite, and
   montmorillonite clays. Bull. Am. Assoc. Pet. Geol., 44, 101-106.
CHILINGAR, G.V. and RIEKE, III, H.H., 1975. Chemistry of inter-
   stitial solutions in undercompacted (overpressured) versus
   well-compacted shales. In: S.W. Bailey (Editor-in-Chief),
   Proc. Internat. Clay Conf., Mexico City, Mexico. Applied
   Publishing Ltd., Wilmette, Ill., pp.673-678.
CHILINGAR, G.V., RIEKE III H.H. and SAWABINI, C.T., 1969.
   Compressibilities of clays and some means of predicting and
   preventing subsidence. In: Symp. Land Subsidence. IASH/UNESCO,
   Tokyo, 89(2), 377-393.
CHILINGARIAN, G.V. and RIEKE III H.H., 1976. Compaction of
   argillaceous sediments. In: W.H. Fertl, Abnormal Formation
   Pressures, Developments in Petrol. Sci., 2, Elsevier, pp.49-
   100.
CHILINGARIAN, G.V., SAWALINI, C.T. and RIEKE III, H.H., 1973.
   Effect of compaction on chemistry of solutions expelled from
   montmorillonite clay saturated in sea water. Sedimentology,
   20, 391-398.
CHILINGARIAN, G.V. and WOLF, K.H., 1975. Compaction of Coarse-
   grained Sediments, I. Elsevier Scientific Publ. Co.,
   Amsterdam, 548pp.
CHILINGARIAN, G.V. and WOLF, K.H., 1976. Compaction of Coarse-
   grained Sediments, II. Elsevier Scientific Publ. Co.,
   Amsterdam, 808pp.
CHILINGARIAN, G.V., WOLF, K.H. and ALLEN, D.R., 1975. Intro-
   duction. In: G.V. Chilingarian and K.H.Wolf (Eds.). Compaction
   of Coarse-grained Sediments, I. Elsevier, Amsterdam, 1-42pp.
COHEN, S.R., 1968. Measurement of the Viscoelastic Properties of
   Water-saturated Clay Sediments. Thesis, U.S. Naval Postgraduate
   School.

COLAZAS, X.C. 1971. Subsidence, Compaction of Sediments and
    Effects of Water Injection, Wilmington and Long Beach Offshore
    Oil Field. Thesis, Univ. South. Calif., 198pp (unpubl.).
COOGAN, A.H. and MANUS, R.W., 1975. Compaction and diagenesis of
    carbonate sands. In: G.V. Chilingarian and K.H. Wolf (Eds.),
    Compaction of Coarse-grained Sediments, I, Developments in
    Sedimentology 18A. Elsevier Publ. Co., 79-166pp.
COOGAN, A.H., 1970. Measurements of compaction in oolitic grain-
    stones, J. Sediment. Petrol., 40, 921-929.
COOK, H.E. and EGBERT R.M., 1981. Diagenesis of deep-sea carbonates.
    In: G. Larsen and G.V. Chilingarian (Eds.), Diagenesis in
    Sediments and Rocks, II. Elsevier (in press).
CURTIS, D.M. (Compiler), 1976. Diagenesis. Soc. Econ. Paleontol.
    Mineral., Reprint Ser. No. 1, 216pp.
DAVIES, T.A. and SUPKO, P.R., 1973. Ocean sediments and their
    diagenesis; some examples from deep sea drilling. J. Sediment.
    Petrol., 43(2), 381-390.
DICKEY, P.A., 1972. Migration of interstitial water in sediments
    and the concentration of petroleum and useful minerals. Proc.
    29th Int. Geol. Congr., Montreal, Que., Sect. 5, 3-16.
DICKEY, P.A. SHRIRAM, C.R. and PAINE, W.R., 1968. Abnormal
    pressures in deep wells of southwestern Louisiana, Science,
    160, 609-615.
DICKINSON, G., 1951. Geological aspects of abnormal reservoir
    pressures in the Gulf Coast region of Louisiana, U.S.A. Proc.
    3rd World Petrol. Congr., 1, 1-17.
DRONG, H.J., 1965. Die Schweminerale des Dogger beta und ihre
    diagenetischen Veranderungen (Vortrag Dtsch. Mineral. Ges.
    Hannover).
EBHARDT, G., 1968. Experimental compaction of carbonate sediments.
    In: G. Muller and G.M. Friedman (Eds.), Recent Developments
    in Carbonate Sedimentology in Central Europe. Springer,
    Heidelberg, 58-65pp.
FATT, L., 1958a. Pore volume compressibility of sandstone
    reservoir rocks. J. Pet. Tech., 10(3), 64-66.
FATT, L., 1958b. Compressibility of sandstones at low to moderate
    pressure. Bull. Am. Assoc. Pet. Geol., 42, 1924-1957.
FERTL, W.H., 1976. Abnormal Formation Pressures. Dev.Petrol.Sci.,
    2, Elsevier, 382pp.
FOTHERGILL, C.A., 1955. The cementation of oil reservoir sands
    and its origin. Proc. 4th World Pet. Congr., Rome. Sect. 1,
    301-314.
FOWLER, Jr. W.A., 1968. Pressure hydrocarbon accumulation and
    salinities - Chocolate Bayou Field, Brazoria County. SPE 2226,
    43rd AIME Fall Meet., Houston, Texas. Sept. 9pp.
FRASER, H.J., 1935. Experimental study of the porosity and
    permeability of clastic sediments. J. Geol., 43, 910-1010.
FRUTH, L.S., ORME, G.R. and DONATH, F.A., 1966. Experimental
    compaction effects in carbonate sediments. J. Sediment. Petrol.,
    36, 747-754.

FUCHTBAUER, H., 1961. Zur Quarzneubildung in Erdollagerstatten. Erdol Kohle, 14, 169-173.

FUCHTBAUER, H., 1967b. Einfluss des Ablagerungsmilieus auf die Sandstein-Diagenes im mittleren Buntsandstein. Sediment.Geol., 1, 159-170.

FUCHTBAUER, H., 1967a. Influence on different types of diagenesis on sandstone porosity. Proc. 7th World Pet. Congr., 2, 353-369.

FUCHTBAUER, H., 1967b. Einfluss des Ablagerungsmilieus auf die Sandstien-Diagenes im mittleren Buntsandstein. Sediment. Geol., 1, 159-179.

FUCHTBAUER, H. and MULLER, G., 1970. Sediments und Sedimentgesteine Schweizerbart, Stuttgart, 726pp.

FUCHTBAUER, H. and REINECK, H.E., 1963. Porositat und Verdichtung rezenter, mariner Sedimente, Sedimentology, 2, 294-306.

GAITHER, A., 1953. A study of porosity and grain relationships in experimental sands. J. Sediment. Petrol., 180-195.

GINZBURG, R.N., 1957. Early diagenesis and lithification of shallow-water carbonate sediments in South Florida. In: R.J. Le Blanc and J.G. Breeding (Eds.), Regional Aspects of Carbonate Deposition. Soc. Econ. Paleontol.Mineral. Spec. Publ., 5, 8-100.

GRATON, L.C. and FRASER, H.J., 1935. Systematic packing of spheres with particular relation to porosity and permeability. J. Geol., 43, 785-909.

HALL, H.M., 1953. Compressibility of reservoir rocks. Trans. Am. Inst. Min. Metall. Eng., 198, 309-311.

HAMILTON, E.L. 1971a. Elastic properties of marine sediments. J. Geophys. Res., 76(2), 579-604.

HAMILTON, E.L., 1971b. Prediction of in situ acoustic and elastic properties of marine sediments. Geophysics, 36(2), 266-284.

HEDBERG, H.D., 1936. Gravitational compaction of clays and shales. Am. J. Sci. 5th Ser. 231(184), 241-287.

HEDBERG, H.D., 1974. Relation of methane generation in under-compacted shales, shale diapirs, and mud volcanoes. Bull. Am. Assoc. Petrol. Geol., 58, 661-673.

HITCHON, B., 1968. Rock volume and pore volume data for plains region of western Canada sedimentary basin between latitudes 49° and 60° N. Bull. Am. Assoc. Pet. Geol., 52(12), 2318-2323.

HITCHON, B., 1969a. Fluid flow in the Western Canada sedimentary basin, I. Effect of topography. Water Resour. Res., 5, 186-195.

HITCHON, B., 1969b. Fluid flow in the Western Canada sedimentary basin, 2. Water Resour. Res., 5, 460-469.

HITCHON, B., 1971. Origin of oil: geological and geochemical constraints. In: H.G. McGrath and M.E. Charles (Eds.), Origin and Refining of Petroleum, Adv. Chem. Ser., 103.

HUBBERT, M.K. and RUBEY, W.W., 1960. Role of fluid pressure in mechanics of overthrust faulting. Bull. Geol. Soc. Am., 71(5), 617-628.

HUMPHRIES, W.K. and WAHLS, H.E., 1968. Stress history effects on dynamic modules of clay. J. Soil. Mech. Found., Div. Am. Soc. Civ. Engrs., 94(SM2), 371-389.

KATZ, D.L. and IBRAHIM, M.A., 1971. Threshold displacement pressure consideration for caprocks of abnormal-pressure reservoirs. 5th Conf. Drill. Rock. Mech., Austin, TX, SPE No. 3222, 113-124.

KISCH, H.J., 1981. Mineralogy and petrology of burial diagenesis (burial metamorphism) and incipient metamorphism in clastic rocks. In: G. Larsen and G.V. Chilingar (Eds.), Diagenèsis in Sediments and Rocks. II. Elsevier, in press.

KNUTSON, C.F. and BOHOR, B.F., 1963. Reservoir rock behaviour under moderate confining pressure. In: C. Fairhurst (Ed.), Rock Mechanics. Pergamon, New York, N.Y., 627-658pp.

LAMBE, T.W., 1958. The structure of compacted clay. J. Soil. Mech. Found. Eng. Am. Soc. Civil Engrs., 84(SM2 Pt. 1), 1654-1711.

LANGES, G.L., ROBERTSON, J.O. and CHILINGAR, C.V. 1972. Secondary Recovery and Carbonate Reservoirs. American Elsevier, New York, N.Y., 303pp.

LAPINSKAYA, T.A. and PROSHLYAKOV, B.K., 1970. Problem of reservoir rocks during exploration for oil and gas at great depths. In: A.V. Sidorenko et al. (Eds.). Status and Problems, 115-121pp.

LARSEN, G. and CHILINGAR, G.V., (Eds.), 1981. Diagenesis in Sediments and Rocks, II, Elsevier Sci. Publ. Co., (in press).

LAUBSCHER, H.P., 1960. Role of fluid pressure in mechanics o overthrust faulting discussion. Bull. Geol. Soc. Am. 71, 611-615.

LINETSKIY, V.F., 1965. Migration of Petroleum and Formation of its Deposits. Naukova Dumka, Kiev, 200pp.

MANUS, R.W. and COOGAN, A.H., 1974. Bulk volume reduction and pressure solution derived cement. J. Sediment. Petrol., 44, 466-471.

MAGARA K., 1968. Compaction and migration of fluids in Miocene mudstone. Nagaoka plain. Japan Bull. Assoc. Pet. Geol., 52, 2466-2501.

MAGARA, K., 1969. Porosity permeability relation of shale. Can. Well Logging Soc. J. 2, 47-74.

MAGARA, K., 1973. Compaction and fluid migration in Cretaceous shales of Western Canada. Geol. Surv. Can. Pap. 72-18, 81pp.

MAGARA, K., 1974. Compaction, ion filtration and osmosis in shale and their significance in primary migration. Bull. Am. Assoc. Pet. Geol., 58, 283-290.

MARTIN, R.J., 1962. Adsorbed water on clay, a review. In: E. Ingerson (Ed.), Clays Clay Miner. Proc. Natl. Conf. Clays Clay Miner., 9 (1960), 28-70.

McBRIDE, E.F. (Compiler). Diagenesis of Sandstone: Cement-Porosity Relationships. Soc. Econ. Paleontol. Mineral., Reprint Ser. No. 9, 233pp.

McGROSSAN, R.G., 1961. Resistivity mapping and petrophysical
  study of upper Devonian inter-reef calcareous shales of central
  Alberta, Canada. Bull.Am. Assoc.Pet. Geologists, 45/4), 441-470.
McCULLOH, T.H., 1967. Mass properties of sedimentary rocks and
  gravimetric effects of petroleum and natural-gas reservoirs.
  U.S. Geol.Surv.Prof. Pap., 528-A, 1-50.
MEADE, R.H., 1963. Factors influencing the pore volume of fine-
  grained sediments under low to moderate overburden loads.
  Sedimentology, 2, 235-242.
MEADE, R.H., 1964. Removal of water and rearrangement of
  particles during the compaction of clay sediments - review.
  U.S. Geol. Surv. Prof. Pap. 497B, 23pp.
MEADE, R.H., 1966. Factors influencing the early stages of the
  compaction of clays and sands - review. J. Sediment. Petrol.,
  36, 1085-1101.
MEKHTIEV, Sh. F., 1956. Questions on Origin of Oil and Formation
  of Petroleum Deposits of Azerbayjan. Izd. Akad. Nauk Azerb.
  S.S.R. Baku, 320pp.
MEYERS, W.J., 1980. Compaction in Mississippian skeletal limestones, Southwestern New Mexico. J. Sediment. Petrol., 50(2),
  457-474.
MITCHELL, J.K., 1956. The fabric of natural clays and its
  relation to engineering properties. Proc. Highway Res.Board,
  35, 693.
MITRA, S. and BEARD, W.C., 1980. Theoretical models of porosity
  reduction by pressure solution for well-sorted sandstones.
  J. Sediment. Petrol., 50(4), 1347-1360.
MUKHIN, U.V., 1968. Evaluation of the amount of pore solutions
  squeezed out from clayey sediments in natural conditions. In:
  G.V. Bogomolov et al. (Eds.), Pore Solutions and Methods of
  Their Study. Nauka i Tekhnika, Minsk, 45-54pp.
MULLER, G. 1967. Diagenesis in argillaceous sediments. In:
  G. Larsen and G.V. Chilingar (Eds.), Diagenesis in Sediments,
  Developments in Sedimentology, 8. Elsevier, Amsterdam 127-
  177pp.
NAZARKIN, L.A. 1979. Influence of Sedimentation Rate and
  Erosional Sections on Oil and Gas Potentials of Sedimentary
  Basins. Izd. Saratov Univ. 336pp.
O'BRIEN, N.R., 1963. A Study of Fissility in Argillaceous Rocks.
  Thesis, Univ. Illinois, Urbana, Ill, 80pp.
PANDEY, G.N., TEK, M.R. and KATZ, D.L., 1974. Diffusion of fluids
  through porous media with implications in petroleum geology.
  Bull. Am. Assoc. Pet. Geol., 58, 291-303.
PHILIPP, W., 1961. Struktur and Lagerstattengeschichte des
  Erdolfeldes Eldingen. Z. Dtsch. Gel. Ges., 112, 414pp.
PHILIPP, W., DRONG, H.J., FUCHTBAUER, H., HADDENHORSTS, H.G. and
  JANKOWSKY, W., 1963. The history of migration in the Gifhorn
  trough (NW Germany). Proc. 6th World Pet. Congr., Frankfurt/
  Main, Sect. 1, Pap. 19, 457-481.

PODIO, A.L., GREGORY, A.R. and GRAY, K.E., 1968. Dynamic prop-
    erties of dry and water-saturated Green River Shale under
    stress. J. Soc. Pet. Eng., 8(4), 389-404.
POL'STER, L.A, VISOVSKIY, YU, A., GUSEVA, A.N., PARNOV, E.I. and
    PLASKOVA, A.G., 1967. Physicochemical Characteristics and
    Hydrogeological Factors of Migration of Natural Solutions.
    (In Relation to Study of Oil- and Gas-Bearing Basins). Nedra,
    Leningrad, 172pp.
POWERS, M.C., 1967. Fluid-release mechanisms in compacting marine
    mudrocks and their importance in oil exploration. Bull. Am.
    Assoc. Pet. Geol., 51, 1240-1254.
PROSHLYAKOV, B.K., 1960. Reservoir properties of rocks as a
    function of their depth and lithology. Nefti Gaza, 41(12),
    24-29. (Assoc. Tech.Serv. Transl. RJ3421.
PROZOROVICH, E.A., 1962. About regularities in compaction of
    clays with depth (illustrated by Sarmatskiy Formation of Pre-
    Caucasus and Azerbayjan). Nov. Neft. i Gaz. Tekh. (Geol.), (4),
    62-64.
PUSCH, R., 1962. Clay particles. Statens Rad Byggforskning
    Handl., 40, 150pp.
PUSCH, R., 1966. Quick-clay microstructure. Eng.Geol., 1(6),
    433-443.
PUSCH, R., 1970. Microstructural changes in soft quick clay at
    failure. Can. Geotech. J., 7(1), 1-7.
QUIGLEY, R.M., 1961. Research on the Physical Properties of
    Marine Soils. Mass Inst. Tech. Soil Eng. Div. Publ., 117.
RAGHAVAN, R., 1974. Consolidation and rebound processes in one-
    dimensional porous columns. J. Geophys. Res., 79, 1687-1698.
RAGHAVAN, R. and MILLER, F.G., 1975. Mathematical analysis of
    sand compaction. In: G.V. Chilingarian and K.H. Wolf (Eds.),
    Compaction of Coarse Grained Sediments. Elsevier Scientific
    Publ. Co., Amsterdam.
RICHARDS, A.F. and HAMILTON, E.L., 1967. Investigation of deep-
    sea sediment cores III. Consolidation. In: A.F. Richards
    (Ed.), Marine Geotechnique, Univ. Illinois Press, Urbana, Ill.,
    327pp.
RIEKE III, H.H. and CHILINGARIAN, G.V. 1974. Compaction of
    Argillaceous Sediments. Elsevier, Amsterdam 424pp.
RIEKE, III, H.H., CHILINGARIAN, G.V. and FERTLE, W.H. 1978. A
    review of the importance of gravitational sediment compaction
    in oil producing areas. Energy Sources, 4(2), 165-193.
ROBERTSON, E.C., 1966. Compaction tests on aragonitic sediment.
    Bull. Am. Assoc. Pet. Geol., 50, 633 (abstr.).
ROBERTSON, E.C., 1967. Laboratory consolidation of carbonate
    sediment. In: A.F. Richards (Ed.), Marine Geotechnique (Int.
    Res.Conf.Mar.Geotech., Monticello, Ill., 1966). Univ. Ill.
    Press, Urbana, Ill., 117-127pp.
ROBERTSON, E.C. 1967. Laboratory consolidation of carbonate
    sediments. In: A.F.Richards (Ed.), Marine Geotechnique, Univ.
    Illinois Press, Urbana, Ill., 327pp.

ROSENQVIST, I. Th., 1958. Remarks on the mechanical properties
    of soil-water systems. Geol.Foren. Stockholm Forh, 80(4),
    435-457.
ROSENQVIST, I, Th., 1959. Physicochemical properties of soils,
    soil-water system. J. Soil Mech., Found. Div. Am. Soc. Civil
    Engrs., 85(SM2), 31-53.
ROSENQVIST, I. Th., 1966. Norwegian research into the properties
    of quick clay - a review. Eng. Geol., 1(6), 445-450.
SAWABINI, C.T. 1971. Triaxial Compaction of Unconsolidated Sand
    Core Samples under Producing Conditions at Constant Over-
    burden Pressures of 3,000 psi and a Constant Temperature of
    140° F. Thesis, Univ. South. Calif., 178pp (unpubl.).
SAWABINI, C.T. CHILINGAR, G.V. and ALLEN, D.R., 1974. Compress-
    ibility of unconsolidated arkosic oil sands. Soc. Pet. Eng. J.
    14, 132-138.
SCHLANGER, S.O. and DOUGLAS, R.G. 1974. The ooze-chalk-limestone
    transition and its implication for marine stratigraphy. In:
    K.J. Hsu and H.C. Jenkyns (Eds.), Pelagic Sediments: On Land
    and Under the Sea. Internat. Assoc. Sedimentol. Spec. Publ.
    No. 1, 117-148.
SHIBAKOVA, V.C. and SHALIMOVA, E.M., 1967. Optimal loading
    pressure and clay texture. Vestn. Moscov. Univ., (4) 90-94.
SIBLEY, D.F. and BLATT, H., 1976. Intergranular pressure solution
    and cementation of the Tuscarora orthoquartzite. J.Sediment.
    Petrol., 46, 881-896.
SINGER, A. and MULLER, G. 1981. Diagenesis in argillaceous sedi-
    ments. In: G. Larsen and G.V.Chilingar (Eds.), Diagenesis in
    Sediments and Rocks. II. Elsevier (in press).
SKEMPTON, A.W., 1953. Soil mechanics in relation to geology.
    Proc.Yorks. Geol. Soc., 29, 33-62.
STETYUKHA, E.I. 1964. Formulas of Correlations between Physical
    Properties of Rocks and Depth of Burial. Nedra, Moscow.
TAYLOR, D.A. 1948. Fundamentals of Soil Mechanics. Wiley, New
    York, N.Y., 700pp.
SMITH, R.E. 1969. Petrography porosity relations in carbonate
    quartz system. Gatesburg Formation (Late Cambrian) Pennsylvania.
    Bull. Am. Assoc. Pet. Geol., 53, 261-278.
Strakov, N.M. 1957. Methodes d'Etudes des Roches Sedimentaires,
    I. Translated by Bureau Recherche Geologie Miniere, 542pp.
STRAKOV, N.M. 1967, 1969, 1970. Principles of Lithogenesis, 1,
    2, 3. Consultants Bureau, New York, N.Y., 1: 245pp; 2: 609pp;
    3: 577pp. Plenum Publ., New York, N.Y.
STRAKOV, N.M., 1960. Principles of Theory of Lithogenesis, I.
    Izd. Akad. Nauk SSSR, Moscow, 212pp.
TAN, T.K., 1957. Report on soil properties and their measurements.
    Proc. Int. Conf. Soil Mech. Found. Eng., 4th, London, 3, 87-89.
TAYLOR, J.M., 1950. Pore space reduction in sandstones. Bull.
    Am. Assoc. Pet. Geol., 34. 701-716.
TAYLOR, S.R., 1968. Geochemistry of andesites. In: L.H.Ahrens (Ed.),
    Origin and Distribution of the Elements. Pergamon Press, Toronto,
    559-583.

TEODOROVICH, G.I. 1961. Authigenic Minerals in Sedimentary
   Rocks. Consultants Bureau, New York, N.Y., 120pp.
TEODOROVICH, G.I. and CHERNOV, A.A. 1968. Character of changes
   with depth in productive deposits of Apsheron oil-gas-bearing
   region. Sov. Geol., 1968 (4), 83-93.
TERMAN, M.J., 1973. Nuclear explosion petroleum stimulation
   projects. United States and USSR Bul. Am. Assoc. Pet.Geol.,
   57, 990-1026.
TERZAGHI, K. and PECK, R.B. 1968. Soil Mechanics in Engineering
   Practice. Wiley, New York, 2nd ed. 84pp.
TERZAGHI, R.A.D. 1940. Compaction of lime mud as a cause of
   secondary structure. J. Sediment. Petrol., 10, 78-90
VAN DER KNAAP, W. and VAN DER VLIS, A.C., 1967. On the cause of
   subsidence in oil producing areas. Trans. World Pet. Congr.
   7th, 3, 85-95
VON ENGELHARDT, W., 1960. Der Porenraum der Sedimente. Springer,
   Berlin, 207pp.
VON ENGELHARDT, W., and GAIDA, K.H., 1963. Concentration changes
   of pore solutions during the compaction of clay sediments. J.
   Sediment. Petrol., 33, 919-930.
VON ENGELHARDT, W. and PITTER, H., 1951. Uber die Zusammenhange
   zwischen Porositat, Permeabilitat und Korngrosse bei Sanden und
   Sandsteinen. Heidelb. Beitr. Mineral. Petrogr., 2(6), 477-491.
WELLER, J.M., 1959. Compaction of sediment. Bull. Am. Assoc.Pet.
   Geol., 43, 273-310.
WELLER, J.M., 1960. Stratigraphic Principles and Practice.
   Harper and Row, N.Y., 725pp.
WELLER, R.R. and MILLS, A.A., 1965. Surface properties and pore
   structure of marine sediments. Deep-Sea. Res., 12, 511-529.
WEYL, P.K., 1959. Pressure solution and the force of crystalliz-
   ation - a phenomenological theory. J. Geophys. Res., 64, 2001-
   2005.
WOLF, K.H. and CHILINGARIAN, G.V., 1976. Diagenesis of sandstones
   and compaction. In: G.V.Chilingarian and K.H. Wolf (Eds.),
   Compaction of Coarse-grained Sediments, II. Elsevier, Amsterdam,
   69-444pp.
WOLF, K.H., CHILINGARIAN, G.V. and BRANAGAN, D.F., 1976.
   Introduction. In: G.V.Chilingarian and K.H.Wolf (Eds.),
   Compaction of Coarse-grained Sediments, II. Elsevier, Amsterdam
   1-40pp.

DIAGENETIC REACTIONS IN SANDSTONES

Knut Bjørlykke,
University of Bergen,
Norway.

INTRODUCTION

Although aspects of sandstone diagenesis were studied earlier as a part of sedimentary petrology, it is only in recent years, particularly through the use of SEM, that it has developed into an important subject.   Increased oil prices have also produced incentives for oil companies to look more carefully into the mineralogical and geochemical nature of sandstone reservoir rocks and to try to understand diagenetic processes. The purpose of this research is:

1. To improve our ability to predict the distribution of porous sandstones to help oil prospecting;
2. To model the distribution of porosity and permeability in sandstone reservoirs as a part of production geology;
3. To obtain a detailed knowledge of the mineralogical and textural composition of sandstone reservoirs, and also of the chemical and physical reactions that may occur between minerals and any fluids that are injected into the reservoir. These are required in secondary and tertiary recovery methods for hydrocarbons.

Diagenesis is also important from a purely academic point of view as it deals with the processes that turn sediments into solid rocks.

Diagenetic reactions in sandstones represent a very complex field of research where disciplines from a wide range of subjects must be integrated in order to obtain some sort of synthesis. The main disciplines are:
1. Mineralogy of clastic and authigenic mineral phases

169

A. Parker and B. W. Sellwood (eds.), Sediment Diagenesis, 169–213.
© *1983 by D. Reidel Publishing Company.*

and their analyses by petrographic microscope,
cathoduluminescence, SEM, X-ray diffraction and microprobe.
2. Physical chemistry of minerals (and amorphous) phases
and their stability in relation to different porewater
compositions.
3. Circulation of porewater in sedimentary basins as a
function of the sedimentary environment, stratigraphy,
basin geometry and subsurface pressures.
4. Plate tectonic controls on sediment composition, basin
types, subsidence, heatflow and porewater flow.

DIAGENETIC PROCESSES IN RELATION TO POROSITY

The primary porosity in sandstones can be reduced by the
following processes:
1. Mechanical compaction of the grains. This process
includes rearrangements and deformation of the grains.
2. Precipitation of cement in the primary pore space.
3. Mineral reactions forming expanded hydrous minerals
like montmorillonite and kaolinite from feldspar and
mica.
4. Pressure solution resulting in increased compaction
and precipitation of cement.

Increased (secondary) porosity is due to:

1. Dissolution of clastic grains.
2. Dissolution of cement.
3. Mineral reactions resulting in loss of water, e.g.
montmorillonite⟶illite, and kaolinite⟶illite
transformations.
4. Fracturing (tectonic or hydrofracturing).

Early diagenesis

Early diagenesis includes all reactions between the mineral
phases of the sandstone and the porewater from the time of
deposition to moderate burial before the temperature is
significantly increased above surface temperature (25°C), and
where also bacterial reactions are important.

A general introduction to early diagenetic processes is
published in a recent textbook by Berner (1980).

Early diagenetic processes are very important in sandstones
because the porosity can be significantly changed at that stage
by early cementation or dissolution and also because these processes
will strongly influence the later diagenetic processes during

deeper burial.

REACTIONS BETWEEN POREWATER AND MINERALS

Immediately after sand is deposited, porewater may start to
react with the grains leading to dissolution of the clastic
minerals and precipitation of new authigenic mineral phases. This
requires, however, that the porewater is significantly out of
equilibrium with at least one of the clastic mineral phases.
Also, if the porewater is stationary, only a volumetrically very
insignificant reaction takes place before equilibrium is re-
established.

The initial porewater in sands may be:
1.   Sea water, for sandstones deposited in a normal marine
environment: porewater composition constant.
2.   Evaporitic porewater in saline basins and sabkha-high ionic
strength.
3.   Meteoric water for continental sandstones - initial porewater composition variable, mostly low ionic strength.
The composition of marine sea water is either in equilibrium with
the most common mineral phases or so slightly removed from the
stability field of the minerals that there is only a low chemical
potential available for equilibration (Garrels and Christ 1966,
p.364). Consequently, very little happens after deposition of
sandstones in the marine environment in terms of mineral reactions
between porewater and the clastic phases. However, shortly
after deposition the initial marine porewater may be exchanged
with porewater of different compositions. This may be either
porewater expelled from underlying sediments or meteoric water
from the continental meteoric water lens. Sea water may also
circulate a few centimetres down into underlying sand and cause
early cementation, e.g. of carbonate or phosphate.

Minerals precipitated as early cement in marine environments

Carbonate cement is the most abundant early diagenetic
cement in sandstones. In the marine environments the circulation of marine surface water, supersaturated with respect to
carbonate mud, precipitates carbonate cement in both carbonate
and clastic beaches forming "Beach Rock" with aragonite or high
Mg-cement (Figure 1).

Upward movements of porewater due to compaction and
diffusion will cause minerals to be precipitated near the sediment-water interface. This includes carbonate, iron oxides and
manganese oxides. Carbonate is the most important of these in
the marine environment since iron and manganese oxides may go
into solution again in a reducing environment.

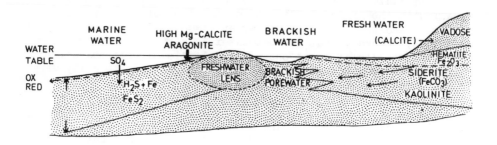

Figure 1.   Schematic presentation of types of early
            diagenetic cement in relation to environment.

The oxidation of organic matter, mostly by aerobic bacteria,
produces carbon dioxide $(CH_2O + O_2 \rightarrow CO_2 + H_2O)$.

Also, nitrate-reducing bacterial produce CO   where the O
concentration is lower $(5\ CH_2O + 4NO_3^- \rightarrow 2N_2 + 2HCO_3^- + CO_2 + 3H_2O)$.

Sulphate-reducing bacteria also contribute to a reduced pH
in the sediments just below the redox boundary by the production
of H_2S (Berner 1980).
$$2CH_2O + SO_4^{--} \rightarrow H_2S + 2HCO_3^-$$

Pyrite will form as an important early diagenetic mineral by
reaction between iron-bearing minerals and H_2S produced from
sulphate in the seawater by sulphate-reducing bacteria.   A long
period of exposure to sea water (slow sedimentation) and an ample
source of iron (iron oxides, chlorites, biotites etc.) will favour
the formation of a high concentration of iron sulphide cement.

Iron-rich chlorites or chamosites may form during early
diagenesis on the surface of the sand grains, and this may inhibit
overgrowth of quartz during deeper burial (Tillman & Almon, 1979).

Glauconite will form above the redox boundary, as it contains

tri-valent iron.   However, it mostly occurs as small pellets
rather than a massive cement, and it rarely reduces the porosity
significantly.

   In the zone of the sulphate-reducing bacteria (less than
0.5 m depth below the sea water -sediment/interface) all iron is
precipitated as sulphide, and other iron minerals such as siderite can only form below this zone (Berner 1971) or in fresh water
by alteration of $CaCO_3$ (Sellwood, 1971).   Clay minerals, feldspar and quartz are essentially stable in marine porewater near
the sediment surface.

   Quartz is not very soluble in marine water (only 6 ppm
SiO_2 at 25°: Siever et al., 1965) and the values are further
reduced in shallow nearshore waters to about 1 ppm SiO_2 (Blatt,
1979) because of precipitation by diatoms.   The silica con-
contration must therefore in such environments be increased 6
times before quartz can precipitate in marine porewater.   There
are several sources of silica.   Sediment containing volcanic
glass or biogenic amorphous silica will be highly supersaturated
with respect to quartz since the solubility of amorphous silica
is 150 ppm at 25 C and pH 7.   Silica may then precipitate as
less stable phases than quartz such as low cristobalite and opal
-CT (Lijima & Tada, 1980) which may later be replaced by quartz.
Porewater in clays vary between 10-60 ppm SiO_2 because of biogenic silica (Siever et al., 1965).

   As pointed out by Blatt (1979) fluvial surface water contains
on average 13 ppm silica, and meteoric water entering marine
basins as groundwater may therefore precipitate silica as it
mixes with marine porewater.   However, the amount of silica
precipitated by each volume of water is only of the order of
10 ppm so that approximately 100,000 volumes of water are required to precipitate one volume of quartz cement. A detailed
study of the Sparta Sand in the southern US shows fresh water
penetration down to about 600 m despite relatively low relief and
a silica concentration of between 10-20 ppm SiO_2 (Payne, 1968).

   Calculations (Sibley & Blatt, 1976; Blatt, 1979) show that
for the average flow rate of ground water, 10-20 m/yr, it will
require $10^7 - 10^8$ years to cement a sandstone by 2% with meteoric
porewater - clearly much too long a time to explain quartz cementation in most stratigraphical sequences.

Precipitation of cement in evaporitic environments

   Both marine and continental evaporitic basins contain water
with increased amounts of dissolved solids, and in these environments we may have precipitation of minerals that are unstable in
normal marine or meteoric water.

As well as carbonates and sulphates, zeolites are also
typical minerals that may form as early cement in evaporitic
environments.   The increased silica and alkali concentrations
also bring such porewater into the stability of smectite (Figure
2).

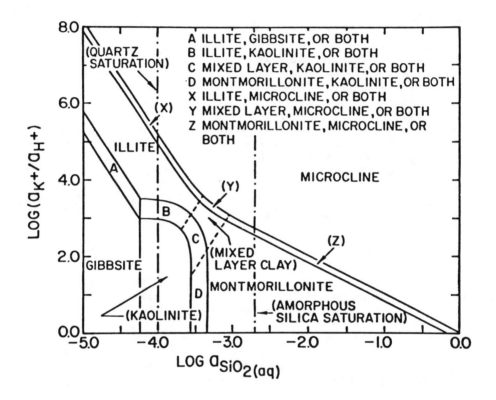

Figure 2.   Schematic presentation of the Gibbs free
energy change associated with the hydrolysis
of K-feldspar (Aagaard, 1979).

The type of early clay mineral cement in sandstones is
therefore to a large extent a function of climate.

In the North Sea basin, Permian and Triassic sandstones,
which were deposited in an arid climate partly as sabkha
(Nagtegaal 1978, 1979), are characterised by abundant smectite,

illite and authigenic feldspar (Waugh, 1978; Kessler, 1978; Hancock, 1978; Glennie et al., 1978).

In the vadose zone of arid environments early carbonate cement (calcite) and also iron and magnese oxides are common.

Precipitation of cement in fresh water

Fresh water is characterized by a low ionic strength and the pH is much more variable than in marine waters.

Meteoric water is initially undersaturated with respect to most mineral phases but circulation of meteoric water in the sub-surface will increase its concentration of solids. Leaching of evaporites drastically changes the composition of meteoric water and may cause reprecipitation of minerals such as gypsum and carbonates.

Weathering processes, such as the dissolution of feldspars and other silicate minerals to give illite, montmorillonite, kaolinite and eventually gibbsite, will all produce silica and its main sources in meteoric water. Early diagenesis in meteoric ground water can be regarded as a "subsurface weathering" where metallic cations (Na^+, Ka^+, Ca^{++}, Mg^{++}) and silica are removed and protons (H^+) are absorbed. One fraction of the silica released through the kaolinitization of feldspar may precipitate in the meteoric environment while more silica will precipitate in the meteoric marine water-mixing zone.

Porewater reactions with feldspar

After quartz, feldspar is quantitatively the most important component in most sandstones except in the lithic sandstones. It is therefore very important to understand the stability of feld-spars in water and how it may dissolve and form new minerals.

Feldspar is essentially stable in marine porewater and also in continental meteoric water with a high concentration of alkalis and silica. Dissolution of feldspar by hydrolysis occurs in acid and neutral water. Rain water is normally slightly acid due to the CO_2 dissolved in it and will dissolve feldspar as a part of normal weathering. Humic acids and CO_2 produced by plants may reduce the pH of meteoric water and increase the rate of feldspar dissolution.

After sandstones have been deposited, feldspar may react with the porewater flowing through the sand.

The rate of feldspar hydrolysis will depend on a complex set of variables:

1.   The composition of the feldspar.
2.   The grain size, degree of fracturing and alteration of
feldspars prior to deposition (specific surface area).
3.   The acidity (pH) of the porewater.
4.   Rate and duration of porewater flow (exposure time).

    Aagaard (1979) has shown how the dissolution of feldspar is
a two-step reaction (Figure 3).   First the alkali ion in the

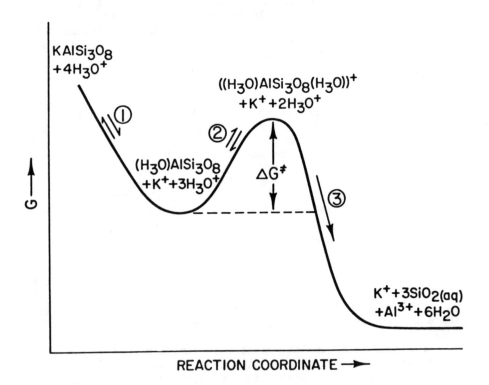

Figure 3.   Activity diagram depicting equilibrium phase
            relations among minerals and aqueous
            solution at 25 , L bar (after Aagaard 1979).

feldspar is exchanged for a hydronium (H_3O^+) ion to form H_3O
$AlSi_3O_8$, and then activation energy is required to break the
tetrahedral Al and Si structure.

In distinctly acid water (pH $<$5) the rate of dissolution is controlled by the activity of protons (pH). For less acid solutions (pH $>$5) the rate of feldspar dissolution is independent of the pH since it is predominantly the H_2O dipole and not H_3O^+ which is responsible for disrupting the feldspar surface structure.

During surface weathering the pH may be acid (pH $<$5) due to the production of humic acid and CO_2 by plants and bacteria. These acids will, however, probably be neutralized rapidly near the surface because their high reaction rate with feldspar and carbonate and other minerals. Subsurface meteoric water will therefore only be slightly acid or neutral (pH 6-7) and the reaction rate with silicate minerals is then slowed down. The rate of feldspar dissolution then becomes a function mainly of potassium (and sodium) concentration, surface area, temperature and time.

As meteoric ground water reacts with silicate minerals, protons will be consumed and alkali ions (K^+, Na^+) will be released into solution. It must be borne in mind that the solubilities of feldspars are very low (e.g. for potash feldspar 3×10^{-7} mole/litre, and for sodium feldspar 6×10^{-7} mole/litre in porewater (Berner, 1978) so that saturation is reached with a very small amount of dissolution). The dissolution will only proceed if the porewater is constantly being renewed by a flow of water which is undersaturated with respect to feldspar.

BURIAL DIAGENESIS

Increased burial has two important effects on diagenetic processes:

1.   The pressure and also the direct stress at grain contacts increase with deeper burial causing more mechanical compaction and pressure solution.
2.   The temperature is increased so that the solubilities for mineral species are different from those at surface temperatures.
3.   Minerals that are stable at surface temperature may become unstable at depth and become transformed to other minerals.

The pressure of the porewater may increase with depth as a function of the weight of the water column up to sea level or ground water table (Figure 4). Pore pressures higher than the weight of the overlying water column are referred to as over-pressures.

The total weight of the overlying rock sequence and water column (from sea level to the bottom) is taken up partly by the porewater and partly by the sand grain framework so that

$$S_g = S_e + pw \quad \text{(Gretener 1978)}$$

$$S_g = \text{the geostatic pressure}$$

$$S_e = \text{grain-to-grain stress}$$

$$pw = \text{porewater pressure}$$

Consequently the grain-to-grain stress or the so-called effective stress is:

$$S_e = S_g - pw$$

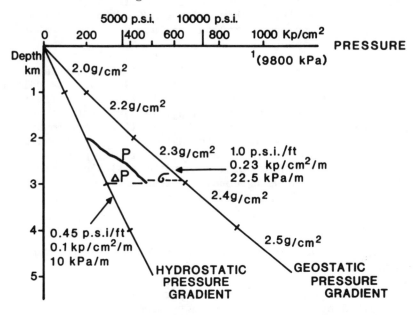

P = PORE PRESSURE σ = EFFECTIVE STRESS

ΔP = OVERPRESSURE

$$1 \text{ Pa} = 1 \text{ N/m}^2$$
$$1 \text{kilo Pascal} = \text{kPa} = 1 \text{ Atm} = 10^5 \text{ N/m}^2$$
$$1 \text{ Mega Pascal} = 1 \text{ N/mm}^2$$

Figure 4. Diagram illustrating the relations between burial depth and pressure for the hydro-static and geostatic load (modified from Gretner (1971)

The effective stress is therefore strongly reduced by increasing overpressure and maximum stress for any depth is obtained when the pore pressure is hydrostatic. When the pore pressure approaches the geostatic pressure the effective stress will approach zero. Both the degree of mechanical compaction of grains, and pressure solution, are directly related to the effective stress in sandstones.

Mechanical compaction includes fracturing and deformation of minerals and also solid-state ductile-grain deformation (McBride, 1978), which may contribute to a denser packing of grains. Such deformation also increases the area of surface contact between the grains and consequently reduces the stress between these grains. Undulose extinction near grain contacts may be a result of such ductile deformation.

Since the solubility of minerals increases with increasing stress of the grain contacts, following Riecke's principle, solution may take place at grain contacts and diffuse out and precipitate on parts of the grain surface facing open pores (Weyl, 1959; Boer, 1972; Boer et al., 1977). Locally, pressure solution may cause very high supersaturation sufficient to form amorphous silica (Boer et al., 1977).

The grain stress which causes pressure solution is the effective stress S_e per surface area of grain contact

$$S_e = (S_g - pw) \frac{100}{k}$$

where k is the percentage of grain contact area relative to the total horizontal cross section of the sandstones. With increasing pressure solution and cementation, the percentage of grain contacts will increase so that the effective stress will be distributed on a larger surface.

The effective stress is determined by the burial depth, rock densities and pore pressure (Figure 4).

At 2 km burial the effective stress is:

$$S_e = 2 \times 10^5 \text{ cm}^2 (2.2 \text{ g/cm}^3 - 1.0 \text{ g/cm}^3) \frac{100}{k} = 2.4 \times 10^5 \text{ g/cm} \frac{100}{k}$$

where the rock density is taken as 2.2 g/cm
if we assume a 10% grain contact (k = 100) the stress at the grain contact is 2.4×10^6 g/cm^2 = 2400 kg/cm^2 (Figure 5).

If the pore pressure is increased up to a pressure equal to a mudweight of 1.8 g/cm^3 the effective stress under otherwise the same condition is:

$$S_e = 2.10^5 \ (2.2 - 1.8) \ \frac{100}{10} \ \text{g/cm}^2 = 8.10^5 \ \text{g/cm}^2 = 800 \ \text{kg/cm}^2$$

in this case the overpressure has reduced the stress at the grain contact by 67% (Figure 5).

GEOSTATIC PRESSURE (0.2 - 0.25 Kp/cm^2/m)

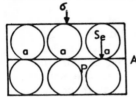

P = PORE PRESSURE

K = PERCENTAGE GRAIN CONTACT = $\frac{a}{A} \cdot 100$

EFFECTIVE STRAIN AT GRAIN CONTACTS:

$$S_e = (\sigma - P) \cdot \frac{100}{K}$$

EXAMPLE I: 2 KM DEPTH, AVERAGE GRAIN CONTACT 10 %

σ = 0.22 Kp/cm^2/m and HYDROSTATIC PRESSURE   0.10 Kp/cm^2

$$S_e = (0.22 - 0.10) \cdot 2000 \ \text{Kp/cm}^2 \cdot \frac{100}{10} = 2400 \ \text{Kp/cm}^2$$

EXAMPLE II: AS EXAMPLE (I), BUT OVERPRESSURE CORRESPONDING TO MUDWEIGHT 1.8 g/cm^3

$$S_e = (0.22 - 0.18) \cdot 2000 \ \text{Kp/cm}^2 \cdot \frac{100}{10} = \ 800 \ \text{Kp/cm}^2$$

THE EFFECTIVE STRAIN S$_e$ IN EXAMPLE (I) CAN BE REDUCED TO

800 Kp/cm^2 BY INCREASING THE GRAIN CONTACT FROM 10-30 %.

Figure 5. A calculated example showing the difference in net stress on grain contact for sandstone with hydrostatic porepressure and overpressure. Overpressure significantly reduces the grain-to-grain stress and therefore also the pressure solution.

In the North Sea the overpressures are high in the Brent Sandstone of the Viking Graben (Chiarelli and Duffan, 1980). The piezometric surface may be up to 3000 m above sea level which

means that the weight of water column of 3000 m must be added
to the hydrostatic pressure to obtain the porepressure. This
means that at 2.5 - 3.0 km depth, approximately 70-80% of the
weight of the overburden is carried by the porepressure and only
20-30% by the sandstone framework.

It is therefore clear that overpressure is a very important
factor in reducing pressure solution. It is highly unlikely
that the Brent Sandstone could have remained loose and friable
at burial depths of 2.5 - 3.0 km if the porepressures had been
hydrostatic.

Effects of increased temperatures as a function of burial depth

The effects of increased temperature with increasing burial
are:
1.   The solubilities of minerals change as a function of
temperature.
2.   The effect of hydration of ions is reduced with increasing
temperature making ions which are strongly hydrated at surface
temperature (i.e. Mg^{++}, Fe^{++}) more available to enter mineral
phases like carbonates.
3.   Hydrous minerals like montmorillonite and kaolinite will
become unstable and form less hydrous mineral phases, expelling
water.
4.   Water will expand slightly with increasing temperatures
raising the subsurface pressure if sealed off by low-permeability
sediments

(1)   The solubility of most mineral phases increased with
increasing temperature. For quartz in seawater it is about
6 ppm and increases to about 100 ppm at 100°C (approximately 3
km depth).   This means that quartz must be precipitated during
upwards porewater flow. It will, nevertheless, take 10,000
volumes of porewater to precipitate one volume of cement in the
entire overlying sediment column.   The solubility of carbonates,
however, will decrease with increasing temperature when the CO_2
partial pressure is equal to that of surface water (0.1 - 0.3
atm CO_2); Sippel and Glover (1964); Bathurst (1975); Blatt (1979,
p.150). A higher partial pressure will significantly increase
the solubility of carbonates.   Carbonates will therefore not
precipitate from upwards-flowing porewater unless it contains a
high concentration of CO_2 (e.g. from kerogen) which comes out
of solution with decreasing overburden.

(2)   Cations which at surface temperatures are strongly
hydrated, such as Mg^{++} and Fe^{++}, are unable to enter the
carbonate mineral lattice at low temperatures in marine porewater.   These ions will become less hydrated with increasing
temperature (60-100°C at 2-3 km) and iron and magnesium

carbonates may form with low Mg^{++}/Ca^{++} ratios (0.1) in the porewater (Usdowsky, 1968).   Late diagenetic carbonates are therefore commonly iron-and magnesium-rich, and dolomite, ankerite or
siderite may form.   We often find these late diagenetic minerals
in rocks such as the Jurassic Brent Sandstone, where euhedral
iron carbonates (ankerite and siderite) are clearly among the
last minerals to form (Bjørlykke et al., 1979).   Iron and
magnesium carbonates at this depth (2-3 km) are probably formed
at the expense of calcite and their concentration is limited
by the availability of Fe^{++} and Mg^{++} which may have come from
silicates such as biotite, chlorite, etc.

      (3) Following Le Chatelier's principle, increased tempera-
ture and pressure will favour denser, less hydrous minerals.
Analyses of cores from sedimentary basins like the Gulf Coast
show that montmorillonite (smectite), and mixed-layer minerals
become unstable between 60-100°C and will be transformed to
illite (Hower et al., 1976) (Figure 6) in the following manner:

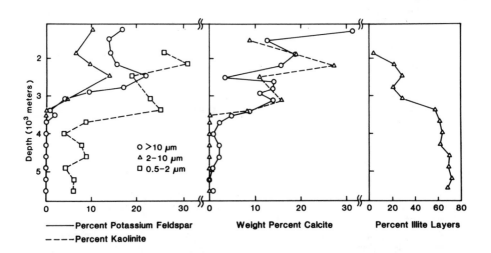

Figure 6. Stability of mineral assemblages in shales
           from the Gulf Coast (Hower et al., 1976).

a) Smectite + K^+ + Al^{3+} ⟶ illite + quartz + H^+
or Smectite + K-feldspar ⟶ illite + chlorite + quartz
Kaolinite becomes unstable between 120° - 150°C (Hower et al.,
1976); (Hoffman and Hower, 1979) and will form illite.
b) $3Al_2Si_2O_5(OH)_4$ + $2K^+$ ⟶ $2KAl_3Si_3O_{10}(OH)_2$ + $2H^+$ + $3H_2O$
         kaolinite   illite
Both these reactions capture cations and expel protons, which
must be neutralized after reactions. The most obvious way is
to dissolve feldspar and form illite and quartz.

c) $3KAlSi_3O_8 + 2H^+ \quad KAl_3Si_3O_{10}(OH)_2 + 6SiO_2 + 2K^+$

Adding up equations b) and c) we obtain:

$$Al_2Si_2O_5(OH)_4 + KAlSi_3O_8 = KAl_3Si_3O_{10}(OH)_2 + 2SiO_2 + H_2O$$

kaolinite   K-feldspar   illite   quartz

Secondary porosity in sandstones may form in this way (Figure 7).

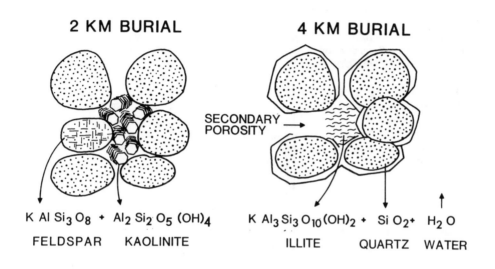

2 KM BURIAL   4 KM BURIAL

SECONDARY POROSITY →

$K\,Al\,Si_3\,O_8 + Al_2\,Si_2\,O_5\,(OH)_4$ $K\,Al_3\,Si_3\,O_{10}(OH)_2 + Si\,O_2 + H_2\,O$

FELDSPAR   KAOLINITE   ILLITE   QUARTZ   WATER

Figure 7.   Reactions between kaolinite and feldspar
            giving illite, owing partly to the
            instability of kaolinite with increasing
            burial depth (>4 km).

The hydrogen ions released from the transformation of
montmorillonite and kaolinite to illite may also be neutralized
by carbonate.   In the cores from the Gulf Coast examined by
Hower et al. (1976), feldspar and carbonate very significantly
disappeared at the same depth as kaolinite ($>$ 4 km).   Poorly
crystalline hydrous illite (IM illite) (Yoder and Eugster, 1955)
is deficient in potassium and will absorb potassium if
available with increasing burial to form well-crystallized (2M)
illite.   For further discussions on the stability of the clay
minerals see Velde (this volume) and Velde (1977).   Both clay
minerals and zeolites can be used as fairly efficient palaeotemperature indicators providing the whole mineral assemblage
is studied carefully (Hoffman and However, 1979; Ghen, 1979).
Zeolites require a high silica concentration in solution to form

and are therefore most abundant in sandstones containing volcanic
and amorphous silica diatoms.   Zeolites will react with quartz
and transform to feldspar at 150-200 C (Ghent, 1979).

SOURCES OF SANDSTONE CEMENT

    As pointed out by Blatt (1979) meteoric water usually has
more silica in solution (average solubility 13 ppm) than sea
water.

    Early diagenetic quartz cement in sandstones may therefore
be introduced through percolation of meteoric water, but since
each volume of water can only precipitate 10 ppm it takes very
large throughflow of meteoric water to supply cement (Figure 8).

    The growth of diatoms in the marine waters will reduce the
silica content to very low values but after deposition in the
marine sediments they will make the porewater supersaturated with
respect to quartz since the solubility of amorphous silica is 60 ppm
at 5°C in sea water (Siever et al., 1965) and this may be another
source of quartz cement.   Kaolinitization of feldspar releases
silica:

$$2KAlSi_3O_8 + 2H^+ + 9H_2O \longrightarrow Al_2Si_2O_5(OH)_4 + 4H_4SiO_4 + 2K^+$$

Because of the low solubility very little of the silica can be
held in solution and must eventually precipitate:

$$2K\ AlSi_3O_8 + 2H^+ + H_2O \longrightarrow Al_2Si_2O_5(OH)_4 + 4SiO_2 + 2K^+$$

    This may be an important source of early diagenetic quartz
cement.   Each volume of feldspar which is dissolve will repre-
cipitate up to 60% vol. of kaolinite and 40% vol. of quartz.

    Carbonate cement may form as an early marine cement or as
cement in the vadose zone in arid environments.   Chlorite may
also form as early pore lining in marine sandstones, possibly
during reworking of barrier bar sandstones, and develop a coat-
ing which will tend to reduce quartz overgrowth (Tillman and
Almon, 1979).

    The origin of quartz cement in sandstones, however, remains
a great problem.   Several authors have noted from the use of
cathodoluminescence (e.g. Blatt, 1979; Sibley and Blatt, 1976)
many sandstones show mostly quartz overgrowths but little
pressure solution of the original clastic quartz grains.

    This implies that most of the silica cement must have been
introduced from outside sources.   Some authors have suggested

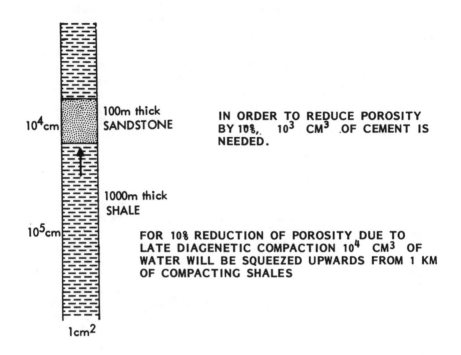

IN ORDER TO REDUCE POROSITY BY 10%, 10^3 CM^3 OF CEMENT IS NEEDED.

FOR 10% REDUCTION OF POROSITY DUE TO LATE DIAGENETIC COMPACTION 10^4 CM^3 OF WATER WILL BE SQUEEZED UPWARDS FROM 1 KM OF COMPACTING SHALES

THE SOLUBILITY OF SILICA MAY BE BETWEEN 10^{-4} AND 10^{-5}. TO PRECIPITATE 10^3 CM^2 OF CEMENT $10^3 \times 10^5 = 10^8$ CM^3 WATER IS NEEDED.

ASCENDING WATER EXPELLED FROM COMPACTING SHALES ARE THEREFORE OF NEGLIGIBLE IMPORTANCE AS SOURCE OF CEMENT FOR SANDSTONE THICKER THAN A FEW METERS, UNLESS THE FLOW OF WATER IS CONCENTRATED OVER A SMALL CROSS SECTION.

Figure 8.   Calculated example showing the vertical porewater flow resulting from compaction of shales. It is clear that water expelled vertically from compacting shales cannot contribute much to the cementation of thicker sandstones except when concentrated or chanelized.

that the quartz cement may have been derived from water squeezed
out of compacting shales (Fuchtbauer, 1978, 1979; Land and Dutton,
1978; Boles and Franks, 1979a, b). It can, however, be shown
that the water contained in the shales, because of the low
solubilities of silica, is much too little to precipitate
significant quantities of quartz cement (Blatt, 1979; Bjørlykke,
1979) (Figure 8). It is, however, difficult to quantify the
amount of silica dissolved by pressure solution, particularly
from stylolites and other irregularly spaced dissolution surfaces,
and pressure solution may easily be underestimated.

FORMATION OF SECONDARY POROSITY

     Evidence of secondary porosity in sandstones has been
observed by several authors (McBride, 1977; Parker, 1974; Loucks,
Bebout and Gallowway, 1977; Lindquist, 1977; Stanton, 1977;
Pittman, 1979; Hayes, 1979; Loucks et al., 1979). Schmidt and
McDonald (1979a, b) have summarised the evidence for the formation of secondary porosity. There is, as pointed out by these
authors, abundant evidence in many reservoir sandstones of the
leaching of minerals (mostly carbonates or feldspars), thus
creating secondary porosity. The main mechanism that has been
proposed for the formation of secondary porosity (Schmidt and
McDonald, 1979a, b) is the late leaching caused by CO derived
from maturation of kerogen.

     Secondary porosity may, however, form by other mechanisms,
and Schmidt and McDonald (1979a, b) do not discuss criteria for
distinguishing secondary porosity formed by leaching in meteoric
water during relatively shallow burial from that developed
during deeper burial leaching.

     The Jurassic Brent Sandstone of the Statfjord field in the
North Sea has been interpreted by Schmidt and McDonald (1979a,b)
as having mostly secondary porosity, assuming that the Brent
Sandstone had extensive early carbonate cement which was later
dissolved during deep burial. It seems unlikely that a deltaic
sandstone like the Brent Sandstone was all carbonate cemented.
The leaching of carbonate cement and feldspar grains may,
alternatively, be explained as being due to leaching by meteoric
water at a much earlier stage of diagenesis and that the preservation of porosity is due to overpressure established in these
sandstones (Bjørlykke et al., 1979). Also, the Brent Sandstone
contains late diagenetic carbonate and feldspar (Blanche and
Whitaker, 1978; Bjørlykke et al., 1979), suggesting relatively
high pH porewater. The occurrence of unaltered volcanic sandstone (Malm, et al., 1979) is difficult to explain if it should
have been subject to intensive leaching.

In attributing most of the secondary porosity formed in sandstones to the leaching effect of CO_2 derived from maturing kerogen, Schmidt and McDonald have, however, made no attempt to quantify the amount of CO_2 yielded from kerogen in source rocks.

The amount of CO_2 released from kerogen, subjected to increasing temperature, is limited by the amount of oxygen in the organic molecules of the kerogen. For example, analyses by Tissot and Welte (1978, p.164) show that in the Paris basin only 0.5% of the carbon atoms in the relatively sapropelic kerogen formed CO_2 between 700 m and 2500 m of burial. Humic kerogen may, however, produce as much as 20 cm^3 CO_2/g kergogen (Tissot and Welte, 1978, p.219). This is equal to about 0.001 mol. (the molar volume for CO_2 being 22.78 1) or 44 mg CO_2/g kerogen. The transformation of lignite to anthracite (250°) may produce as much as 75 cm^3/g or 140 mg of CO_2/g (Hunt, 1979) (Figure 9).

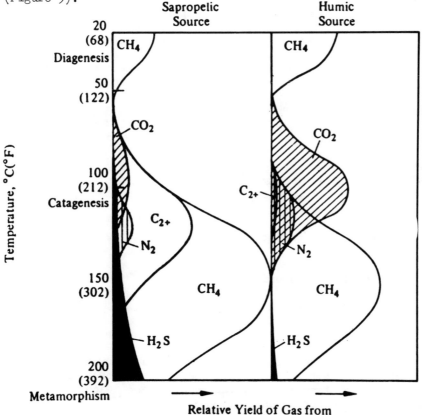

Figure 9. Generation of gases with depth from kerogen (from Hunt, 1979).

It should therefore be possible to calculate the amount of CO_2 produced from a given concentration of kerogen in a sedimentary sequence, as will be discussed below.

Let us consider a 1 km thick shale with an average kerogen content of 1%. This will, in a column of 1 cm^2 by 1 km, contain 2×10^3 g of kerogen, from which 1% to 10% CO_2 can be produced depending upon the kerogen type. Since each mole of CO_2 may dissolve one mole of carbonate by the following reaction:

$$CO_2 + H_2O + CaCO_3 = Ca^{++} + 2HCO_3^-$$

then 1 g CO_2 will dissolve approximately 2.3 g of $CaCO_3$, or almost 1 cm^3 (0.89 cm^3) (Figure 10).

In the case of feldspar the volume dissolved by CO_2 is larger but kaolinite is normally precipitated.

Assuming that 10% of the kerogen is converted to CO_2, 2×10^2 g of CO_2 may dissolve 200 cm^3 of carbonate cement and increase the porosity by 2% in a 100 m thick carbonate cemented sandstone. In the case of sapropelic kerogen the figure will be 0.2% or less. This is based on the assumption that CO_2 moved vertically up the sedimentary column without being concentrated. It follows from the above that the production of CO_2 in most sedimentary sequences will be insufficient to produce significant secondary porosity.

More generally, the amount of secondary porosity that can be leached from carbonate cemented sandstone and shales in the same sequence is:

$\emptyset$ = 100 (shale/sandstone ratio) · (kerogen/shale ratio)·
     K . C

where $\emptyset_s$ = % secondary porosity formed

K =   (kerogen $\longrightarrow$ CO_2 factor)

C = % CO_2 formed from kerogen during maturation

This is assuming vertical upward flow of porewater. If the flow is concentrated through a smaller cross-section up-dip or through tops of anticlines, faults etc., the leaching capacity will be increased correspondingly.

The values calculated here represent absolute maximum possible leaching capacities for given concentrations of kerogen. A very significant part of the CO_2 produced by the maturation

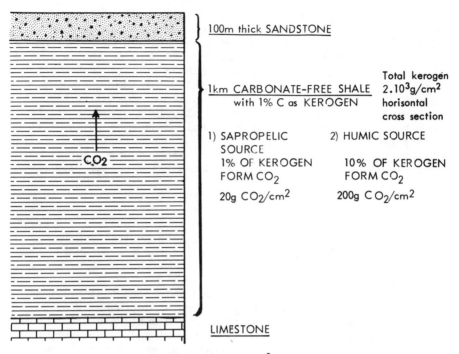

1g CO_2 WILL DISSOLVE 2.3g OF $CaCO_3$ ~ 1cm^3 $CaCO_3$

EXAMPLE: 1) WILL RESULT IN A LEACHED POROSITY OF 0,2 %

2) WILL RESULT IN A LEACHED POROSITY OG 2,0%

Figure 10.   A calculation of the amount of CO_2
generated from 1 km of shale with 1%
kerogen.   We can see that in the case of
vertical porewater flow the leaching
capacity of the CO_2 will only be able to
produce minor amounts of secondary porosity
in a 100 m thick overlying sandstone.
The shale is here assumed to contain no
carbonate or feldspar that could neutralize
any of the CO_2 produced.   In most cases
the amounts of CO_2 that can dissolve
carbonate or feldspar cement in the sandstone will be much less than in the
calculated example.

of kerogen must be expected to be neutralized by reactions with
minerals such as feldspar and carbonate in the shales.

Secondary porosity from maturating kerogen should therefore
only be considered in cases where sandstones are underlain by
carbonate-free source rocks.   In the North Sea, source rocks
like the Kimmeridgian shale normally do contain some carbonate.
Coal beds contained within the reservoir rock, as in the Brent
Sand, may also contribute to the leaching but are probably not
very significant.   In the lower Tertiary of the Gulf Coast,
however, the conditions for the formation of secondary porosity
are much more favourable (Loucks et al., 1977) because of the
more humic kerogen and the higher shale/sandstone ratio.   It is
also possible for expelled porewater to concentrate by water flow
through anticlines or by up-dip channeling of porewater from
basinal shales through sandstone bodies (Figure 11a, b).

    The calculations above suggest that in most basins the
CO_2 produced from maturing kerogen is insufficient to create
very significant secondary porosities at deep burial.   A very
large percent of the CO_2 released will be neutralized by
carbonate and feldspar in the shales (source rocks) where the
CO is generated.

    In the Gulf Coast Tertiary, from which the best evidence of
formation of secondary porosity at greater depth has been described, we do find that all feldspar and carbonate are
dissolved at 3-4 km depth (Hower et al., 1976).   However, as
pointed out above, clay mineral reactions may be more important
in producing acids than CO_2 from kerogen.   To understand the
formation of secondary porosity by leaching we must look into
all situations where particularly carbonate and feldspar may be
dissolved.   This includes:
1.   Leaching by meteoric water
2.   Leaching by CO_2 from maturing kerogen
3.   Transformation of smectite and kaolinite to illite
In the Gulf cores described by Loucks et al. (1977), however,
kaolinite seems to have formed at relatively great depth (4 km),
equivalent to where kaolinite is disappearing in the shales
studied by Hower et al. (1976).   This must be controlled by
low pH and low alkali concentrations.

CONTROLS ON DIAGENETIC REACTIONS CAUSING CHANGES IN POROSITIES
OF SANDSTONES

    Many well-sorted sandstones with a high primary porosity
will be cemented and lose most of their primary porosity at
relatively moderate burial depth.   This may occur prior to the
time of oil migration, which means that the sandstone loses its
potential as a reservoir rock.   It is therefore important to
try to model diagenetic processes in the subsurface in order to
be able to predict the distribution of porous sandstones at

depth.

Cementation of sandstones

Sandstones may be cemented in two principally different ways (Figure 12):
1. By net precipitation of cement introduced in solution from outside the sandstone. This implies that the sandstones retain constant volume.
2. By dissolution (pressure solution) of grains and reprecipitation of cement.
Net cementation by ions in solution from outside the sandstone requires a flow of porewater from an area where the solubilities of the minerals are higher than in sandstone. Only a fraction of the ions held in solution can be expected to precipitate. It therefore requires very large quantities of porewater to supply the necessary cement for a significant reduction in porosity.

For quartz the solubility at 5 km depth ($150°C$) is 100 ppm (Moray et al., 1962), and assuming a 10% efficiency of precipitation 10^5 volumes of water is required to precipitate one volume of cement. We must therefore look into the porewater circulation of sedimentary basins.

Diffusion is important to the short range transport of ions in porewater. The diffusion factor in loose sediments is of the order of 10^{-6} cm^2 $^{-1}$ (Berner, 1980), but it is reduced to 10^{-7} - 10^{-8} cm^2 $^{-1}$ in more consolidated sediment (Manheim, 1970). The rate of diffusion is related to the concentration gradient, and diffusion is therefore most important where there is an abrupt change in solubilities such as near a redox boundary where iron and manganese are trapped.

In sandstones at greater depths diffusion is probably most important in the short range transport of ions, in particular where pressure solution builds up a local concentration.

Downward flow of meteoric water

Since this requires a pressure head above sea level, downward movements of porewater are normally restricted to meteoric water. The driving force is the pressure head from a fresh water column above sea level. A flow will occur perpendicular to equipotential lines and will meet the porewater expelled from the basin in a hydrological boundary (Galloway, 1977). In the case of confined aquifers fresh water may extend deep into the basin until isostatic equilibrium is obtained between the column of meteoric water and the saline porewater. The depths of meteoric water penetration in metres below sea level is

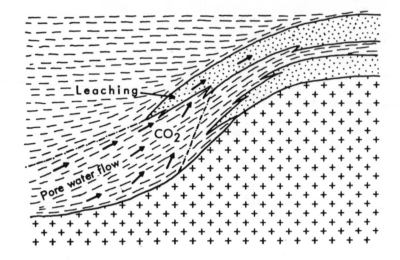

Fig. 11a.

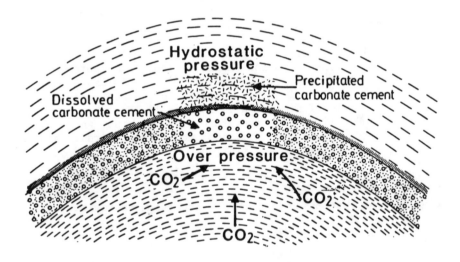

Generation of CO₂ from kerogen

Fig. 11b.

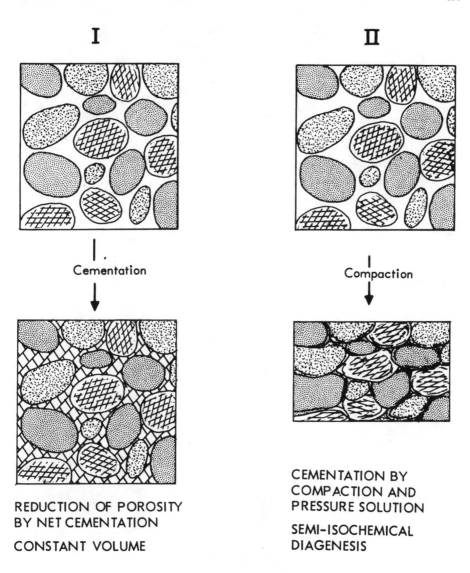

Figure 12.   Two ways to cement well-sorted sandstones:
             1) requires a large throughflow of
             supersaturated porewater, 11) is
             compaction and cementation by pressure
             solution (does not require a high
             porewater flow).

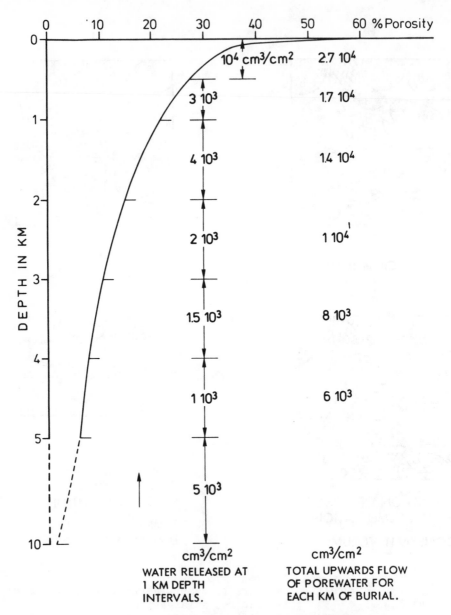

UPWARDS FLOW OF POREWATER FROM
COMPACTIONAL DEWATERING OF CLAYS AND
SHALES. CALCULATED FROM POROSITY GRADIENTS.

Figure 13.   Calculated upwards porewater flow
resulting from compaction.

theoretically 40 times the hydraulic head of the water table
assuming fresh and marine porewater densities of 1.00 g/cm^3
and 1.025 g/cm^3 respectively.   The depth of penetration of
meteoric water will normally be considerably less than this
figure due to higher pore pressure in the basin related to restricted circulation.   Fresh water may in some cases, however,
extend hundreds of metres or more below sea level and also flush
intercratonic basins (Chiarelli, 1978; White et al., 1963).

Fresh water aquifers have been reported on as far as 120 km
west off the Florida coast in Tertiary carbonates (Manheim,
1967).   Periods of regression, like the ones associated with
the Quaternary, provide an efficient pumping mechanism for
meteoric flow by the lowering of sea level.

Since the solubility of quartz is low, meteoric water
quickly reaches saturation with respect to quartz, but it has a
much larger capacity to leach feldspar and carbonates.

Porewater flow resulting from compaction (dewatering) of sediments

The main effect of compaction in sediments is loss of water,
and for a given porosity/depth relationship the total upwards
flow of porewater in a sedimentary basin can be calculated.

Its is very difficult to find a basis for estimating rates
of porewater flow in sediment basins as attempted by Blatt (1979).
The total flow of porewater upwards in a sedimentary basin is,
however, related to compaction and cannot exceed the total
volume of water contained in the underlying sequence as porewater and as molecular water in minerals.

Convection currents may be set up by temperature differences in the sedimentary column, but these are probably of
minor importance because of low permeability of certain layers,
and pressure gradients higher than hydrostatic pressure will
prevent water from flowing down back against this pressure
gradient.   Porewater is therefore probably not recycled to any
large extent in vertical loops, particularly at depth.

The total upwards flow of porewater during the entire
burial history due to compaction is in the order of 10^4 cm^3 /
cm^2 to 10^5 cm^3/cm^2 (Figure 13).   This means the porewater
can only precipitate about 1 cm   of quartz in the entire column
of overlying sediment.

Since the flow of porewater at depth in a sedimentary basin
is too small to transport large quantities of cement over long
distances it must be assumed that the diagenetic processes are
essentially isochemical except over short (< 10 m) distances.

At greater depth some of the mineral phases formed during early diagenesis, such as kaolinite and smectite, become unstable. These minerals have formed through the uptake of protons from meteoric water, thus releasing metallic cations.   At 3-4 km depth these protons are released again causing acid products (Figure 14).   Also, kerogen releases CO_2 during maturation to hydrocarbons and both these acidic producs must be neutralized by dissolving other minerals, particularly carbonate and feldspars.

Diagenetic processes in relation to basin types and plate tectonic setting

The above discussion shows that diagenetic processes depend primarily on the following factors:
1.   Primary composition of the sandstone grains related to source rock weathering and transport.
2.   Sedimentary facies and stratigraphic relationships in the sedimentary basin.
3.   Flow of meteoric water related to elevated ground water table around the basin.
4.   Subsurface pressure and sediment dewatering.
5.   Subsidence history and heatflow.

It is possible to relate these factors to basin types developed in different plate tectonic environments.

Intercratonic basins

Epicontinental basins are characterised by a very low relief both in the basin and in the source area (Shaw, 1964). Sedimentation is slow and carbonates, shales and orthoquartzites are the most typical lithologies.   During periods of regression the sediments may be exposed to meteoric water, which causes leaching of carbonate rocks.   However, the groundwater table in the exposed area and around the basin is not likely to be very much elevated above sea level, which would tend to prevent penetration of meteoric water to great depths.   Interlayering of low permeability clays and shales also discourages vertical movements of porewater.

The possibility of large scale upwards porewater flow from deeper strata is also reduced by the fact that the total thickness of the sedimentary sequences will normally be limited to 2-3 km unless there is no further thinning of continental crust (Kinsman, 1975).   Epicontinental sediments therefore have a low degree of mobility of porewater and are likely to undergo relatively isochemical diagenesis.

Sandstone in epicontinental sequences are usually thin and

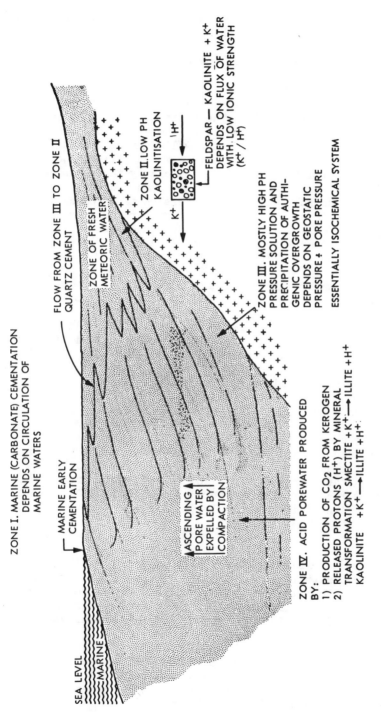

Figure 14. Schematic representation of diagenetic reaction and environments.

mineralogically mature which reduces the potential for secondary porosity through feldspar leaching.

Basins along passive (Atlantic) margins

Sedimentary basins forming along passive margins are fed by major rivers from the continent and prograde into deep water underlain by oceanic crust. The basins are asymmetrical by the very nature of their formation since the direction of sediment supply is constant. Progradation will form offlapping units which are inclined seaward, a feature which will be accentuated by increased compaction of the distal clayey facies and also by isostatic responses to loading of the oceanic crust. Only major regression followed by transgression may produce onlapping relations. This may be on the shelf or particularly in the slope system (Vail et al., 1977). Good examples of onlapping sands in submarine canyons have been described off Brazil (Brown and Fisher, 1977).

Meteoric water diagenesis may be strong in the deltaic sand deposits. The offlapping sand limits will form good conduits for updip water expulsion by low permeability mud (Figure 15). Growth faults are the most important barrier for updip migration of water or oil. Because of the low geothermal gradients in these basins maturation occurs relatively late, and where the sandstones are not over-pressured, substantial pressure solution may occur before oil mingration.

Since the sediment thickness of those basins may be very high, as in the Mississippi delta, the total upwards porewater flow due to compaction may be substantial and contribute to secondary leaching and the development of secondary porosity. The sediments deposited in a passive margin basin, like the Mississippi, are derived from a low relief source area of older sedimentary rocks or from weathered basement and will therefore have a relatively mature mineralogy.

Rift basins

Rift basins are characterised by sandstones with a low mineralogical maturity and feldspar-rich sandstones or arkoses are common. This is because the sediment may be derived from uplifted horsts or platforms along the rift where erosion has pierced the sedimentary cover and eroded into the crystalline basement.

The Middle Jurassic Brent Sandstone and equivalent Jurassic sandstones in the North Sea are good examples of this where erosion has reached into Precambrian basements and produced arkosic and micaceous sandstones (Figure 16).

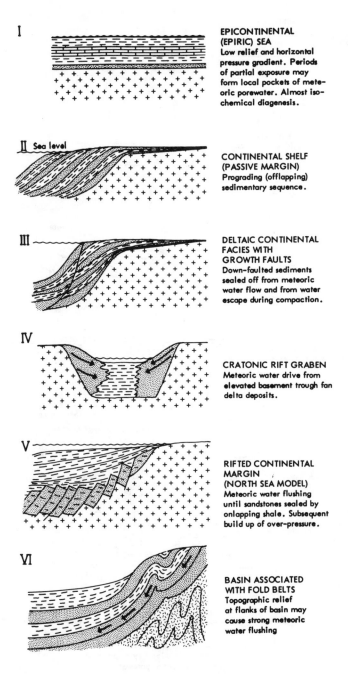

I

EPICONTINENTAL (EPIRIC) SEA
Low relief and horizontal pressure gradient. Periods of partial exposure may form local pockets of meteoric porewater. Almost isochemical diagenesis.

II Sea level

CONTINENTAL SHELF (PASSIVE MARGIN)
Prograding (offlapping) sedimentary sequence.

III

DELTAIC CONTINENTAL FACIES WITH GROWTH FAULTS
Down-faulted sediments sealed off from meteoric water flow and from water escape during compaction.

IV

CRATONIC RIFT GRABEN
Meteoric water drive from elevated basement trough fan delta deposits.

V

RIFTED CONTINENTAL MARGIN (NORTH SEA MODEL)
Meteoric water flushing until sandstones sealed by onlapping shale. Subsequent build up of over-pressure.

VI

BASIN ASSOCIATED WITH FOLD BELTS
Topographic relief at flanks of basin may cause strong meteoric water flushing

Figure 15. Schematic representation of basin types with respect to meteoric porewater flow.

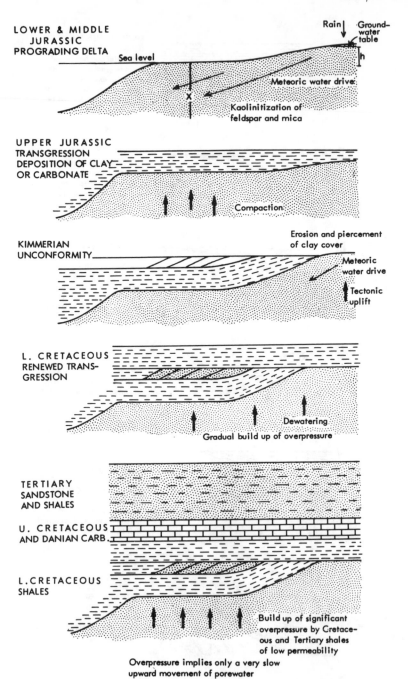

LOWER & MIDDLE
JURASSIC
PROGRADING DELTA

Rain Ground-
water
table

Sea level

h

Meteoric water drive

x

Kaolinitization of
feldspar and mica

UPPER JURASSIC
TRANSGRESSION
DEPOSITION OF CLAY
OR CARBONATE

Compaction

KIMMERIAN
UNCONFORMITY

Erosion and piercement
of clay cover

Meteoric
water drive

Tectonic
uplift

L. CRETACEOUS
RENEWED TRANS-
GRESSION

Dewatering

Gradual build up of overpressure

TERTIARY
SANDSTONE
AND SHALES

U. CRETACEOUS
AND DANIAN CARB.

L. CRETACEOUS
SHALES

Build up of significant
overpressure by Cretace-
ous and Tertiary shales
of low permeability

Overpressure implies only a very slow
upward movement of porewater

Figure 16.   Interpretation of the diagenetic history of
the Jurassic sandstones of the North Sea.

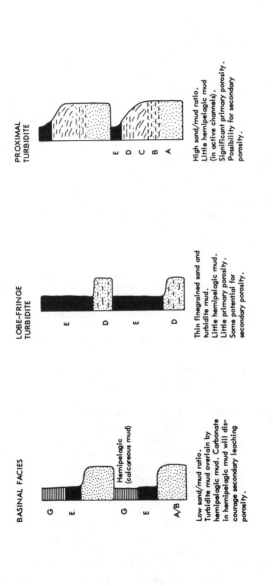

Figure 17. Potentials for primary and secondary porosity in different types of turbidite basins.

When the basement contains micaceous gneisses like the
Shetland Platform this sandstone becomes very rich in mica.
Mica (particularly biotite), feldspar and also certain cements
may be affected by flushing by meteoric water through delta
complexes causing leaching of carbonate cement and kaolinitization of feldspar.

When the period of active rifting is ended, the horst and
platform around the rift basin will subside owing to lower geothermal gradients and cause a transgression with a deposition of
an onlapping sequence (Figures 14, 15).   The feldspathic sandstones deposited during the active rifting stage are therefore
likely to be overlain by more distal shaley facies which,
because of onlapping relationship, will form a low permeability
barrier for water released during compaction.   The resulting
overpressure therefore is a typical feature of rift basins of
which the North Sea is a good example (Chiarelli and Duffaud,
1979).   As pointed out earlier an overpressure strongly reduces
pressure solution.   Relatively high geothermal gradients will
produce maturation of organic matter, and the possibility for
oil migration into reservoir rocks at an earlier stage before
a strong porosity reduction by pressure solution.

Turbidite basins

Deep water turbidite sandstones may be deposited in a wide
range of basin types (passive margins, wrench basins and basins
associated with subduction or continent collision zones).

From a diagenetic point of view the most important factors
of deep water are:
1.   They have never been flushed with meteoric water during
shallow burial and therefore normally contain little diagenetic
kaolinite.
2.   With the exception of proximal turbidites and sand
deposited by traction currents the sand will be poorly sorted
and will lose primary porosity rapidly by compaction.
3.   Basinal turbidites will often be intercalated with pelagic
carbonate and will be subjected to early carbonate cementation,
and the calcareous shales may prevent the formation of secondary porosity by leaching (Figure 17).

Sedimentary basins associated with oceanic island arcs

Sedimentary basins fed totally or partly from exposed
ensimatic island arc complexes or obducted oceanic crust have
an initial composition which is fundamentally different from
that derived from continental crust.   The sand grains from
basalt or gabbro are highly unstable both chemically and
mineralogically.   The primary porosity may therefore be

reduced rapidly due to formation of clay rims or coats and
mechanical squashing of the grains (Galloway, 1979).   It is therefore probable that the primary porosity and permeability may be
reduced to uneconomical levels before the sandstones reach the
depth of oil migration.

RELATIONS BETWEEN DIAGENESIS AND POROSITY AND PERMEABILITY

    We have discussed diagenetic processes above in relation to
porosity.   It is however very important to try to relate diagenetic processes to permeability, which may be highly variable
in sandstones.   We often find that there is an extremely poor
correlation between porosity and permeability (Sommer, 1978).

    Rim cement and porelining cement reduce the permeability by
very small amounts for a given porosity reduction (Pittman, 1977).
Porefilling cement like kaolinite has a very marked effect on
the permeability and the pores between kaolinite crystals will
either not be filled with oil or oil cannot be produced from
such pores.   Typical kaolinite crystals 3μm x 3μm x 1μm have a
specific surface of 1.5 m^2/g which is important in relation to
the surface properties of the sandstone (Figure 18a).

Figure 18a.   Magnification of kaolinite aggregates
from Figure 18f.

Low permeability/porosity ratios in the Jurassic sandstones
of the North Sea can be related to high kaolinite contents.

Fibrous or flaky illite often bridging the pores also offers
a high resistance to fluid flow through the sandstones.   This
type of illite is particularly common in the Permian sandstones
(Nagtegaal, 1979; Glennie et al., 1978; Hancock, 1978).   Fibrous
illite is also typical of some Triassic sandstones (Figure 18b).

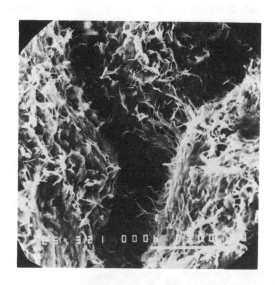

Figure 18b.   Fibrous illite from Triassic sandstones.

In Jurassic sandstones fibrous illite is much less common
but fibrous and flaky illite has been formed in the Brent Sandstone as one of the last diagenetic stages.   Sommer (1979) has
interpreted the formation of this late diagenetic illite as
formed during oil migration and K-Ar dating of the illite suggests
a time of migration of 45-55 my.   Late-stage illitization in
the Brent Sand strongly reduces the permeability in these sandstones (Sommer, 1978).

METHODS USED IN THE STUDY OF SANDSTONE DIAGENESIS

The petrographic microscope is still the most useful tool
in the study of sandstones.   By impregnating the sample with
stained (blue) Araldite under vacuum, the pores stand out much
more clearer (see pictures Schmidt and McDonald, 1979, p. 814).
It is, however, difficult to obtain full penetration of Araldite
into the smaller pores.

Porecasts can be obtained by impregnating a small cube of
the sandstone with Araldite or other resins, and dissolving out
the grains by hydrofluoric acids.   An Araldite cast of the pores
can then be studied in normal incident light microscope and by
SEM.   A pore cast gives a good impression of the characteristic
pore size and pore geometry, particularly pore throats etc.

Scanning electron microscopy has become the most valuable
tool for the study of the micro-textures, particularly of
authigenic minerals which have formed in the pores of sandstones.
Many of the authigenic minerals can readily be recognized by
their characteristic crystallography and habits.   These are mostly
typically quartz, feldspar, carbonate and clay minerals such as
kaolinite, chlorite and illite.

Some authigenic minerals, however, may develop an anhedral
surface during stages of mineral growth and may therefore be
difficult to recognise.   In some cases, stages of growth (e.g.
of kaolinite and feldspar) may resemble dissolution (Figures
18c, 18d).   Semiquantitative element analyses should therefore
always be used in such studies to aid mineral identification.
Comparisons with XRD analyses of the same samples should also be
made.

Figure 18c.   Feldspar from the Mid-Jurassic sandstones
               from the Statfjord field.   Note the
               growth forms which may resemble dissolu-
               tion textures. The pinnacles grow up to
               form perfect crystals indicating that
               leaching has not occurred.

Clastic clay minerals, which are particularly abundant in
less well sorted sandstones, are often very difficult to
recognize in the SEM because they occur in aggregates without
characteristic textures.

Figure 18d.   Irregular knobby kaolinite which is
              interpreted as a further growth stage of
              kaolinite (Stafjord Fm., Statfjord Field).

The microprobe offers the possibility for analyses of both
clastic and authigenic minerals in thin sections.   This method
is particularly useful for the study of mineral transformations
(e.g. transformations of mica to kaolinite), since profiles of
chemical analyses can be run across mineral boundaries.   The
microprobe also makes it possible to analyse thin authigenic
mineral overgrowths on clastic minerals, e.g. on feldspar
(Bjørlykke et al., 1979).   The composition of these authigenic
mineral phases reflects the porewater chemistry at different
stages of burial (Figures 18e, f, g, h).

Cathodoluminescence is a useful tool in the study of sandstones (Sippel, 1968).   The difference in cathodoluminescence
intensity between the original clastic grains from metamorphic
or igneous rocks and low temperature overgrowths (e.g. on quartz)
can be distinguished.   This helps the evaluation of the degree
of pressure solution and overgrowth in sandstones (Sippel and
Blatt, 1976).   It should, however, be borne in mind that sand
grains which have been recycled from older sedimentary rocks

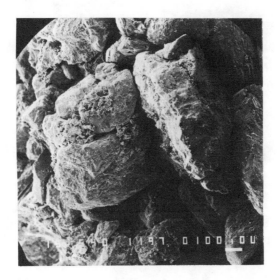

Figure 18e. Example of Jurassic sandstone from the
Statfjord field.   Statfjord Fm. porosity
20.8% permeability 424 md.   Low kaolinite
content.

Figure 18f. Example of Jurassic sandstone from the
Statfjord field. Brent Fm. porosity
20.3% permeability 3.7 md.   Porosity
blocked by kaolinite.

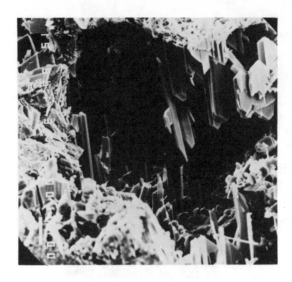

Figure 18g. Authigenic albite growing into mould formed
             by earlier leaching of a feldspar grain:
             Triassic sandstone (3300 m depth), Northern
             North Sea.

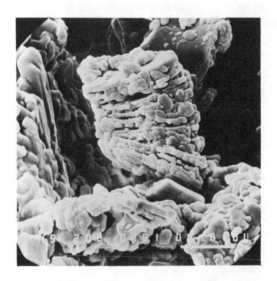

Figure 18h. Authigenic kaolinite in the Statfjord Fm.,
             North Sea. The irregular surface is
             interpreted to represent stages of growth
             of kaolinite.

may already have overgrowths.   If earlier overgrowths are not
distinguished from those formed after deposition, one will obtain
a misleading impression of the ratio between overgrowths and
pressure solution.

X-ray diffraction analyses are necessary for correct identification and semiquantitative analyeses of the authigenic clay
mineral phases in sandstones.   It is useful to analyse the bulk
sample for the identification of feldspars and carbonates and also
to separate out and concentrate the clay mineral fraction ($< 2\mu$m).

Stable isotopes are also of great importance for sandstone
diagenesis.   The stable isotopes S^{34}/S^{32}, O^{18}/O^{16} and $C^{13}/^{12}$
can be used to determine the origin of porewater precipitating
different types of sandstone cement.

CONCLUSION

Although diagenetic textures are studied on a microscopic
scale, diagenetic processes are primarily controlled by large-
scale sedimentological stratigraphical and structural relations.
It is important to have a semiquantitative approach to diagenetic
reactions to understand what is feasible in this scale.

Because of the low solubility of most mineral phases, net
cementation or dissolution requires large-scale porewater flow
through the sandstones.

Compactional isochemical diagenesis (pressure solution) is
related to the effective stress in sedimentary basins and is
significantly reduced by overpressure.   Secondary porosity may
form by leaching of carbonate and feldspar by undersaturated
meteoric water.   At greater depth CO_2 released from maturing
kerogen may leach carbonate and feldspar.   Calculations show,
however, that this process may be important only under very
specific conditions where we have humic kerogen, high shale/sand
ratio and a low carbonate content in the shales.

The acids released from the transformation of kaolinite and
montmorillonite to illite are also important with respect to the
leaching of carbonate and feldspar.

Diagenetic processes can to a large extent be related to
basin types in a plate tectonic setting.   To a large extent, the
circulation of porewater in sedimentary basins controls diagenetic
processes, and these can be related to stratigraphic and tectonic
relationships.   Sands covered by shales deposited as part of a
transgressive onlapping sequence will provide a seal for both
porewater flow and oil and overpressures will develop.   Off-

lapping progradational units usually require sealing faults or updip pinchouts for overpressure to occur.

Fluvial sandstones and in some cases shallow marine sandstones will be flushed by meteoric water after deposition forming authigenic kaolinite which can reduce the permeability. Also, fibrous fine-grained illite is effective in reducing the permeability/porosity ratio.

ACKNOWLEDGEMENTS

The present paper is to a large extent based on projects on sandstone diagenesis in the North Sea basin. Support from the NTNF, Statoil and the Oil Directorate is gratefully acknowledged.

Dr. W.E. Galloway is thanked for discussions on the diagenesis of the Tertiary of the Gulf Coast. Dr. Paul Ryan has kindly corrected the English manuscript.

REFERENCES

AAGAARD, P. 1979. Thermodynamic and Kinematic Analysis of Silicate Hydration at Low Temperatures and Pressures. PhD Thesis, University of California, Berkeley.

BATHURST, R.G.C. 1975. Carbonate Sediments and their Diagenesis. Developments in Sedimentology 12. Elsevier (2nd Ed.), 658pp.

BERNER, R.A. 1971. Principles of Chemical Sedimentology. McGraw-Hill, New York, 240pp.

BERNER, R.A. 1978. Rate controls of mineral dissolution under Earth surface conditions. Am.J.Sci., 278, 1235-1252.

BERNER, R.A. 1980. Early Diagenesis. Princeton University Press, 237pp.

BJØRLYKKE, K. 1979. Cementation of sandstones. J.sedim.Petrol., 49, 1358-1359.

BJØRLYKKE, K., MALM, O, and ELVERHØI, A. 1979. Diagenesis in the Mesozoic Sandstones from Spitsbergen and the North Sea. Geol. Rdsch., 68, 1151-1171.

BLANCHE, J.B., and WHITAKER, J.H. MCD. 1978. Diagenesis of part of the Brent Sand Formation (Middle Jurassic) of the northern North Sea Basin. J.geol.Soc., London, 135, 73-82.

BLATT, H. 1979. Diagenetic processes in sandstones. SEPM Special Publication, 26, 141-157.

BOER, R.B. DE 1972. On the thermodynamics of pressure solution-interaction between chemical and mechanical forces. Geochim. Cosmochim. Acta., 41, 249-256.

BOER, R.D. DE., NAGTEGAAL, P.J.C. and DUYVIS, E.M. 1977. Pressure solution experiments on quartz sand. Geochim. Cosmochim. Acta., 41, 257-264.

BOLES, J.R. and FRANKS, S.G. 1979a. Clay diagenesis in Wilcox
  Sandstones. J.sedim.Petrol., 49, 55-77.
BOLES, J.R. and FRANKS, S.G. 1979b. Cementation of sandstones: a
  reply. J.sedim.Petrol., 49, 1362.
BROWN, L.F. and FISHER, W.L. 1977. Seismic-stratigraphic Inter-
  pretation of Depositional Systems: Example from Brazilian Rift
  and Pull-Apart Basins in Payton. Mem.Am.Ass.Petrol.Geol. 26.
CHIARELLI, A. 1978. Hydrodynamic framework of Easter Algerian
  Sahara - influence on hydrocarbon oceanics. Bull.Am.Ass.Petrol.
  Geol., 62, 667-685.
CHIARELLI, A. and DUFFAUD, F. 1980. Pressure origin and distri-
  bution in the Viking-Basin (United Kingdon-Norway). Am. Ass.
  Petrol.Geol., 64, 1245-1266.
FUCHTBAUER, H. 1978. Zur Herkunf des Quartzements. Abschistzung
  der Quartzauflosung in silt und sandsteinen. Geol.Rdsch., 67,
  991-1008.
FUCHTBAUER, H. 1979. Die Sandsteindiagenese in Spiegl der neuren
  literatur. Geol.Rdsch., 58, 1125-1151.
GALLOWAY, W.E. 1977. Catahoula Formation of the Texas Coastal
  Plain. Depositional Systems, Composition, Structural Development,
  Groundwater Flow History and Uranium distribution. Bur.econ.
  Geol., Univ.Tex., N87.
GARRELS, R.M. and CHRIST, C.L. 1966. Solutions, Minerals and
  Equilibria. Harper and Row, London, 450 pp.
GLENNIE, K.W., MUDD, G.C. and NAGTEGAAL, P.J.C. 1978. Depositional
  environment and diagenesis of Permian Rotliegendes sandstones
  in Leman Bank and Sole Pit areas of the U.K. Southern North Sea.
  J.geol.Soc., Lond. 135, 25-34.
GHENT, E.D. 1979. Problems in zeolite facies geometry and fluid
  composition. SEPM, Special Publication, 26, 81-88.
GRETENER, P.E. 1978. Pore pressure: Fundamentals, general
  ramifications and implication for structural geology. AAPG
  Continuing Education Course Note, Series 4.
HANCOCK, N.J. 1978. Possible causes of Rotliegendes sandstone
  diagenesis in northern W. Germany. J.geol.Soc.Lond., 135, 35-40.
HAYES, JOHN B. 1979. Sandstone diagenesis - the whole truth.
  SEPM,Special Publication, 26, 127-139.
HOFFMAN, J. and HOWER, J. 1979. Clay mineral assemblies as low
  grade metamorphic Geomometers: Application to the thrust-
  faulted distrubed Belt of Montana - U.S.A. SEPM, Special
  Publication, 26, 35-79.
HOWER, J., ESLINGER, E.V. HOWER, M.E. and PERRY, E.A. 1976.
  Mechanism of burial metamorphism of argillaceous sediments.
  1) Mineralogical and chemical evidence. Bull.geol.Soc.Am.,
  87, 725-737.
HUNT, J.M. 1979. Petroleum Geochemistry and Geology. Freeman
  & Co., San Francisco, 716pp.
IIJIMA, A. and TADA, R. 1981. Silica diagenesis of Neogene dia-
  tomaceous and volcanolcastic sediments in northern Japan.
  Sedimentology, 28, 185-200.

KESSLER, L.G. 1978. Diagenetic sequences in ancient sandstones deposited under desert climatic conditions. J.geol.Soc.Lond., 135, 41-49.

KINSMAN, D.J.J. 1975. Rift Valley Basins and the Sedimentary History of Trailing Continental Margins. In: Fischer and Judson (eds.): Petroleum and Global Tectonics. Princeton University Press, 83-128.

LAND, L. and DUTTON, S.P. 1978. Cementation of a Pennsylvanian deltaic sandstone: Isotope data. J.sedim.Petrol., 48, 1167-1176.

LINDQUIST, S.J. 1977. Secondary porosity development and subsequent reduction in over-pressure, Frio Formation Sandstone (Oligocene), South Texas. Gulf Coast.Assoc.Geol.Soc., 27, 99-107.

LOUCKS, R.G., BEBOUT, D.G. and GALLOWAY, W.E. 1977. Relationship of porosity formation and preservation to sandstone consolidation history - Gulf coast Tertiary, Frio Formation. Bur.econ. Geol., Univ.Tex., 77, 109-120.

LOUCKS, R.G. DOUGE, M.M. and GALLOWAY, W.E. 1980. Importance of secondary leached porosity in Lower Tertiary Sandstone Reservoirs along the Gulf Coast. Geol.Circ.Bur.econ.Geol., Univ. Tex., 80-2.

MALM, O.A., FURNES, H. and Bjørlykke, K. 1979. Volcanoclastics of Middle Jurassic age in the Statfjord oilfield of the North Sea. N.Jb.Geol.Palaont., 10, 607-618.

MANHEIM, F.E. 1967. Evidence for submarine discharge of water on the Atlantic continental slope of the Southern United States and suggestions for further research. Trans.N.Y.Acad.Sci., 2.29, 839-853.

MANHEIM, F.T. 1970. The diffusion of ions in unconsolidated sediments. Earth Planet.Sci.Lett., 9, 307-309.

MATSUMOTO, R. and IIJIMA, H. Origin and diagenetic evolution of Ca-Mg-Fe carbonates in some coalfields of Japan. Sedimentology, 28, 237-261.

McBRIDE, E.F. 1977. Secondary porosity-importance in sandstone reservoirs in Texas Transactions - Gulf Coast. Association of Geological Societies, 17, 121-122.

McBRIDE, E.F. 1978. Porosity loss in sandstones by ductile grain deformation during compaction (Abstr.). Bull.Am.Ass.Petrol. Geol., 62, 1261.

MOREY, G.W., FOURNIER, R.C. and ROWE, J.J. 1962. The solubility of quartz in water in the temperature interval from 25°to 300°C. Geochim.Cosmochim.Act., 26, 1029-43.

NAGTEGAAL, P.J.C. 1979. Sandstone-framework instability as a function of burial diagenesis. J.geol.Soc.Lond., 135, 101-105.

NAGTEGAAL, P.J.C. 1979. Relationship of facies and reservoir quality in Rotliegendes desert Sandstones, Southern North Sea Region. J.Petrol. Geol., 2, 145-158.

PARKER, C.A. 1974. Geopressures and secondary porosity in the deep Jurassic of Mississippi. Gulf Coast Assoc.Trans., 29, 69-80.

PAYNE, J.N. 1968. Hydrological Significance of the Lithofacies
  of the Sparta sand in Arkansas, Louisiana, Mississippi and
  Texas. Prof.Pap.U.S. geol.Surv., 569-A.
PITTMAN, E.D. 1979. Porosity, diagenesis and productive capabil-
  ity of sandstone reservoirs. SEPM, Special Publication, 26,
  159-173.
SCHMIDT, V. and MCDONALD, D.A. 1979a. Texture and recognition of
  Secondary porosity in sandstones. SEPM, Special Publication,
  26, 209-225.
SCHMIDT, V. and MCDONALD, D.A. 1979b. The role of secondary
  porosity in the course of sandstone diagenesis. SEPM, Special
  Publication, 26, 175-207.
SELLWOOD, B.W. 1971. The genesis of some sideritic beds in the
  Yorkshire Lias (England). J.sedim.Petrol., 41, 854-858.
SHAW, A.B. 1963. Time in Stratigraphy. McGraw-Hill, N.Y. 365pp.
SIBLEY, D.F. and BLATT, H. 1976. Intergranular pressure solution
  and Cementation at the Tuscarora orthoquartzite. J.sedim.Petrol.,
  46, 881-891.
SIPPEL, R.F. and GLOVER, E.D. 1964. The solution alternation of
  carbonate rocks, the effects of temperature and pressure.
  Geochim. Cosmochim.Acta., 28, 1401-1417.
STANTON, G.C. 1977. Secondary Porosity in Sandstones of the
  Lower Wilcox (Eocene) Karnes County, Texas. Gulf Coast Ass.
  Geol.Soc., 27, 197-207.
SIEVER, R.,BECK, K.C. and BERNER, R.A. 1965. Composition of
  interstitial waters of modern sediments. J.Geol., 73, 39-73.
TILLMAN, R.W. and ALMON, W.R. 1979. Diagenesis of Frontier
  Formation offshore bar sandstone. Spearhead Ranch field,
  Wyoming. SEPM, Special Publication, 26, 337-378.
TISSOT, B.P. and WELTE, D.H. 1978. Petroleum Formation and
  Occurrence. Springer-Verlag, Berlin, 538pp.
USDOWSKI, H.E. 1968. The formation of dolomite in sediments.
  pp.21-32 in: Miller, G. & Friedman, G.M. (eds.). Recent
  Developments in Carbonate Sedimentology in Central Europe.
  Springer-Verlag, Berlin. 255pp.
VAIL, P.R., MITCHUM, R.M.J. TODD, R.G. WIDMER, S. THOMSEN,
  SANGREE, J.B. BUBB, J.N. and HATFIELD, W.G. 1977. Seismic
  stratigraphy and global changes of sea level. Mem.Am.Ass. Petrol.
  Geol., 26, 47-212.
VELDE, B. 1978. Authigenic K-feldspar in British Permo-Triassic
  Sandstones. J.geol.Soc.Lond., 135, 41-49.
WEYL, P.K. 1959. Pressure solution and the force of crystalliza-
  tion - a phenomenological theory. J.geophys.Res., 64, 2001-
  2025.
WHITE, D.E., HEIN, J.D. and WARING, G.A. 1963. Chemical Composi-
  tion of Subsurface waters. U.S.Geol.Surv., 44F.
WILSON, M.D. and PITTMAN, E.D. 1977. Authigenic clays in sand-
  stones: Recognition and influence on reservoir properties and
  palaeoenvironmental analysis. J.sedim.Petrol., 47, 3-31.

DIAGENETIC REACTIONS IN CLAYS

Bruce Velde
Ecole Normale Superieure
Laboratoire de Geologie
46, rue d'Ulm
75230, Paris, France

INTRODUCTION

In the study of phyllosilicate diagenesis two problems must
be immediately considered:

    1 - The chemical composition of the phases concerned and
    2 - The possibility of metastable reactions as well as
        slow reaction rates among the phases present.

In the first case, one is immediately hindered by small grain
size (< 2 microns is the definition of a clay mineral as far as
sedimentology is concerned) and by frequent occurrence as a
mixture of mineral species which combine to make chemical identification most often impossible. One must rely upon infrequent
natural occurrences of mono-mineralic material which may or may
not typify the phase chemically. Initial work using an electron
microprobe provides some hope of solving the problem of mineral
chemical composition in the case of sandstone diagenesis. Several
examples will be given further on.

In attempting to rationalise clay mineral assemblages into a
system of phase diagrams (which imply an approach to phase
equilibria), a major problem concerns the determination of which
minerals are stable in sedimentary rocks that have undergone burial.
There are two basic possibilities. The first is that the phases
present are being transformed into another, stable, assemblage but
the rate is so slow that geological events overtake the reaction
and it is stopped in its course thus leaving an assemblage which
does not represent a specific set of P-T conditions which it has
experienced. This is an unreacted assemblage. In principle, too

215

A. Parker and B. W. Sellwood (eds.), Sediment Diagenesis, 215–268.
© *1983 by D. Reidel Publishing Company.*

many phases will be present following the phase rule.   The second
possibility is that a reaction does occur in a reasonable time
span, but it does not produce the most stable assemblage for the
P-T conditions experienced.   This is a metastable reaction which
produces a metastable assemblage.   This aspect will be discussed
in more detail.

Given these two basic stumbling blocks to a reasonable
approach to the "truth" concerning the petrology of clay minerals
in diagenesis we can proceed in the following manner with caution.
First of all, let us assume that nature knows what it is doing,
at least in the measure that the natural laboratory will behave
according to the same laws of chemistry and physics found in the
"scientific" laboratory.   What I mean by this is that natural
mineral assemblages formed during diagenesis will follow a certain
restrained number of patterns which can be interpreted with some
competence using simple rules of chemistry and physics - so-called
thermodynamics.   We will take the point of view that if a given
assemblage is found often enough in similar geologic conditions,
its presence and contents will reflect a given set of physical and
chemical parameters which were active during its formation.
Generalization based upon reported mineral assemblages is the
foundation of our approach.   We want to know what minerals occur
together in which geologic environments.

Once the associations or clay mineral "facies" are defined,
we must fix the components in a chemiographic framework to see if
it is logical and to see what were the chemical limits acting to
produce the phases.   This means that one must determine the limits
of chemical variability of a clay mineral species (solid solution)
and one must fix the appropriate chemical parameters to describe
the mineral adequately.

The last step is to correlate, when possible, differences in
physical conditions of formation with the types of clay mineral
assemblages that are found for a given general chemical environment.   Here we can use the laboratory experiments which have been
performed on synthetic and natural clay minerals as well as
observations on natural assemblages.   Since we know that reaction
kinetics are a great problem, it is difficult to assign a given
reaction observed in the laboratory to a given temperature because
the duration of the experiment will most likely not be long enough.
However, it is possible to observe the sequence of assemblages
formed as, say, temperature increases and then to correlate this
information with observations on natural rocks.   This allows one
to verify if a given series of facies represents change in physical
conditions or rather variations in chemical parameters.

Thus our approach will be in three parts:
    Chemistry of the phases

    Stability of the phases
    Evolution in P-T-x space.

CHEMICAL COORDINATES FOR PHYLLOSILICATE ASSEMBLAGES

    The basic representation of the clay mineral compositions has
been chosen in the coordinates M^+R^3 - $2R^3$ - $3R^2$. SiO_2 and H_2O
are considered components in excess, i.e. these oxides are always
present in nearly pure monophase form so that no reaction is pro-
hibited or promoted through a variation in their relative abundance in the chemical system. In this general system the follow-
ing identities are assumed: R^2 = Mg^2, Mn^{2+}, Fe^{2+}; R^3 = Fe^{3+},
Al^{3+}, Ti^{4+}; M^+ = Na^+, K^+ and 2 x Ca^{2+}. An M^+R^3 pole represents
the bulk composition of feldspars, commonly found associated with
clay minerals. They are the most alkali-rich minerals present
(excluding zeolites). Calcium is considered to fulfil a chemical
role similar to the alkali ions in most clay minerals. It is
notably not present in micas stable at low temperatures (Velde,
1971; Hemley et al., 1971) and hence does not compare directly to
potassic systems. It is also possible that the calcium concentration in a given aggregate of phyllosilicates could be largely controlled independently of them through its activity in aqueous
solution in equilibrium with carbonates. Therefore its presence
is difficult to interpret in a purely silicate framework. As a
result, calcium is largely ignored in the analysis of clay minerals
here. It is assumed that high calcium concentration will not
provoke the crystallization of a specific clay mineral.

    The poles $2R^3$ and $3R^2$ were chosen because such ions are
present in these proportions in the clay minerals found to be
composed of R^{2+} or R^{3+} ions plus SiO_2 and H_2O. Kaolinite and
pyrophyllite are found at the $2R^3$ pole and serpentine or talc at
the $3R^2$ pole. Further, the choice of two R^3 ions is useful in
that clay minerals either combine two ions in octahedral co-
ordination, or three ions (ideally R^{2+}). The former are called
dioctahedral and the latter trioctahedral minerals. The major
advantage of combing the constituent elements in this way is the
resulting convenient distortion of a more simple M^+ - R^2 - R^3
geometry. By spreading the components out, one can see more
precisely the region of interest which contains the majority of
clay minerals. They are found in between the compositions of
feldspar-kaolinite-serpentine.

    The chemiographic relations of these phases are related to
the chemical coordinates in Figure 1.

    There are two obvious problems that come to mind when such
a method of analysis is proposed - does Fe^{3+} = Al^{3+} in all of
the possible phases and does Mg^{2+} = Fe^{2+} as well? As we will see

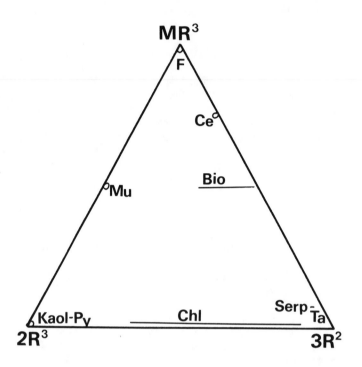

Figure 1. Representation of the ideal compositions of the more
            important phyllosilicate phases found at low
            temperatures (Serp = serpentine; Ta = talc; Chl =
            chlorite; Kaol = kaolinite; Py = pyrophyllite; Mu =
            muscovite; Bio = biotite; Ce = celadonite; F = feldspar).
            Coordinates are MR^3 = (K + Na + 2Ca) al; $2R^3$ = Al less
            value from MR^3 + Fe^3 divided by two; $3R^2$ = Mg + Fe^{2+}
            divided by three.

later on, probably the Mg^{2+} - Fe^{2+} identity is effectively correct
for most clay minerals and related phases. However, Fe^{3+} certainly
does not always substitute with equal facility for aluminium.
First we know that iron feldspars are quite rare in nature for the
potassium forms and unknown for sodium species. The same is true
for the micas. However, aluminous and ferric clay minerals do
exist. It is apparent then that some point of divergency will be
reached where aluminous systems will contain one set of phases,
and ferric systems, under the same physical conditions, will form
another. Here usage is the best proof - we will see that only
one major instance is evident where Fe^{3+} concentrations provides
the crystallization of a phase which would not be present in an

aluminous system.

Further it is not enough to assign Fe^{3+} content a fixed value in a given rock and then to forget about it in the evolution of the sediment during burial and metamorphism. Fe^{3+} is different from Al^{3+} in its ability to change oxidation state and one should therefore change it identity as a coordinate in a simplified system such as is proposed here. Let us consider the situation in which the loss of Fe^{3+} by chemical reduction to $Fe2+$ suppresses a phase - the so-called kaolinite-pyrophyllite "anomaly". This problem is an illustration of the uses and problems involved with representing rocks and clay minerals in the MR^3 - $2R^3$ - $3R^2$ coodinates.

The detailed arguments and geologic setting of the example are given by Velde (1968). Essentially, the geologic observation which has been made is that pyrophyllite is rare in sedimentary and epi-metamorphic pelitic rock assemblages. The question is posed - why does an aluminium silicate appear at low pressure-temperature conditions and disappear at higher ones (frequently to reappear as kyanite or andalusite upon strong metamorphism)? The answer can lie in the interplay between changing bulk compositions of sedimentary mineral assemblages and the constituent minerals stable at various physical conditions. The following chemiographic analysis is possible.

In general, minerals and rocks contain less ferric iron as metamorphism becomes more apparent. As a result, the composition of a rock in the MR^3 - $2R^3$ - $3R^2$ system is displaced toward the $3R$ pole through chemical reduction of iron as pressure and temperature are increased. Figure 2 shows the relation of different clay mineral assemblages before and after this displacement. Simply stated, one moves from a three phase-kaolinite (aluminium silicate)-bearing assemblages to a two phase (non-aluminium silicate-bearing) assemblage: illite + chlorite. The chemical reduction in the rocks is assumed to take place through the agency of hydrogen-rich gases, produced by the diagenesis of organic matter. If we now consider a kaolinite-bearing rock which initially contains relatively little ferric iron and which is rich in organic matter, the reduction effect will not take place during metamorphism. Concerning the aluminium-silicate mineralogy in such a rock, pyrophyllite should appear at higher metamorphic grades through the reaction kaolinite + quartz = pyrophyllite. Thus we see that mineralogic evolution of a given rock in P-T space is not necessarily represented by fixed values of chemical variables. Even though little or no material is taken from or added to the rock, it can change its bulk composition in the chemical coordinates which have been chosed through oxidation or reduction of iron. As a result of this change a new mineral assemblage can be produced. One must conclude that the problem of iron oxidation state will be omnipresent.

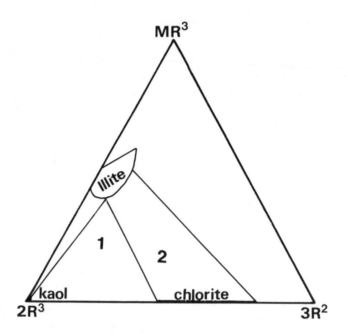

Figure 2. Illustration of the possible displacement in bulk
composition of a rock which contains Fe^{3+} as metamorphic
grade increases and the iron is reduced to Fe^{2+}.
Chemical coordinates as in Figure 1.   Composition 1 is
the initial and 2 final bulk composition where the
initial assemblage is three-phase and contains kaolinite
while the final assemblage is two-phase containing no
aluminosilicate.

STABILITY AND REACTION STATE

One critical problem encountered in observing and interpreting
natural mineral assemblages and experimental results is the
determination of the relative stability of different phases or
assemblages of phases.   The object of most experimentation is to
establish which is the stable assemblage at given P-T-x conditions,
i.e. that which will remain at infinite periods of time when
maintained under the same physical and chemical conditions.   This
criterion is especially important for geologists in that time can
be an important factor in their interpretations.   Observations
such as those made by Weaver (1959) or Bartholomé (1966)

concerning the clay mineralogy of sedimentary rocks as a function
of their age, i.e. certain parageneses are typical of rocks of a
certain age or older, indicate the potential importance of
reaction kinetics in clay mineralogy. However the time scale
involved in these studies ($1.3 - 2.3 \times 10^7$ years) probably
excludes consideration for most diagenetic problems.

Experiments by Eberl and Hower (1976) suggest that a time
span of 10^6 years is necessary at 60°C to transform smectite to a
mixed-layer mineral with low expandibility. However, if this
reaction occurred at low temperature, 20°C, it would require 10^8
years to be completed. These values begin to interest the
geologist. In this sequence, it is certain that when a sediment
has reached 80°C, it will be reacting "instantaneously" in a
geologic sense (10^5 years) and one can consider that the phases
present represent a near approach to equilibrium. However, if
the sediments never were brough to temperatures above 50°C, the
problem of the significance of the mineral assemblages found must
be considered. In the Gulf Coast (U.S.A.) sections, it is not
uncommon to find sediments at 1.5 km depths at temperatures less
than 50°C. These materials must then be considered as likely to
contain much argillaceous material which has not recrystallized.

In another example, Mitzutani (1970) has demonstrated by
laboratory experiments that the transformation of amorphous silica
into quartz, the most stable polymorph, passes through a
cristobalite phase before forming quartz. All of this takes 10^9
years to be accomplished at 0°C, and 10^6 years at 100°C. Observations of silica polymorphs in near-surface, deep-sea sediments
by Calvert (1971) confirmed these laboratory studies. Cristobalite
is replaced by quartz in late Cretaceous sediments. It is thus
evident that the chemical properties of amorphous silica, then
cristobalite will enter into reaction equilibria throughout most
of sediment deposition and early burial history; whereas the
properties of quartz can be ignored in thermochemical considerations.

One might make a slight digression here to explain what is,
or should be, meant by stable and metastable phases as they relate
to geological problems. First of all the two terms relate to
one another and are in fact a measure of a thermodynamic parameter
which indicates a tendency toward transformation. A stable
phase is one which will persist under all variables of time or
energetic configuration. The tendency toward change is zero if
pressure, temperature and bulk composition are constant. Metastable indicates a state other than a stable one at given physical
conditions and thus there is inherent in a metastable state a
potential for change. However, this potential is not always
realized, somewhat as the explosive potential of a mixture of
methane and air is not always realized. If the energetic

conditions necessary for change are not fulfilled, the metastable
material can well persist for semi-infinite periods of time. In
fact virtually all metamorphic and magmatic mineral assemblages
found at the Earth's surface are metastable where they are found.
This does not prevent geologists from studying them and applying
thermodynamic principles of reasoning to their existence and
formation.

Let us consider a simple example. Figure 3 shows the

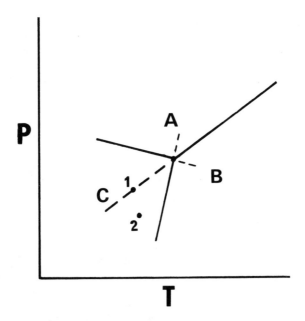

Figure 3. Graphic illustration of a metastable reaction: A⇌B,
in P-T space where the stable phase is C. At point 1
both A and B coexist metastably. At point 2 phase B
is metastable with respect to C, phase A is metastable
with respect to B and C.

stability fields of phases A, B and C, all of the same composition,
as they occur in pressure-temperature space. Let us assume that
the transformation of A into C and B into C is very difficult,
requiring a large amount of excess energy in order to rearrange
the constituent atoms into a complex configuration. However, the
A-B transformation involves little difference in atomic configuration and one finds that the reaction occurs with ease. If

we take phase A at point 1 it will transform partially into B because this is easier energetically than the A → C transformation. The A → C transformation will not occur at point 2 either, but the A → B transformation will.   Above the dashed line A is metastably present and below it B is metastably present.   A is _more_ metastable than B at point 2.   All that interests us from a practical point of view is that A becomes B at point 2 in a short time span while the stable phase C will not appear until a great period of time has elapsed.   Therefore, all of the phases present which accompany A or B in nature will be in equilibrium with and their appearance will be controlled by the particular thermodynamic properties of A or B instead of those of phase C.

Taking a concrete example, as mentioned previously amorphous silica appears more readily than does quartz as a newly formed phase at or near the Earth's surface.   The solubility of amorphous silica is much greater than quartz.   Then most surface aqueous solutions can contain more dissolved silica than would be permitted by a quartz-water equilibrium and they will not precipitate the stable silica phase quartz.   Further, other phases such as zeolites, whose composition is silica-rich, will form at low temperatures where the assemblage feldspar-quartz is the most likely stable assemblage.   These zeolite phases will persist until the system becomes energetically capable of producing quartz plus sodium feldspar (near 100°C).   Zeolites are then metastable phases but their persistence in nature is such that we must deal with them as reality over geologic time spans.   Their apparent stability field must therefore be limited to conditions where quartz does not crystallize in P-T-time space.   However when quartz does not crystallize, the rigours of the phase rule will apply just as they would to a more stable set of phases (see Kujawa and Eugster (1966) for a more learned discussion of these concepts).   Thus metastable phases can be dealt with using the phase rul and thermochemical constraints as well as stable phases can.   The trick is to recognize which phase will occur for a given set of conditions.

PELITIC CLAY MINERALS - COMPOSITION AND OCCURRENCE

We will now look briefly at the common clay minerals found in pelitic argillaceous rocks and arenaceous sandstones.   The first step is to ascertain their compositional range in the framework of the MR^{3+} - $2R^{3+}$ - $3R^{2+}$ coordinates.

Kaolinite

We will begin with the easiest phase-kaolinite, which has an almost invariable composition, with Al = Si in atomic proportions. The importance of kaolinite in chemiographic space is its polar

position, that of a pure R^3 component.   Since there is no solid
solution with other phases, the presence of kaolin in a rock will
suggest that the accompanying phases will have their highest
possible Al_2O_3 content under the prevailing P-T conditions if
they have been formed stably together.   The range of P-T
conditions of kaolinite stability is greater than those attributed
to diagenesis, 25°- 300°C at one atmosphere to 2 kb pressure
(Velde and Kornprobst, 1969).   However, its persistence in
sedimentary rocks is most often a factor of combination with
other phases more than its thermal stability (Velde, 1977).   Thus
we can consider the presence of kaolinite as a helpful chemiographic marker but it will not be useful as a palaeotemperature
indicator.

Chlorites

    Relatively little is known about the chemical composition of
chlorites found in soils, sediments or sedimentary rocks.   It is
generally admitted that authigenic chlorites in low temperature
environments are iron-rich (Grim, 1968), but their exact Fe^{2+},
Fe^{3+}, Mg and Al content cannot be determined by the most common
indirect means such as X-ray diffraction or infra-red absorption
(Brown, 1961; Velde, 1973).   Other classical methods cannot be
used since chlorites can rarely be separated from their readily
mineral environment.   Classical chemical analyses are therefore
not readily available.   One exception to this statement is the
7 Å pelletal chlorites called berthierines.   It is known however
that chlorites are the minerals which eventually concentrate the
iron and magnesium in pelitic rocks upon their metamorphism in
the greenschist facies.

    Chlorites are present in sedimentary rocks throughout the
geologic column (Weaver, 1959).   The typical assemblage in
evolved or old sedimentary rocks is illite + 14 Å chlorite +
quartz (Millot, 1964; Dunoyer de Segonzac, 1969; Weaver, 1959).
Generally speaking, the importance of chlorite increases in a
sequence of rocks to the detriment of expaning phases (van Moort,
1971; Dunoyer de Segonzac, 1969).   Chlorite has been produced
under experimental conditions of 2 kb pressure, 300°- 400°C, from
natural illites and expandable mixed layered minerals, both
dioctahedral and trioctahedral types (Winkler, 1964; Velde and
Bystrom-Brusewitz, 1972; Velde, 1977).   These experiments
reinforce the observations made on sedimentary rocks which lead
to the conclusion that chlorite can be an early product of
diagenetic or epimetamorphic change in pelitic rocks.   Although
the gradation between sedimentary and metamorphic rocks can be
defined by the polymorph and composition of the illite-muscovite
association, no such distinction can be made for chlorite.   The
14 Å polymorph is present in association with diagenetic or
authigenic minerals and remains present into medium grades of
metamorphism.

The change between 7 and 14 Å has been noted to occur at incipient diagenesis (Schoen, 1964; Mitsui, 1975) for the chamosite-chlorite series containing 60% Fe^{2+} in the octahedral position. There is apparently a slight enrichment in iron during the 7-14 Å transformation.   Less is known for compositions which are more magnesian.   14 Å chlorite has been noted as the first diagenetic mineral in an undeformed turbidite (Lovell, 1969).   Present-day temperatures of 100°C and coexistence with montmorillonite attest its precocity.   The 7-14 Å polymorphic transition probably occurs at rather low temperatures, somewhere near 100°C (Velde et al., 1974; Mitsui, 1975) for berthierines, i.e. chlorites isolated from other phyllosilicates in the rocks.

Although as already mentioned the information on the chemical composition of natural low temperature chlorites in sedimentary rocks is limited, some new data have been gathered using micro-probe analysis of grain mounts or rock thin sections.   The samples studied come from rather different geographic areas - western Montana, Algeria and the Franco-Italian Alps and African off-shore Atlantic coast shelf sediments.

First we will consider the chlorite known as berthierine. These are the pelletal ocean bottom 7 Å chlorites.   Those reported in the literature (Velde et al., 1974; Leone et al., 1975), and several new analyses from Alpine samples, indicate a homogeneous composition throughout the pellet.   These samples have undergone metamorphism and the minerals now have a 14 Å polymorph. This may well explain their relative compositional homogeneity.   The analyses done on many single grains (with microprobe) show some scatter of compositions but they all lie within the 7 Å chlorite range determined experimentally.

Samples reported by Schellmann (1969) from the sediments in the Gulf of Guinea were analyzed at grain edges and centres (Carmoze et al., 1978).   These pellets have not completed the evolution toward a single-phase 7 Å mineral.   They are iron-rich, containing no detectable phyllosilicate other than berthierine but they contain goethite.   Grain edges are variable in composition whereas grain centres tend to group near a berthierine composition comparable to those of the meta-berthierines.   The grain centre compositions tend to lie close to those of synthetic 7 Å chlorites in the $Si-Al-R^{2+}$ coordinates (Velde, 1973.   Their composition is indicated by zone 3 in Figure 4 using Fe-Mg-Al coordinates.

The most important character of all of the berthierine compositions is their low silica content.   The variation of compositions found for pellets from the recent sediments appears to be the result of the crystallization of a chlorite structure with full octahedral occupancy.   The meta-berthierines fall within the limits deduced for synthetic magnesian 7 Å chlorites,

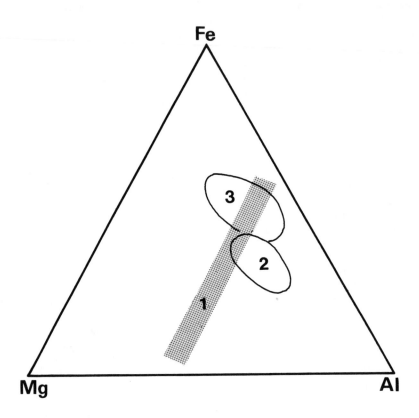

Figure 4. Representation of chlorite compositions in Fe^{2+} - Mg -
    Al coordinates.   Grey zone (1) shows compositions of
    metamorphic chlorites found with muscovite for all
    facies (Velde and Rumble, 1977). Area 2 shows compo-
    sitions of some chlorites found in rocks of the mixed
    layered minerals paragenesis and zone 3 shows compo-
    sitions of sedimentary 7 Å chlorites (berthierines).
    All analyses were obtained using a microprobe.

limits which are also near full octahedral occupancy.   The samples
studied are richer in iron than in magnesium and they show a
limited variation in relative contents of these two ions.

Figure 4 indicates the compositional spread of chlorites from six rocks in the illite-montmorillonite mixed-layer mineral facies and from the illite-chlorite zone in the French Alps (Velde, 1977). The grains analyzed with the microprobe are chlorites replacing isolated grains of detrital mica or are newly formed grains. They are usually 15 microns in the smallest dimension. These chlorites are usually richer in silica than berthierines or magnesian 7 Å synthetic minerals. The compositions are quite variable ($\pm$ 5 ionic percent Si or Al), but nevertheless the compositions determined from grain to grain in one sample do not exceed the limits defined for synthetic 14 Å chlorites (Velde, 1973). Those chlorites associated with mixed-layer clay minerals are most silica-rich and have the greatest compositional variations for grains in a single thin section; they tend to be iron-rich and aluminous. One chlorite vein was found to transect a glauconite pellet. This chlorite was quite iron-poor indicating attainment of a local chemical equilibrium between chlorite and mica upon its crystallization.

A brief study by Velde and Rumble (1977) which compiled available microprobe data for chlorites from metamorphic rocks shows that alumina content of chlorites becomes a fixed quantity in most assemblages where white mica is present. Al atomic percent is 28 $\pm$ 2 for all facies. Thus the only important chemical variation in chlorites from pelitic bulk compositions under metamorphic conditions is between Fe and Mg, which must depend upon the bulk composition of the rock, coexisting phases and temperature conditions. However, this information should confirm the usefulness of chlorites as indicators of palaeometamorphic environments since partitioning of Mg-Fe must be only a factor of P-T and not bulk composition, i.e. alumina content will not interfere with the partioning of the divalent ions between different phases. The composition of these chlorites is shown by zone 1 in Figure 4.

It would seem possible to summarize the chlorite composition data here in the following manner. The chemistry of chlorites is a function of the general geological conditions under which they form. P-T values control the alumina content of chlorites. Local chemistry controls their Mg-Fe content.

Thus only Al_2O_3 concentration in a chlorite can be used to indicate its origin if it formed at diagenetic conditions, i.e. below those of a metamorphic rock. Mg-Fe content might be used in metamorphic assemblages as a means of estimating P-T conditions if the chlorites are in equilibrium with another ferro-magnesian phase. In this case, the percent Mg (partitioning coefficient) in the ferro-magnesian phase is likely to be a function of the physical conditions of equilibrium.

Illite

    Nineteen well-characterized illites are plotted in the MR^3 - $2R^3$ - $3R^2$ coordinates (Figure 5). The main features of their formulae are traditionally compared to muscovite, the mineral closest to illite compositions. Where muscovite has the ideal

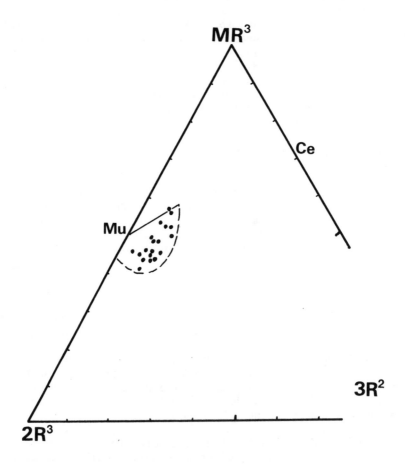

Figure 5. Compositions of well-characterized illites after Velde
(1977). Solid line shows solid solution between
muscovite and celadonite mica (phengites), dashed line
shows provisional illite composition boundary which
will be used in an indicative manner throughout the
text.

formula $K Al_2 Si_3 Al O_{10} (OH)_2$, illite can be represented by $M^+_{.6-.9}$ $(R^2 R^3)_{2.00} (Si Al)_4 O_{10} (OH)_2$. Illite does not have a total lattice charge equal to 1.00, and the total silica ions in the tetrahedral sites exceeds 3.00. The charge, distributed between octahedral and tetrahedral sites, is usually compensated by potassium and sodium ions, more questionably by Ca^{2+} or H_3O^+ (Hower and Mowatt, 1966). Two main compositional trends are evident in looking at this information. The first is the tendency to have a low lattice charge relative to mica, therefore a montmorillonite-type compositional solid solution, the second is the R^{3IV} = Si substitution typical of phengite micas. These two trends are apparent in Figure 5. The scatter of points indicates a combination of these two substitutions simultaneously and thus, although illite has essentially a mica structure (i.e. 10 Å non-expandable), it most often has a bulk composition not found in high-temperature micas. This extension of the illite chemical domain is very important in determing the relations of illite or sedimentary "mica" with other phyllosilicate phases. It should be noted here that the solid solution at low P-T conditions is complete between muscovite and extremes of illite compositions; all phases are stable at these P-T conditions. As a result the readjustment of detrital muscovite to chemical conditions of sedimentation will be made slowly if at all and sedimentary illite will contain newly crystallized material as well as older, inherited detrital micas which have not come to equilibrium with the new chemical conditions, i.e. the average composition of the illite does not represent the phase in equilibrium with other neo-formed silicates.

Figure 6 indicates the necessary change in composition which a muscovite would need to become stable under conditions in a sedimentary rock where chlorite is present (x to y). The solid solution for mica-illites is delimited by the shaded area which represents a much larger variation than is possible under meta-morphic or igneous conditions. The detrital muscovite (composition x) is in itself stable if the bulk composition of the sediment as projected into the coordinates is found at x. The ΔG between the assemblage of muscovite + chlorite at compo-sition y and illite of this is likely to be relatively small and the tendency to recrystallize the muscovite from x to y compo-sitions will be small at sedimentary conditions. However, as more thermal energy is added to the rock system, under conditions of deeper burial, the recrystallization will proceed more rapidly as temperature is increased. An indication of this situation can be given by the data presented in Figure 7: microprobe analyses of detrital mica grains (circles) and recrystallized silicate material which forms overgrowth (dots) in rocks from several sedimentary rocks of high diagenetic grade. The detrital grains show higher sodium contents (> 0.10 atom per $O_{10} (OH)_2$ formula unit) and tend to be grouped along the muscovite-phengite

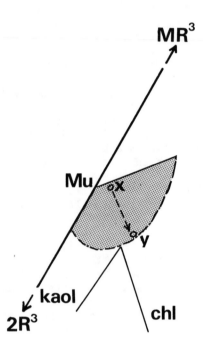

Figure 6. Illustration, using MR^3 - $2R^3$ - $3R^2$ coordinates of
change in composition necessary for the transformation
of a detrital muscovite (x) into an illite (y) which
would be stable with chlorite under conditions of
burial diagenesis. kaol = kaolinite; chl = chlorite,
Mu = muscovite.

mica line in the figure. These represent what are probably the
varieties of micas found in the magmatic or metamorphic rocks
which were sources of the sediments. New micaceous phases have
variable compositions but they tend to be grouped towards the
edges of the illite compositional field.

It is apparent that the new micaceous minerals have been
crystallized to form illitic phases approaching a composition
closer to an illite which would be in equilibrium with other
clays. Thus we can see the effect of increasing diagenetic
grade on the extent of equilibration in the composition of
micaceous material of different origins.

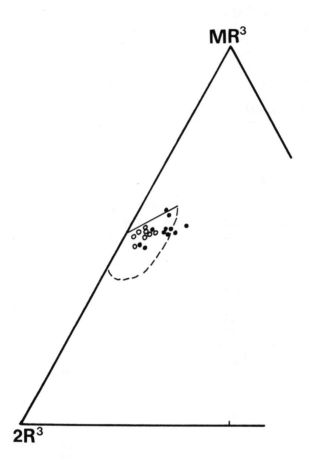

Figure 7. Composition of some detrital, sodic micas (circles)
          and their low sodium overgrowths (dots). The tendency
          is for the new micaceous phases to have an illite compo-
          sition (field delimited by dashed line) which can
          coexist with chlorite which occurs in the rocks. Un-
          published microprobe analyses are by E.Nicot and the
          present author.

Glauconites

    Much work has been done on these green pelletal minerals

which are so often found in sandstones. A general idea of their compositional variation can be found in Velde (1977). If we plot natural glauconite compositions in our $MR^{3+} - 2R^{3+} - 3R$ co-ordinates, they fall somewhat below a line between muscovite (M) and celadonite (Ce) in Figure 8. In this respect they resemble illites which also fall below this line but to the $2R^{3+}$ side of the diagram. The continuity in composition is only apparent and it is due to the assimilation of Al and R^3 as being the same chemical "component". In fact, there is most likely a discontinuity between illites and glauconites in their $Al-Fe^{3+}$ contents. This seems evident from microprobe as well as bulk chemical analyses of multi-grain separates (Velde, 1976).

Going a step further, the phase diagrams (P-T-x variations) for glauconite- and illite-smectite associations are very different (Velde and Odin, 1975). This leads one to believe that there is no direct continuous connection between the mineral groups illite and illite-smectite mixed-layer minerals and glauconites. If we consider the typical geologic origin of illite and glauconite, they are quite different also. Illite can be produced through many chemical reactions at a range of P-T conditions which covers most of what can be attributed to diagenesis. By contrast glauconite is uniquely a low temperature phase due most often to chemical migration in a localized volume at the sediment-sea interface. Some glauconite can originate from the destabilization of glauconitic rock (Courbe and Velde, 1981) during weathering but this is of a very limited nature. There is petrographic evidence that glauconitic materials can be mobilized during burial also. But again, this latter case is minor in amount.

Of greater importance is the fact that glauconites once formed in the sedimentary environment will not change their composition much during most diagenetic conditions of burial metamorphism. The effect of pressure and temperature are small concerning glauconite solid solution or chemical range. Thus fixed in the sediment, glauconite tends to remain rather unchanged until temperatures about 300°C are reached. For this latter reason, glauconite is "uninteresting" because one cannot derive conditions of burial diagenesis from it. Further, since its composition is derived from sediment-sea water chemical interchange where most of the initial solid material is completely transformed, the composition of glauconites does not give much information about the conditions of sedimentary diagenesis either.

All in all, glauconites will not give much useful information concerning diagenesis and for this reason they will not be discussed further here.

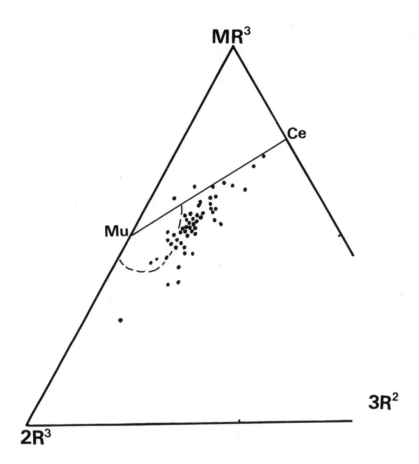

Figure 8. Celadonite and glauconite compositions in MR^3 - $2R^3$ -
           $3R^2$ coordinates. Mica compositions lie between
           muscovite (Mu) and celadonite (Ce) while illitic or
           low charge micas lie below this line. An apparent
           continuous composition series exists between illite
           and celadonite - glauconite due to Al - Fe^{3+} being
           used as one component. However, illites and
           celadonite - glauconites are separated in compositional
           space when Fe^{3+} is distinct from Al^{3+} (data from Velde,
           1977).

Smectite

Compiled data for dioctahedral smectites are mixed-layer
minerals are shown in Figures 9 and 10 (the coordinates are
based upon calculated formulas for the 2:1 layers and subsequent
assignment to the M^+ as equivalent to the total charge on the
structure. This eliminates the problem of H^+ interlayer ions
and others, which frequently arises in formula calculations for
smectites.   Data are those used in Velde (1977)).

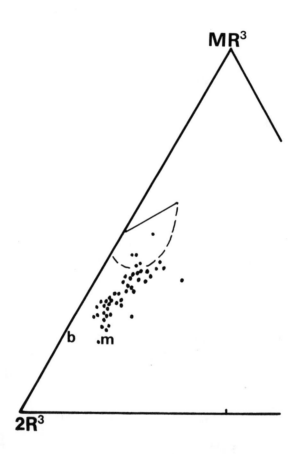

Figure 9. Composition of natural dioctahedral smecticies: b =
          ideal beidellite; m = ideal montmorillonite (data from
          Velde, 1977).

The field of illite composition is indicated in each figure. It
is evident that these smectites, designated as fully expandable
phases by each author, cover a large field of compositions - from
theoretical montmorillonites $(M^+)_{0.7}$ $(R^{2+}R^{3+})_2$ $(Si\ Al)_4$ $O_{10}(OH)_2$.
It can be surmised that even though X-ray data indicated only
expandable material there must be significant interlayering with

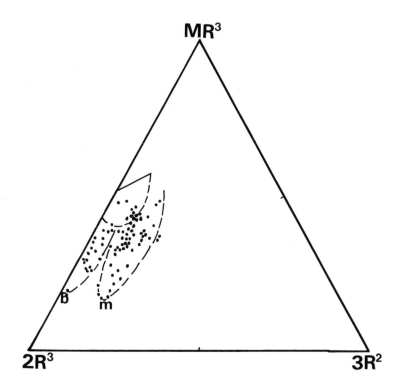

Figure 10. Composition of natural mixed-layer minerals; b =
          ideal beidellite, m = ideal montmorillonite fully
          expandable phases. Dashed lines indicate illite and
          the two mixed-layer mineral compositional series
          which can be deduced from these data (Velde, 1977).

illite or other non-expandable mica-like phases such as glauconite-celadonite in order to give such a high structural charge imbalance. If not, one wonders why illites, with a similar chemical formula, are not expandable as well.

Basically, little can be said about the smectite data. The compositional dispersion is quite large. Some of this variation can be explained in part by an overpopulation of octahedral sites due to mis-assignment of magnesium ions as proposed by Foster (1951). In such a case, a portion of the magnesium present is exchangeable and interlayer ions which are mis-assigned to octahedral sites will decrease the apparent interlayer charge. The overall effect is to place the bulk mineral composition toward the $3R^2$ pole in the diagram. This is borne out to a certain extent by the scatter of the compositions toward the $3R^2$ pole beyond the celadonite-montmorillonite line (Figure 9) which represents maximum theoretical $3R^2$ content.

If one considers the ideal formulae for the two end member minerals, beidellite and montmorillonite (Ross and Hendricks, 1945), it appears that the substitutions which create a net charge of deficiency in the 2:1 structure are found either in the octahedral site-montmorillonite $M^+_{0.25}$ $(R^3_{1.75} R^2_{0.25})$ Si_4O_{10} $(OH)_2$. nH_2O or in the tetrahedral site-beidellite $M^+_{0.25} R^{3+}_2$ $(Si_{3.75}$ $Al_{0.25})$ $O_{10} (OH)_2 . nH_2O$. .   The compositions represented by b and m in the $MR^3-2R^3-3R^2$ system (Figure 10) obviously are rarely found in nature. Thus natural dioctahedral smectites appear to have other than ideal or theoretical mineral compositions.

If we now consider the bulk compositions of the aluminous mixed-layer minerals which contain both expandable and non-expandable layers, two series are apparent, one between theoretical beidellite and illite and one between theoretical montmorillonite and illite (Figure 11). The intersection of the lines joining muscovite-montmorillonite and beidellite-celadonite (i.e. expand-able mineral to mica) is a point which roughly delimits the apparent compositional fields of the two montmorillonite-illite compositional trends for the natural mixed-layer minerals (Figure 10). That is, the natural minerals appear to show a compositional distribution due to solid solutions between each one of the two smectite types and the two mica types - muscovite and celadonite. There is apparently a gap in solid solution between the two highly expandable (80% montmorillonite) beidellitic and montmorillonitic end members. The point of intersection of the theoretical substitutional series beidellite-celadonite and muscovite-montmorillonite is located at about 30-40% expandable layers - 70-60% illite. This interlayering is similar to that in the "mineral" allevardite. It appears that as the expandability of the mixed layered series decreases, the total range of solid solution (as a function of $MR^3-2R^3-3R^2$ components increases.

The general compositional dispersion of fully expandable
dioctahedral smectite bulk compositions is then replaced by two
substitutional trends as expandability begins to decrease (i.e.
mixed layering is apparent).   It can be noted that there is a
tendency toward apparent excess or high $3R^2$ content evident for
the mixed-layer minerals as was seen for the fully expandable
smectites.   This is again probably due to interlayer Mg^{2+}

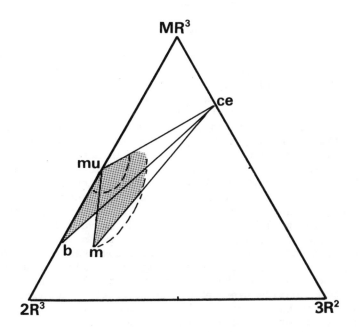

Figure 11. Comparison of natural mixed-layer mineral compositions
           (dashed lines from figures 5 and 10) with area defined
           by the intersection of muscovite-beidellite (mu-b),
           muscovite-montmorillonite (mu-m), muscovite-celadonite
           (mu-ce), celadonite-beidellite (ce-b) and celadonite-
           montmorillonite (ce-m) compositional joins as they
           project in MR^3 - $2R^3$ - $3R^2$ coordinates.

which is calculated as octahedral ions. It is neverless quite
possible that the analyses of the fully expandable smectites do
show a valid chemical variation and not just analytical error of
one sort or another. A remarkable point in comparing the mixed
layered and fully expandable bulk compositions is that the former
defines two compositional series while the latter is found just
between these two series. If indeed, this is not the result
simply of analytical errors, the relations would suggest that
the fully expandable series are mixtures of the two extreme
compositional types beidellite and montmorillonite. Since
neither these nor the two forms are found alone, one would suspect
the above deduction to be true. The possibility of the coexistence of two fully expanding phases has important implications
in the phase relations as we will see.

Paragenesis diagrams for dioctahedral smectites and illite

     As we have seen in the previous section, the bulk chemical
compositions of smectites taken from the literature are dispersed
over the field of fully expandable, mixed-layer and even extreme
illite compositions. Just what are the limits of true smectite
composition cannot be established at present. We can, nevertheless, as a basis for discussion, assume that the ideal compositions
of beidellite with 0.25 charge per 10 oxygens and of montmorillonite
with the same structural charge do exist in nature and that they
form the end-members of smectite solid solutions. Using this
assumption one can suppose either solid solution between these
two points, or inimate mixtures of these two theoretical end-
member fully expandable minerals. In either case, the observable
phase relations will be similar, since it is very difficult if
not impossible to distinguish between the two species by physical
or chemical methods should they be mixed together. As the bulk
chemistry of the expandable phases suggests a mixture of two
phases, we will use this hypothesis, and it will be assumed here
that the two smectite types are compositionally independent and
that a tie-line exists between them. When fully expandable
dioctahedral minerals are stable, a phase diagram such as Figure
12 will be applicable. Most important in this figure is the tie-
line beidellite-montmorillonite. One of the consequences of
these phase relations is that where both beidellite and montmorillonite are present kaolinite and illite do not coexist stably.
Compositions more rich in divalent ions will fall in the chlorite
(or expandable trioctahedral phase)-montmorillonite field.

     As has been deduced from the compositions of natural
dioctahedral mixed-layer phases, two series of interlayered
minerals are present, based upon beidellite or montmorillonite
end-members and illite. As the fully expandable phase becomes
unstable, the coexistence of mixed-layer phases and kaolinite will
be possible. This assemblage is commonly observed in deep

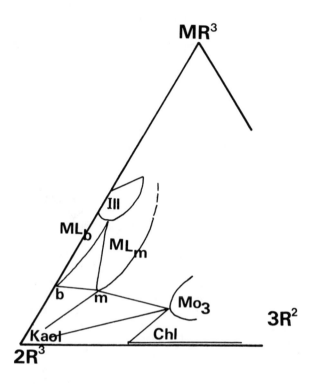

Figure 12. Deduced phase relations for aluminous clay minerals
at low temperatures. ML_b = beidellitic mixed
layered minerals, ML_m = montmorillonitic mixed
layered minerals, Ill = illite, Kaol = kaolinite,
Chl = Chlorite, Mo_3 = expanding trioctahedral phase.

boreholes and other series of deeply buried sedimentary rocks.
However, the assemblage of kaolinite mixed-layer phase and illite
is even more common in these rocks. In order for this assemblage
to be stable, the chemical continuity of interlayered phases
between illite and smectites must be interrupted at some points;
i.e. the tie-line kaolinite-illite should exist. Looking again
at the compositions of natural mixed-layer minerals, the existence
of this tie-line is not particularly evident, i.e. there is no
obvious gap in the compositions of the mixed-layer minerals near
the illite field (Figure 12). However, chemical analyses are
not accurate enough to show such a discontinuity, nor are mineral

separations normally precise enough to eliminate all foreign
material.   The only real indication of the existence of an
illite-kaolinite tie-line is the common and persistent assemblage
of kaolinite, illite and mixed-layer mineral in sedimentary rocks.
Since these minerals are otherwise sensitive to physical changes
(Perry and Hower, 1970) it is unlikely that this assemblage would
persist so often if it were metastable, especially since kaolinite
is a mineral formed at low temperatures and, therefore, less
likely to remain metastable under physical conditions of deep
burial as we will see in the following section.   If one accepts
these arguments, intuitive as they may be, a possible phase
diagram for moderate temperatures and pressures can be drawn as
in Figure 13.   The amount of solid solution (interlayering)
between expandable and micaceous layers is continually restricted
with rising temperatures.   This was seen to be the case for
synthetic aluminous minerals (muscovite-K-beidellite) and natural
Mg-Fe forms treated hydrothermally and it is known for those
found in deeply buried sediments as we shall see later.

    The most important assumption used to construct Figure 13 is
the existence of the kaolinite-illite tie-line.   When illite
mixed-layer series are no longer continuous in the 0-25% mixed-
layer range, two-phase fields (compositional tie-lines) of illite-
mixed-layer phases appear, joining illite with expandable phase
in the mixed-layer series.   It is now necessary to fix the
relations between kaolinite, illite and the restricted range of
mixed-layer minerals (allevardite).   There are two possibilities:
joining kaolinite to all mixed-layered compositions by compositional
tie-lines or joining only the most expandable phase stable to
illite and kaolinite, thus having coexistence with the full range
of expandable mixed-layer minerals which are stable.   The only
basis available at present for choosing between the two alterna-
tives is the assemblage found in nature.   As we know, illite is
commonly associated with montmorillonite and highly expandable
phases or with kaolinite.   In the studies of deep boreholes,
this situation persists as the maximum expandability of the
mixed-layered mineral allevardite.   This strongly suggests a full
range of illite-expanding mineral tie-lines as shown in Figure
13.   An important result of such a geometric disposition is the
gradual enlarging of the compositional zone where the three
phases illite, kaolinite and mixed-layer mineral assemblage will
exist as P-T conditions increase (with burial depth).   This
three-phase assemblage will contain the interlayered phase of
maximum expandability which is stable under the prevalent P-T
conditions.

    The two series of phase relations deduced above result in,
at a first approximation, two "facies" for the expandable
dioctahedral minerals - that of low temperature where fully
expandable minerals exist and where the tie-line or association

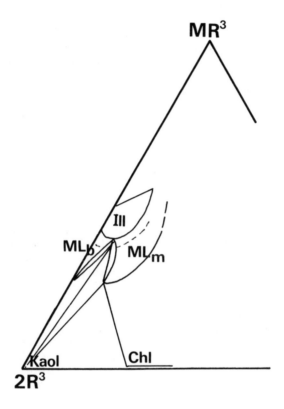

Figure 13. Deduced relations at diagenetic or very low grade
metamorphic conditions where no fully expandable
dioctahedral smectite is present and the mixed layered
mineral series is limited in compositional range.
Symbols as in Figure 12.

beidellite-montmorillonite persists.   More elevated conditions
produce a kaolinite-illite tie-line characteristic of sequences
of buried rocks.

Illite, Montmorillonite and Mixed-layer Minerals in Sequences of
Buried Rocks (P-T Space)

It has been known for some time (Burst, 1959) that
argillaceous rocks frequently present a continuous sequence of
clay mineral assemblages as depth of burial increases. Weaver
(1959) attempted to duplicate these transformations by applying
high pressure to natural montmorillonites at room temperature.
Since these studies were made a number of investigations of
deeply buried sediments have been completed; all show similar
parageneses (Dunoyer de Segonzac, 1970; Mitsui, 1975). The most
valuable information gathered in such studies is that of
mineralogical determination as a function of both depth and
temperature. With such data one can correlate the mineral facies
observed in the context of a geothermal gradient and ultimately
delimit the existence of critical assemblages as a function of
pressure and temperature. Eventually a grid of diagenetic facies
can be produced for argillaceous rocks.

The most important prerequisite in selecting information of
this sort is that there be no major unconformity in the sequence
of sediments and sedimentary rocks studied and that they be recent
in age. A disconformity is even worse but more difficult to
detect in drill hole cuttings. It is absolutely necessary that
the depositional sequence has experienced only the geothermal
gradient which is at present imposed on the silicate mineral
assemblages. If the sediments experienced a previous cycle of
burial and subsequent uplift, creating an unconformity, the
mineral assemblages below this surface cannot be treated as
continuous with those above. This is simply normal geological
reasoning since it is well known that rocks tend to maintain the
silicate assemblages formed at the highest physical conditions.
Retrometamorphism is a slow process at low temperatures.

Basically, the available information is of two kinds: that
from sequences of rocks in sedimentary basins where the geothermal
gradient is on the order of 25°C/km and that from so-called hydrothermal areas where the gradient is much higher, 130°C/km.
The very close correspondence of mineral transformations under
both gradients leads one to believe that the parageneses observed
are both a function of depth (pressure) and temperature. It
should be noted that in areas of high geothermal gradient,
rapidly circulating fluids frequently increase the temperature
locally, and most notably produce hot springs at the surface. In
these cases a thermal gradient is superimposed locally upon the
geothermal gradient. Schoen and White (1965) and Steiner (1968)
have observed that in alteration zones around rock fissures a
new mineral assemblage is produced adjacent to the site of
circulating fluids. However, the general mineralogy of the rock
series appears to conform to a general trend which is dependent

upon the geothermal gradient for the area.

It is important to distinguish between the different observed types of interlayering which derive from the regularity of the repetition of the types of layer units.   We can distinguish four types among the X-ray diffractograms published.   This classification is based upon consideration of Reynolds and Hower (1970) and those found in Brindley and Brown (1980).   First one must consider a situation where no ordering or sequential repetition of mica and montmorillonite layer will occur.   This gives the disordered structure which has a slightly less than 10 Å reflection upon glycollation and an 18 Å reflection also.   Ordering of the layers results in a single reflection between 10 and 18 Å upon glycollation. Short range ordering - sequences of mica-montmorillonite repeated - can give a 27 Å reflection, sometimes called a superstructure reflection.   Allevardite presents this structure and is composed of equal numbers of mica and expandable layers.   It is possible to have an ordering reflection with other proportions of mica and montmorillonite.   Presence of a 27 Å reflection distinguishes what is called here an allevardite-type mineral.   A fourth type of ordering gives an ordering peak approximately at 47 Å due to a three layer ordering unit (10 + 10 + 10 + 17 Å).   This gives a 10 Å and near 11 Å reflection when glycollated;   this is the IMII or four-layer ordering which can occur in minerals with less than 20% expanding layers.

It is also important to consider that identifications of fully expandable smectite are difficult to make.   The range of mixed-layering between 100 and 70% expandable layers gives very similar X-ray diffraction patterns (see Scrodon, 1980) if one considers only the reflections near 17 Å.   It is probably necessary to assume that phases reported as fully expandable could range in composition from 100 to 70% expandable.

In a general way unordered interlayered minerals are succeeded by ordered ones as physical conditions become more intense. Browne and Ellis (1970), Muffler and White (1969), Steiner (1968), Schoen and White (1965) and Eslinger and Savin (1973) report the following clay minerals in rocks of varying origin - volcanic tuffs to normal terrigenous sediments - which are found in "hydrothermal" areas where the geothermal gradient is high.

1 - Montmorillonite exists up to a temperature of 100°C at 100 m depth.   Calcic montmorillonites may persist to 150°C or so at more shallow depths.

2 - Mixed-layer clays, most often ordered, are present up to temperatures near 200°C at depths of 500 to 1500 m.   The minerals form in two distinct zones.   At shallow depths (between 100° and 200°C) mixed layering is between 90 and 0% montmorillonite.

Above 200°C or so no expandable minerals are present. In the
second zone (1.5 km depth) one finds order-interlayering showing
the superstructure reflection near 27 Å glycollated.

3 - Above 200°–220°C only illite or sericite is found, usually
with chlorite.   No dioctahedral mixed-layer phase is present.
This is the "illite-chlorite" zone.

The occurrence of kaolinite is generally erratic but in the
terrigenous sediments (Muffler and White, 1969) it can apparently
react with dolomite to form the assemblage calcite + chlorite
between 120°–180°C.   Expandable chlorite was noted in shear zone,
and iron-rich chlorite is common in most of the rocks becoming
more evident at greater depths.   In the terrigenous rocks
observed, the apparent alumina content of chlorite decreases with
depth.   Alkali zeolites have been observed at temperatures up
to 100°C in the deeply buried rocks.

Iijima (1970), Perry and Hower (1970, 1972), Dunoyer de
Segonzac (1969), Weaver and Beck (1971) and Hower et al., have
studied the sequences of clay minerals found in deeply buried
sediments, both terrigenous and tuffaceous.   Their data are
similar to those obtained from "hydrothermal" areas in that the
three major assemblages are present containing (1) montmorillonite,
(2) highly expandable mixed-layer minerals and (3) low expandability minerals, frequently with an allevardite and eventually
IMII-structure.   The upper limit of apparently fully expandable
montmorillonites appears to be slightly variable at different
depths but it is generally below 100°C and above 80°C.   The
range of mixed layering between 60 and 30% is restricted to
temperatures between 80° and about 120°C, depending upon the
depth and precision of the temperature observations.   The
allevardite-type mineral which has 30-40% expandable layers is
present over a much larger temperature range.   In fact, it exists
up to the greatest depths in the rocks studied.   A last and ill-
defined field is that of the IMII-type ordering where 15-5%
smectite is present. This represents the highest temperature
range of mixed-layer dioctahedral minerals and, by consequence,
the lowest amount of expandable mineral layers present. Kaolinite
seems to be stable to the greatest depths in some cases.   Chlorite
is commonly present but no expandable trioctahedral minerals have
been reported.   Zeolites are found in tuffaceous rocks (Iijima,
1970) presenting the three assemblages: alkali zeolite, analcite,
and analcite-albite as a function of increasing depth.

It is important to note that in all cases the bulk composition of the argillaceous samples was determined no major
systematic bulk compositional variation was observed as a function
of depth (Perry and Hower, 1970; Weaver and Beck, 1971; Dunoyer de
Segonzac, 1969; van Moort, 1971; Hower et al., 1976). This is

especially true for alkalis.   Thus the occurrence of illite or
mica is apparently not a function of bulk composition but one of
pressure-temperature conditions.   A second important observation,
made by van Moort, is that the sequence of mineral changes does
not appear to be related to the age of the rocks, i.e. younger
rocks do not appear to be richer in montmorillonite than older
rocks for given P-T conditions and fully expandable minerals can
be found in pre-Cambrian shales.   In the deep drill holes studies
of more recent sediments (Tertiary, late Cretaceous), the ages of
the rocks vary enough so that it is obvious that physical conditions
are the predominant factors in forming the mineral assemblages.
If there is a kinetic factor involved it is not obvious in the
range of temperatures between $80°$ and $200°C$.

     Figure 14 compiles the above information on a temperature-
depth plot.   The montmorillonite (fully expandable phase), mixed
layer between 80 and 50% montmorillonite and the ordered
(allevardite type) 30-40% expandable minerals have definite zones
of occurrence.   However, information is not abundant for the
IMII-illite boundary at great depths and low geothermal gradient,
i.e. the zone where no partially expandable dioctahedral phase
can exist in a pelitic mineral assemblage.   It has been suggested
by Hower et al. (1976) that this limit is nearer to $200°C$ at
greater than 6 km depths.   Boles and Franks (1979) observed the
IMII-phase at $210°C$, 4 km and Steiner (1968) found it at $230°C$,
2 km depth.   Notable in this diagram is the enlarged field
for allevardite at great depths.   This suggests that the initial
stage of mixed layering, 70-30% expandable, will be much reduced
at great depths and low temperatures.   Such an effect has been
observed by Weaver and Beck (1971).   There is an obvious influence
of pressure on the appearance of the superstructure ordering
reflection.

     It is instructive to compare the natural mineral assemblages
with those in the simplified system which has been studied
experimentally.   The system muscovite-pyrophyllite is useful
because it contains phases analogous to natural minerals.   The
system is dominated by two factors:

          1 - the stability of the expandable phase (potassic
              beidellite) either alone or as a mixed layered phase;
          2 - the stability of kaolinite in the pressence of quartz.

It appears from the data presented by Velde (1969) that the
temperature of transformation from mixed-layer to an ordered superstructure 30% smectite mixed-layer phase (allevardite-type) is
controlled by the stability of kaolinite plus quartz.   The slope
of the reaction kaolinite + quartz = pyrophyllite is considered
to be slightly positive between 1 and 2 kb near $300°C$ (Velde and
Kornprobst, 1969).   This temperature is much higher than those

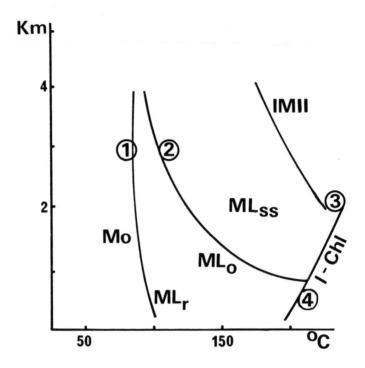

Figure 14. Representation of depth-temperature plot for mixed-
layer mineral occurrence in shale and tuffaceous
sequences of recent sediments. Mo = phases identified
as fully expandable, ML_o = ordered mixed-layer structures,
ML_{ss} = allevardite-type ordering with super structure
reflection; IMII = four-layer ordered, 15-5% expandable
structure, I = illite, Chl = chlorite. Numbers refer
to phase field boundaries found in Figure 15.

where natural superstructure minerals are found to occur.
Further, the slope of the transformation in natural minerals is
negative as is that for the stability of smectite in the synthetic
system. The major phase fields and their disposition as a
function of composition in the muscovite-pyrophyllite system seem
to correspond with the information obtained from the natural
mineral assemblages. In the synthetic system muscovite is the

member of a field of low charge, silicic potassium deficient mica
- a phase assimilable to illite.   At low temperatures complete
solid solution in the form of mixed layering between illite and
beidellite is possible.   Under higher temperature conditions,
there is a gap in solid solution between illite and mixed-layered
minerals.   Also, the maximum extent of mixed layering with an
expandable phase decreases rapidly as temperature increases.
Above $300°C$, the mixed layering becomes approximately constant
($\sim 30\%$ smectite) and ordering is more apparent.   This situation
is maintained until the expandable phase becomes unstable.   The
ultimate stability of beidellite-mica (illite) solid solution is
dependent upon pressure as well as temperature, the maximum
thermal stability decreasing as pressure is greater.   However, in
the synthetic aluminous system, the upper stability is much
higher ($300°-400°C$) than that observed in clay mineral assemblages
from argillaceous rocks at low pressures ($\sim 200°C$).   The relations
between phases are very similar and the sequence of changes is
identical in natural and synthetic systems.

Velde (1977b) has performed a number of experiments on
natural materials which have various types of interlayering and
various non-expandable phases present such as illite, kaolinite
and quartz.   The starting materials were chosen in order to
determine the effect of composition as well as physical conditions
upon the transformation of expandable minerals.   This expands
the information from the synthetic system described above by
adding the chemical variables Mg and Fe to those of Si, Al and K.

Briefly, the samples chosen verified the construction of
the phase relations proposed in Figure 15.   The only difference
between what can be surmised from the data on natural minerals
and the experimental results is the production of a corrensite-
like ordered trioctahedral mixed-layer mineral at conditions
approaching those which produce ordering in the dioctahedral
mixed-layer minerals.   We will discuss these results here only as
they affect the sequence of ordered minerals found as P-T
conditions increase.   Figure 15 summarizes the experimental
results as a type of phase diagram where P-T conditions effect
different arrangements in the mica-expanding minerals present
between muscovite and a chlorite composition.   We see that
the lowest temperatures would give disordered interlayer minerals.
The second step with metamorphism is a reduction in the expanding
layers present in the structure and an ordering of the units.

Since most sedimentary rocks contain several clay minerals
at the same time, the phase relations that they reflect are those
of the phase fields to the right of the figure.   If a rock contained a dioctahedral mixed-layered mineral only, it would be
represented by the central portion of the diagram, the ML zone.
Such rocks do exist; they are called metabentonites.   In some

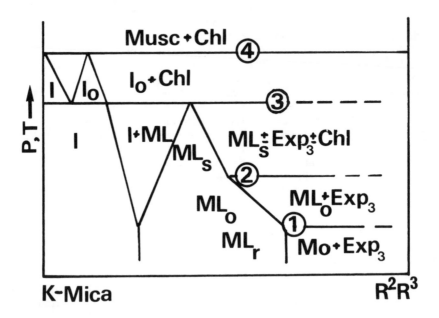

Figure 15. Generalized phase diagram of relations in non-binary
          chemical coordinates between muscovite-phengite
          (K-mica) and aluminous minerals such as chlorite
          (R^2R^3). Exp_3 = expandable trioctahedral minerals,
          Io = IMII-type phase. Other symbols as in Figure 14.
          Numbers refer to major changes in mixed-layer mineral
          type as observed in deep drill holes. The phase
          diagram is based upon experimental results using
          natural and synthetic starting materials.

instances, there is a variation in chemistry of the rocks between
the edge and centre of a metabentonite bed (J.Hower, A-M Bystrom-
Brusewitz, personal communications). One can see what the effect
of chemical variations has on the system at constant P-T conditions.
This is especially important for the types of interlayering as
they are affected by the bulk composition of the phase. It appears
that the low potassium content mixed layered phases are randomly
interstratified from 100 to about 50% expanding layers. Then one
finds ordered minerals and then minerals with a superstructure
reflection in the 30-20% expanding layer compositions. In the
experiments, the type of ordering produced is a function of the

physical conditions employed which change the proportions of phases present by changing the composition of the mixed layered phase. Only ordered phases were encountered near the 30% expandable region.   However, this is not true for all natural mineral assemblages.   In the less than 15% expanding phase samples, ordering is of the type IMII, designated as I in the figure. Such minerals are found at high temperatures in the experiments and in the later stages of diagenesis in rocks.   In most assemblages of clay minerals in pelitic rocks, the degree of ordering and composition of the mixed layered phase will be dependent upon the physical conditions of crystallization.

Now let us consider the mica-like component in the system. In looking again at the phase diagram for the aluminous system (Figure 15) the compositional field of the mica-like phase (an aluminous illite) is seen to decrease as temperature increases. Although the experimental data are imprecise, it can be expected that some 10% "chlorite" component is lost between $300°$ and $400°C$. Interestingly, a similar effect has been observed in sequences of deeply buried sediments and sedimentary rocks (Dunoyer de Segonzac, 1969; Dunoyer de Segonzac et al., 1968).   Using a crystallinity index or sharpness ratio proposed by Weaver and redefined by Kubler (1968), these authors have observed that the shape of the illite reflection becomes more sharp and narrow-based as depth and temperature increase.   It can be assumed that better crystallinity can be assimilated to a more micaceous composition of the mica-like phase, i.e. one with + 1.0 charge satisfied by an alkali ion in the interlayer position (per $O_{10}(OH_2)$). This would correspond to the narrowing of the compositional range for the micaceous phase in the synthetic system.   Thus as depth and temperature increase the illitic and mixed-layer phases become more well defined, ordered and better crystallized; the association illite (possibly of the $2M_1$ polymorph) and allevardite (ordered mixed-layer illite$_{70}$- montmorillonite$_{30}$) is the stable assemblage. The illite will contain more potassium and less silica, iron and magnesium than at low temperatures.

Thus the sharpening of the basal reflection is probably due to the increasing K_2O content of the phase, as suggested by Dunoyer de Segonzac (1969) and Weaver and Beck (1971), which permits a more ordered strucutre, i.e. a more ideal mica form.   It is known that illite is usually of the 1Mx polymorph, where there is disorder in the stacking sequence in the $c \sin \beta$ crystallographic direction.   This would be expected where less than ideal occupancy of the interlayer ion site would inhibit orderly arrangement of the layering sequence producing stacking faults.   It is also possible that some of the low total structural charge could be due to a small admixture of expanding low charge montmorillonite layers which broaden the basal reflection to low angles (Reynolds and Hower, 1970).   A greater frequency of $2M_1$ polymorphs, as

noted by Dunoyer de Segonzac (1969), in the illites with a
sharpened peak is most likely the result of mica compositional
minerals which can become ordered in the stable polymorph near
the muscovite composition (Velde, 1965).   This is indicated in
Figure 15.

     Considering the compositions of the mixed layered minerals
found in sedimentary rocks (Figure 13) it is obvious that magnesian-
iron expandable dioctahedral minerals will be in equilibrium not
uniquely with kaolinite but also in many instances with a magnesian-
iron phase - either chlorite or an expanding trioctahedral mineral.
In such a situation the slope in P-T space of the reaction mixed-
layer $\rightarrow$ allevardite + phyllosilicate will be controlled not by
the aluminum silicate phases, as in the muscovite-pyrophyllite
system, but by the production of chlorite.   The reaction most
likely to occur in the range of conditions concerned is the
transition between a trioctahedral expandable phase and 14 Å
chlorite + quartz.   In the experimental system $MgO-Al_2O_3-SiO_2-$
H_2O investigated by Velde (1973), this transition has been observed
to maintain a negative slope in P-T space.   Of course, the temperatures at which this phase change occurs in the synthetic system
are well above those found in the natural system, due mainly to
the presence of iron in nature, but the implication remains that
such a reaction "fits" the slope defined by the data for natural
assemblages.   If the parallel between aluminous and magnesio-
aluminium systems can be assumed, kaolinite stability and that of
expandable chlorite can be considered analogous in their influence
upon the amount of interlayering present in the illite-montmorillonite structures present.   If we look back to the experimental
studies on natural expandable minerals at high pressures, it can
be recalled that the production of a chlorite phase occurred
when interlayering in the natural dioctahedral mineral had reached
about 30% interlayering.   It is possible that below this transition only expandable phases are present for most magnesium-iron
compositions;   one is dioctahedral, the other would be trioctahedral.   Thus, at temperatures below the transition to an ordered
allevardite-type phase, dioctahedral mixed-layer minerals will
coexist with expandable chlorites or vermiculites as well as
kaolinite.   The distinction between these two phases is very
difficult because both respond in about the same manner when
glycollated.   There can also be interlayering in both di- and
trioctahedral minerals.   The temperature of mineral transition
to non-expanding minerals will be a function of the Mg-fE^{2+} content
in the sysmtem as a whole.   It follows then that magnesian 14 Å
chlorite in sediments will be in disequilibrium until allevardite
is stable.   It can be suggested then that most of the 14 Å
chlorite in sediments is of detrital origin.   The data of van
Moort (1971), Perry and Hower (1970) and Hower et al., (1976)
suggest such a conclusion in that chlorite content appears to
decrease until the allevardite zone is reached, where chlorite

content begins to increase.

Let us now consider more closely the conditions under which
allevardite, or IMII-super-structure expandable dioctahedral
phases, disappear from sedimentary rocks.   There are several key
assemblages which have been reported that distinctly limit the
possible physical conditions for this transition.   Most notable
is the assemblage allevardite (actually IMII mixed layering)-
pyrophyllite, reported by Dunoyer de Segonzac (1969) in a series
of shales found in the French pre-Alps and in shales of a
Palaeozoic basin in northern Africa, and by Frey (1970, 1974) in
Mesozoic sediments of the Swiss Alps, for Palaeozoic rocks in
northern France, Esquevin (personal communication) and for
Palaeozoic rocks of the French Pyrenees (Dunoyer de Segonzac and
Heddbaut, 1971).   The mineral pair has also been reported in
supposed hydrothermally altered shales in Utah (Ehlmann and Sand,
1959).   The stability of pyrophyllite relative to Kaolinite +
quartz has been determined to be near $300°C$, at 1 and 2 kb
presssure (Velde and Kornprobst, 1969).   This is a minimum value
compared to other experimental studies made previously which
indicate significantly higher temperatures.   Because the
association of interest is certainly low temperature, we will
use the lowest plausible values in the discussion.   Given then
the minimum of $300°C$ for the stability of pyrophyllite at 2 kb,
the assemblage pyrophyllite-IMII mixed layered mineral must exist
above this temperature.   Since illite + chlorite replace
allevardite in low pressure "hydrothermal" sequences (Figure 14)
we know that the slope of the curve IMII = illite + chlorite must
be positive as pressure increases in order to attain a temperature
above $300°C$ at 2 kb total pressure.   If this is accepted, the
assemblage IMII + pyrophyllite would be stable at this temperature
for depths greater than 7 kilometres, assuming an average rock
density of 2.8 and $P_{total} = P_{lithostatic}$.   Further, the experimental results on pyrophyllite stability indicate a $300°C$ minimum
temperature at 1 kb or depths of 3.5 km.

Discussion of the mixed-layer mineral facies depth-temperature
diagram

There are, as we have said, two ways of looking at clay
mineral assemblages and the reactions which took place to form
them.   One is to say that under given geological conditions,
most assemblages or different constituents of an assemblage are
not the stable forms.   This supposes that reaction rates are too
slow, at a given temperature, for the true equilibrium assemblage
to form from the materials initially present.   If a range of
temperatures is observed in a sequence of rocks, different
approaches to the equilibrium assemblages will be observed
depending upon the lenght of time which the temperature persisted
and the rate of the reaction at a given temperature.   This point

of view is held by Eberl and Hower (1976) where their experiments
were used to demonstrate that aluminous smectite is unstable under
geologic conditions.   It will be replaced by an assemblage of
mica + hydrous aluminosilicate when either sodium or potassium
ions are available.   Thus aluminous smectites found in deeply
buried rocks should be considered to be metastable.   Similar
experiments were performed by the present author (Velde, 1969)
from which he concluded that smectite is stable at low temperatures
but the amount of smectite which can exist in a mixed-layer
structure (solid solution between smectite and mica) will diminish
as temperature increases.   Both series of experiments were
performed at temperatures above those at which a fully expandable
smectite would be stable according to the experiments of the
present author and according to the thermal conditions in nature
where smectite occurs in pelitic sediments.   The conclusions of
this study are that aluminous smectites in argillaceous rocks can
represent an approach toward equilibrium.

    Here we are confronted by a classic problem: what do the
laboratory experiments mean compared to the observations made
upon geological materials?   It is probable that both sets of
experiments (Eberl and Hower, 1976; Velde, 1967) can be interpreted in two ways since the results are similar using similar
materials.   However, if we turn to natural materials as the final
referee, we are still confronted to a certain extent by a time-
temperature problem (kinetic).   This is possibly due to a basic
geologic configuration and a lack of geological information.   If
we consider the source of information used to construct depth-
temperature-facies plots (Figure 14), we see that in recent basins,
those which are deepest (i.e. have the thickest series of sediments)
are those with the lowest geothermal gradients.   Those with
lower sedimentation rates have higher geothermal gradients.   Thus
for a given depth, the rocks will be hotter and older in shallow
basins than those in deep basins.   It seems that there is a
general pressure (depth)-temperature-time surface for sedimentary
basins.   This relation creates a problem in distinguishing between
time, temperature and pressure effects as independent variables.
How then can one go about sorting out the importance of each factor
concerning the stability of mixed-layer phases in sedimentary
rocks?   At the moment there is probably no decisive argument
since we do not know how long a geothermal gradient is maintained
during a period of an orogenic cycle.   How then should one proceed
in interpreting clay mineral assemblages?   If we consider the
simple depth-temperature plot of the mixed-layer mineral facies
(Figure 14), we see that the limit for high expandability mixed-
layer minerals (100-80% smectite) seems to be somewhat independent
of depth, and one can consider it to lie at $80°C \pm 15°C$.   There is
a possibility that this limit decreases in temperature as depth
increases.   This limit does not appear to depend on the age of
the sediments.   The data of Perry and Hower (1972) are possibly

the best supporting evidence for this.

The next facies limit, that where about 30% expandable layers are present in a superlattice structure (allevardite-type interlayering), seems to be more variable in time-temperature -depth space. Here, the reaction is not the same as that of a simple reduction of smectite layers which are transformed into illite layers. The alternation of illite-smectite must be respected in allevardite which suggests a selective organization of ions in the tetrahedral and octahedral sites in each layer. The kinetics of this reaction should be different from those of a simple, random change of smectite to illite layers as the material recrystallizes (i.e. the first facies boundary) and as a result one would expect different relations on the time-temperature plane. Since the reaction involves a relatively large release of water (see Perry and Hower, 1972), it is likely that pressure and hence depth will become an important variable in the kinetics of the reaction. In sedimentary basins of low geothermal gradient, this change appears to occur between 120°-150°C. In areas of high geothermal gradient, this reaction occurs at temperatures above 200°C, but we do not know how long these high temperatures have persisted in the areas investigated (New Zealand and California). If one considers the laboratory experiments, the transformation of smectite to mixed-layer minerals of low expandability (< 20% smectite) occurs at 250°C in several months' time. If illite plus kaolinite (or chlorite) were the stable phases (Eberl and Hower, 1976), they would surely have already formed in the hydrothermal altered rocks. But we do not know what the assemblage in these rocks would be 10^6 years from now at constant temperature. As a result, we can tentatively place an upper limit for the reaction to form allevardite in time-temperature space at low pressure as being somewhat above 200°C, in fact at 220°C in New Zealand (Steiner, 1968). In hydrothermal areas, it appears that the illite-chlorite facies (i.e. no interlayered aluminous minerals are present in pelitic rocks) follows closely after the allevardite-bearing rocks. At lower pressures, less than 2 km depth, the allevardite minerals are not present and one passes from low expandability non-super-structure minerals to illite-chlorite assemblages. Again according to the kinetic studies of Eberl and Hower (1976) which were performed in this temperature range, the reaction times are short, in geological terms, for this reaction and one would expect that the clays present in these rocks would represent an approach to equilibrium.

In this way, we can put outer limits on the mixed layer mineral facies between 80°C and 200°C at shallow depths. There remain now the more normal geothermal gradients which occur in most sedimentary basins. Here, the problem of reaction kinetics will undoubtedly be important in determining the precise clay

mineral assemblage in a shale. The data of Boles and Frank
(1979) indicate that Eocene Gulf coast sedimentary rocks attain
assemblages of lower smectite content at temperatures below
those of Oligocene and Miocene rocks (Perry and Hower, 1972;
Hower et al., 1976). It appears that, if present day geothermal
gradients had been maintained throughout the history of the Gulf
Coast basin, 20 to 40 million years could produce a difference
of 20°C in the apparent temperature for the appearance of
allevardite-type mixed layered minerals of 30-20% expandable
layers at 3-5 km depths. There is significant overlap in
transition temperatures for a series of different ages which
might suggest that some assumptions of constant temperature over
the time period or possibly differences in reaction rates due to
differences in starting materials varies the resulting observed
assemblages.

One must conclude that the allevardite facies limit cannot
be determined with precision at present for deeply buried rocks.

For the last stage in the evolution of mixed layered minerals
the IMII interlayered form is perhaps easier to interpret but we
have less depth-temperature data as to its occurrence. Steiner
(1968) reports it to occur at 220°C, 2 km depth and Hower
(personal communication) has identified the phase at 5 km depth
in rocks which have experienced temperatures of 180°C. Boles
and Franks(1979) report its occurrence in samples at 210°C.
Since it has been reported in rocks containing pyrophyllite and
chloritoid, the IMII-type interlayering must exist over a large
range of temperatures. However, one must correlate supposed
reaction rates and criteria for equilibrium between these phases
and clay minerals.

The most important aspect of the information on buried
sediments is the apparently similar sequence of mineral assemblages
found in deep and in shallow series. For high and low geothermal
gradients, in pelitic or acidic volcanic materials, zeolite-
bearing or calcareous, all rocks containing a predominantly dioctahedral phyllosilicate mineral assemblage are found to have con-
cordant mineral assemblages. Further, there is no evidence that
the bulk rock composition changes significantly during these transformations and the elements most likely to vary (alkalis) remain
more or less constant at depth increases although minor changes
certainly occur (Leikine, 1980). The sequence of mineral
stabilities - montmorillonite, random mixed-layer, allevardite-
type, IMII, and uniquely illite - can be used as an index to the
pressure-temperature conditions which a rock has undergone. This
gives a useful facies grid for a large range of physical conditions which encompass most sequences of deeply buried rocks.
It should be remembered that the assemblages containing kaolinite
should indicate the composition of mixed-layer minerals with the

maximum amount of expandable layers stable at give P-T conditions.
Another very important observation to be made is whether or not a
high-spacing ordering peak is present indicating the stability of
the allevardite-type phase.   With such information available, a
reasonable assessment of mineral facies can be made.

1 - The mica-montmorillonite cycle in weathering and
diagenesis. An interesting use of phase diagrams can be made to
explain the apparently paradoxical relationship between dioctahedral micas and montmorillonites in weathering and diagenesis.
The observation has been made (Millot, 1964; Weaver, 1959;
Dunoyer de Segonzac et al., 1970; among others) that illite and
dioctahedral micas are "degraded" or altered to form montmorillonites during weathering. Upon burial and initial diagenesis,
the montmorillonite fraction decreases, being replaced apparently
by illite.   There is an apparent symmetry observed in weathering
and diagenetic processes.   This observation has led to the idea,
or reinforced it, that montmorillonite represents a metastable
form of mica which returns to its initial form when exposed to
sea water sedimentation and burial; it becomes reconstituted
(Weaver, 1959) upon diagenesis.

Let us consider the first aspect of these statements, the
weathering-diagenesis identity.   If we use the T-x diagram of
the muscovite-pyrophyllite compositional join as an approach to
the behaviour of the natural montmorillonites associated with
illites, both weathering reactions and diagenesis can be
portrayed.   Figure 16 shows two arrows representing possible
paths taken by individual phases during the two processes.   One
arrow shows the evolution of the mica component as alkalis,
principally potassium, are removed from a rock via weathering
processes and as the bulk composition of the system allows the
formation of first a mixed layered phase then a montmorillonite.
This sequence has been observed in many weathering profiles.   The
removal of alkalis from the mica is not the only change necessary
to produce a smectite.   In order to maintain charge balance, the
octahedral or tetrahedral ions must become globally more positive
in charge, either by ionic substitution (divalent or trivalent
ions) or by oxidation of the material.   As the total alkalis
diminish beyond the limit of montmorillonite-bearing assemblages,
kaolinite becomes apparent in the clays.   Should the weathering
be rapid, that is, should alkali removal be rapid, kaolinite could
appear with only a small amount of the compositionally intermediate montmorillonite phase present in a narrow horizon.
However, in slower processes an apparent degradation sequence can
be traced on the phase diagram of Figure 16 as a function of
decreasing alkali content in each horizon of the profile.

Let us now take a phase of montmorillonite composition (fully
expandable at low temperature) and subject it to an increase in

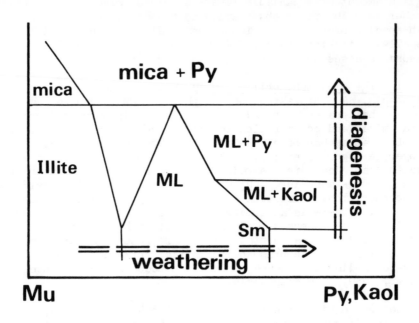

Figure 16. Schematic diagram used to illustrate possible paths
           in P-T-x space in a weathering-diagenesis cycle,
           where the low P,T trajectory is one of chemical
           change and the diagenesis trajectory effects phase
           changes.   The aluminous, potassium-bearing phase
           will show first a loss in K content (production of a
           smectite, Sm) then will recrystallize to form a
           multiphase assemblage where the alkali-bearing phase
           contains more potassium as temperature increases.
           Thus one sees a mica-to-mica cycle.

temperature and pressure in an argillaceous system.   The following transformations take place: the initially expandable phase
changes composition to become an interlayered mineral with a
mica-like component, and kaolinite is produced as well as quartz.
Higher temperatures effect an apparent reconstitution of the mica-
like component which is accompanied by the continued production
of kaolinite then pyrophyllite.   If one looks only at the
smectite-illite part in the assemblage, it takes the reverse
course to that observed during weathering.   Thus the degradation
of a mica due to weathering is in appearance reversed: illite
becomes illite again.   However, several other transformations
have occurred during the processes.   In the first instance
alkalis have left the system and the proportion of Al relative

to the other elements has been changed from that in the initial
2:1 mica structure. The result is the production of a
montmorillonite. In the second transformation, a new non-mica
phase was produced as the expandability of the dioctahedral
mineral decreased while producing more of the illite component.
By looking only at the alkali-bearing 2:1 mineral in the weathering-
diagenesis cycle an apparently circular process might be observed.
However, this is accomplished through the removal of material and
the subsequent production of new phases in the weathering cycle
and crystallization of new phases in the burial metamorphic
(diagenetic) cycle. The whole process can be envisioned as a
normal series of phase equilibria attained under varying chemical
and physical constraints where minerals recrystallize according
to P-T-x variations during geologic processes.

    2 - Recrystallization in sandstones. One can imagine other
configurations that should be commonly encountered in sedimentary
rocks. Take, for example, a shale layer and an adjacent sandstone which have been buried to depths where allevardite (30%
expanding layers) is present. As the orogenic process ends the
geothermal gradient decreases, and a new charge of less saline
waters flows into the sandstone. Lower temperatures allow the
crystallization of a mixed-layer mineral with more expandable
layers. This reaction is hastened by a lower chemical potential
of alkalis, notably potassium, in solution. We have moved to
lower temperatures in our diagram (Figure 16) and we have moved
away from its mica side. However, the shale which is relatively
impermeable has not had an incentive to recrystallize under
constant chemical conditions and it retains its allevardite
mineralogy. Now if we drill-core the two rocks, we will find
two different clay mineral assemblages side by side using X-ray
diffraction techniques. It would be necessary to see where the
new phases occur in the sandstone in order to ascertain their
origin - selectively placed in or near pores or within the clay
mass of the rock. In order to obtain these data, one must use
petrographic and microprobe methods.

DETAILED ANALYSIS OF CLAY SYSTEMS: THE ELECTRON MICROPROBE

    Up until now we have tried to generalize about clay mineral
compositions and clay assemblages. Our aim was to simplify.
In this way, several general patterns of clay mineral occurrences
can be developed and explained and the evolution of mixed-layer
minerals during burial diagenesis of argillaceous rocks can be
taken as an example. However, if general statements about clay
mineral assemblages and phase equilibria have been made, they
signify a simplification towards an ideal system. We know very
well that natural clay minerals are frequently mixtures of new
and old phases which reflect various origins. This knowledge has

led clay mineralogists to work with the finest fraction of an
assemblage hoping that small grain size will lead to greater
reaction rate.   Hower and his coworkers have found this method
to be very useful as have others.   However, if we deal only with
the finest materials present, we naturally ignore what happens in
the coarse fraction.   We do know that as a general rule, kaolinite
for example tends to form in larger crystals than smectites even
when both are the result of the same geological processes.   Thus
it is evident that eventually one must look at the whole rock in
order to understand the diagenetic processes which takes place.

It is significant to use the term "look at" the whole rock
because modern methods of analysis do in fact now allow us to
use a microscope in the investigation of clay minerals and one
can become a sedimentary petrologist in their study.   The electron
microprobe should become a tool of routine investigation of clay
minerals in heterogeneous rocks.   This is possible largely due to
the use of solid state "energy-dispersive" detectors which allows
very low beam currents to be used on the relatively unstable clay
masses.   Although these techniques have been available for several
years now, little work has thus far been done upon clay mineral
problems.   However, several examples of the types of studies
possible will be given as a suggestion as to the possibilities
that this method present.

1 - Coarse-grained phases.   Recent work by Nicot (1981) and
the present author shows that, at least in some samples, detrital
micas can persist in sedimentary rocks even though new micaceous
phases crystallize.   One phenomenon observed is an overgrowth of
a new mica-like phase on a detrital mineral.   The new mineral is
less sodic (< 0.10 atom/O_{10} (OH)$_2$ formula) than the detrital centre,
contains more silicon, magnesium and generally more iron.   The
general tendency is an under-filling of the alkali ion site
(< 0.85 ion).   Figure 17 indicates these relations in MR^3-$2R^3$-
$3R^2$ coordinates.   The overgrowth zone is from 10 to 20 microns
wide.   Other micaceous zones occur where detrital feldspars or
other minerals have been replaced by illite.   These are also
plotted in the figure.

If it is possible to analyse the new micas, it is also
possible to analyse new chlorite grains when they are 5-10 microns
wide as has been reported in the section on chlorites.   Since
kaolinite has an invariable composition it need not be analysed
but at times replacement textures of detrital minerals such as
feldspars and white micas are not diagnostic enough to allow one
to establish when kaolinite is present.   Use of the microprobe
gives a quick and accurate identification on a microscale.   In
this way the three clay minerals most likely to occur in the
> 2 micron size fraction can be identified as to their chemical
composition: variations in their composition indicate to what

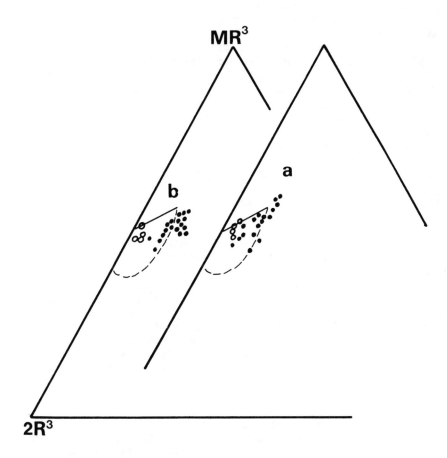

Figure 17. Plot of microprobe analyses of micaceous clays for
          two sandstones from the North Sea. a = sample
          containing clays with higher expandability than
          sample b. Circles show sodic micaceous minerals
          (detrital) and dots show low sodium diagenetic
          minerals.

extent the detrital phases have persisted and to what extent local
chemistry has played a role in determining the composition of the
new phases.   This is a measure of the approach to chemical and
thus thermodynamic equilibrium on the scale of the rock sample
(centimetric scale in the case of thin section observations).

   2 - Replacement and pore filling. It is possible in sandstones to find areas of newly formed phases which seem to replace
previous aluminous minerals or fill pore zones. These areas are
less coloured than the matrix which is a very finely crystallized
clay mineral assemblage.   Two samples are presented which show
that the composition of these clayey zones seems to vary as does
the overall clay mineralogy (Figure 18).   The sample with about
20-25% expandable layer mixed-layer mineral present in the < 2
micron fraction shows a spread of clayey zone composition around
the edge of the "illite" sector" in the figure.   The other sample,
which contains as < 2 micron clay fraction with a 15% expandable
mixed-layer phase shows clayey zone compositions within the
illite zone or just at its edge.   We see that, on the average,
there is a difference in composition of these zones from one
sample to the other which could well be correlated with the
general clay mineralogy of the samples.   It is also evident that
there is a scatter of compositions of this material indicating
chemical heterogeneity on a local scale (assuming the material
to be mono-phase).

   3 - Matrix compositions. Now, if we look at the matrix or
groundmass clay compositions of the same samples (Figure 19), it
is apparent that there is a mixture of phases present. In one
case the groundmass is a mixture of a mixed-layer phases (plus
possibly illite also) and chlorite while the compositions in the
other sample lie between a mixed layered phase and kaolinites.
In this latter sample, chlorite is expressed in individual grains.
Both samples have larger crystals of kaolinite present. It is
interesting to note that no compositions fall into the three-
phase area between mixed-layer kaolinite and chlorite phases.

Microprobe analyses and phase diagrams

   The most obvious result shown by the plots of the microprobe data on clays from sandstones is that from point to point
the same phase has different compositions.   This immediately
suggests non-equilibrium of the phases on the scale of the thin-
section.   There are two reasons for this - one is that the phases
have not reacted sufficiently to produce the single, stable phase
of a fixed composition which is determined by the bulk composition
of the system and its pressure-temperature conditions; the other
explanation is that the reacting agents which produce the phase
are part of a system smaller than that represented by the thin
section.   In this manner very local chemical conditions are
responsible for the composition of the phases present.   Further,
it is quite possible to consider that both causes of mineral
inhomogeneity are operative. What then should be done to reconcile
the observed "facts" of mineral heterogeneity and the assumption
of phases at equilibrium which has been used to formulate the
preceding phase diagrams to this text?

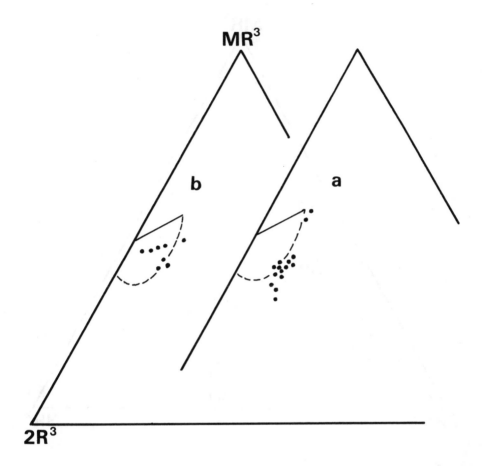

Figure 18. Plot of microprobe analyses for the samples shown
            in Figure 17.   Here argillaceous patches were
            analysed which appeared either to fill pore spaces
            or replace detrital minerals. These are identified
            with the expanding mineral present in the rock.

        In order to assess the meaning of phase diagrams and their
application to geological problems we will have to go back to a
few thermodynamic definitions.   What is a phase?   According to
the most strict definitions, a phase is separable physically
from other phases.   This means that one will see optic, structural
and compositional discontinuity between types of segregation into

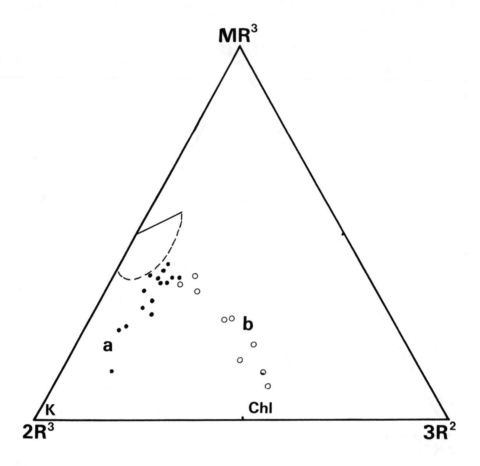

Figure 19. Plot of microprobe analyses of the clay matrix for the
           North Sea sandstones. Compositions are aligned between
           the expanding phase and kaolinite (k) for the sample
           a and chlorite (chl) for sample b of Figures 17 and 18.

phases.   Second, each phase is physically and chemically homogeneous, so that when at equilibrium the chemical potential
differences are null between different phases, and thus elements
will not migrate between phases.   The structural state of a
phase in a system as well as its chemistry are the same throughout.
These are the definitions used since the 19th century.   However,
geologists know that natural metamorphic silicate phases containing more than three cationic elemental species are rarely

homogeneous.   Does this mean that almost all minerals in rocks
have been formed metastably with respect to their neighbours? Does
this mean that if the rock had been held at given P-T conditions
for a longer period of time, it would contain another mineral
assemblage?   Could one go so far as to say that all zoned garnets
are metastable phases?

The great success of metamorphic petrology in the twentieth
century has been based upon the use of the phase rule to explain
mineral assemblages.   This has largely been done despite the
fact that metamorphic minerals are quite inhomogeneous in a given
thin-section sized sample.   As it turns out, one can estimate the
number of phase which occur or were formed together on the basis
of optical criteria - it is usually fairly obvious when minerals
are being recrystallized.   This allows one to establish compatible
phase assemblages according to the norms of thermodynamic
equilibrium using the phase rule.   However, phase homogeneity is
frequently not considered to be important.   The phase rule "works"
in a general way.   Of course certain minerals present problems
such as those of the aluminosilicates-kyanite, sillimanite and
andalusite which are very slow to react.

I think that one can consider clay minerals under diagenetic
conditions, say those above 100°C, to be the equivalent of
metamorphic rocks as far as the phase rule is concerned.   We
can apply the same methods of reasoning to the mineral assemblages
present.   What then will we see?   There will be basic assemblages
of newly formed clay minerals which will represent a given set of
P-T-x conditions for systems closed to significant chemical
migration.   Within these minerals groups one will find variations
in mineral composition which reflect either slow reaction rate to
produce the equilibrium composition for the phase or which represent
variations in local chemical composition of the systems.   Thus one
will first attain an equilibrium which determines the types of
phases present, then one will attain the homogeneity of composition
if reaction rates are rapid enough.   This sort of a general
process has been observed by the author and his coworkers for
systems at low temperatures involving weathering of intrusive
rocks and in studies of glauconite genesis and weathering.   It
would seem reasonable to expect that such a system would apply
to diagenetic reactions which are between weathering and metamorphic
conditions.

It is likely that, as temperatures or the duration of a burial
event are greater, one will find a greater homogeneity in the
phases present.   Such information will be geologically important
if we can generalize using well established cases to fix our points
of reference.   Much work needs to be done.

Another important aspect of microprobe work on silicates in

sandstones is that, depending upon the petrographic or textural site, one can determine what are the compositions of pore filling phases as opposed to those either in the groundmass or those which form from or on detrital phases. It could thus be possible to distinguish between primary diagenetic clay minerals and those which formed at a later time. Again, much work has yet to be done on the subject before we can make serious generalizations. However, these last aspects of the study of clay minerals can lead us to new applications of the determination of clays as they relate to their environment of formation.

CONCLUSION

In the preceding pages, we have seen what could be considered to be the exposition of a method of reasoning concerning clay mineral assemblages in pelitic or aluminous sedimentary rocks. This method is none other than that used in metamorphic petrology for many years. However, its application to clay mineralogy has not been commonly used until now due to a lack of precise data. It is now possible to summarize the chemical data available in order to fix the compositional value of the major clay mineral species in chemical space. These data, based upon global analyses of pure or purified clay minerals, seem to be confirmed when microprobe determinations on appropriate minerals have been made. Thus a basis for determining the relations between the minerals has been established.

The next step has been to determine the P-T limits for the different clay minerals. This can be done in a general way when the bulk composition of the sample does not limit the occurrence of a phase. For example we have seen that the apparent stability of kaolinite could well be due to reactions among some phases which are not normally present and which do not normally intervene in the pelitic clay mineral assemblages. A correlation of depth-temperature occurrence with chemical composition data for the phases allows one to establish general paragenesis diagrams which delimit the mineral stability zones in P-T space.

Problems of identifying local kinetic or chemical equilibrium will undoubtedly arise when one uses X-ray diffractograms of whole rock samples. It is believed that, when possible, electron microprobe analyses will be useful to establish the precise phase relations in a given rock. In fact such observations might well lead to the establishment of different scales of duration for the P-T paroxysms of orogenic events which cause clay mineral recrystallization. Further, it should be possible to isolate different parts of a porous rock which have reacted under different chemical or physical conditions during the burial and diagenesis of a sedimentary rock. Thus two aspects of sedimentary petrology which have seen thus far rather elusive - kinetics and chemical domains - might well be accessible in the near future.

The phase diagrams presented above must, of course, be taken as being a preliminary attempt to establish a P-T gride for clay mineral facies in sedimentary rocks.   It is hoped that continued work will establish such diagrams on a more firm basis and allow one to use them in the analysis of a wide range of geologic conditions.

REFERENCES

BARTHOLOME, P. 1966. Sur L8abondance de la dolomite et de la sepiolite dans les series sedimentairs. Chem.Geol., 1, 33-48.
BRINDLEY, G. & BROWN, G. 1980. Crystal Structures of Clay Minerals and their X-ray Identification. Mineralogical Society, London. 495 pp.
BOLES, J.R. & FRANKS, J.G., 1979. Clay diagenesis in Wilcox sandstones of Southwest Texas: implications of smectite diagenesis in sandstone cementation. J.sedim.Petrol., 49, 55-70.
BROWN, G. (ed.). 1961. The X-ray Identification and Crystal Structure of Clay Minerals. Mineralogical Society, London, 544 pp.
BROWNE, D.R.L. & ELLIS, A.J. 1970. The Ohaki-Broadlands hydro-thermal area, New Zealand: mineralogy and related geochemistry. Am.J.Sci., 269, 97-131.
BURST, J.E. 1959. Postdiagenetic clay mineral environmental relationship in the Gulf Coast Eocene. Clays Clay Miner., 6 327-341.
CALVERT, S.E. 1971. Nature of silica phases in deep sea cherts of the North Atlantic. Nature (Phys. Sci.), 234, 133-134.
CARMOUZE, J.P., PEDRO, G. & VELDE, B. 1978. Pelletal montronite formation in recent sediments of Lake Chad. Chem.Geol., 23, 139-149.
COURBE, C. & VELDE, B. 1981. Weathering of glauconites. Reversal of the glauconitization process in a soil profile in Western France. Clays Clay Miner., (in press).
DUNOYER DE SEGONZAC, G., FERRERO, J. & KUBLER, B. 1968. Sur la cristalinite de l'illite dans la diagenese et l'anchimetamor-phisme. Sedimentology, 10, 137-143.
DUNOYER DE SEGONZAC, G. 1969. Les mineraux argileux dans la diagenese. Passage au metamorphisme. Mem.Serv.Carte Geol. Alsace-Lorraine, 29, 320 pp.
DUNOYER DE SEGONZAC, G. 1970. The transformation of clay minerals during diagenesis and low-grade metamorphism. A review. Sedimentology, 15, 281-346.

DUNOYER DE SEGONZAC, J. & HEDDEBAUT, C. 1971. Paleozoique anchi-
    metamorphism a illite, chlorite, pyrophyllite, allevardite et
    paragonite dans les Pyrenees basques. Bull.Carte serv.Geol.
    Als.Lorr., 24, 277-290.
EBERL, D. & HOWER, J. 1976. Kinetics of illite formation. Bull.
    geol.Soc.Am., 87, 1326-1330.
EHLMANN, A.J. & SAND, L.B. 1959. Ocurrences of shales partially
    altered to pyrophyllite. Clays Clay Miner., 6, 368-391.
ESLINGER, E.V. & SAVIN, S.M. 1973. Mineralogy and oxygen
    isotope geochemistry of the hydrothermally altered rocks of
    the Ohaki-Broadlands, New Zealand geothermal area. Am.J.Sci.,
    273, 240-269.
FOSTER, M.D. 1951. The importance of the exchangeable magnesium
    and cation exchange capacity in the study of montmorillonitic
    clays. Am.Miner., 36, 717-730.
FREY, M. 1970. The step from diagenesis to metamorphism in
    pelitic rocks during alpine orogenesis. Sedimentology, 15,
    261-179.
FREY, M. 1974. Alpine metamorphism of pelitic and marly rocks
    of the central Alps. Schw.Petr.Mitt., 54, 489-506.
GRIM, R.E. 1968. Clay Mineralogy. 2nd Ed. McGraw-Hill, New York.
HEMLEY, J.J., MONTEYA, J.W., NIGRINI, A & VINCENT, H.A. 1971.
    Some alteration reactions in the system CaO-Al O -SiO -H O.
    Soc.Mining Geol. Japan Spec.Issue, 2, 58-63.
HOWER, J. & MOWATT, T.C. 1966. The mineralogy of illites and
    mixed-layer illite/montmorillonites. Am.Miner., 51, 825-854.
HOWER, J., ESLINGER, E.V., HOWER, M.E. & PERRY, E.A. 1976.
    Mechanism of burial metamorphism of argillaceous sediment:
    1 - Mineralogical and chemical evidence. Bull.geol.Soc.Am.,
    87, 725-737.
IIJIMA, A. 1970. Present day zeolitic diagenesis of the Neogene
    geosynclinal deposits in the Niigata Oil field, Japan. Am.
    Chem.Soc., 2nd Int. Zeolite Conf.
KUBLER, B. 1968. Evaluation quantitative du metamorphisme par
    la cristallinite de l'illite. Bull.Cent.Rech.Pau. 1, 259-278.
KUJAWA, F.B. & EUGSTER, H.P. 1966. Stability sequences and
    stability levels in unary systems. Am.J.Sci., 264, 620-642.
LEONE, M., ALAINO, R. & CALDERONE, S. 1975. Genesis of chlorite
    pellets from Mesozoic bedded cherts of Sicily. J.sedim.Petrol.,
    45, 618-628.
LEIKINE, M. 1981. Influence des faibles metamorphismes sur les
    variations du chimisme global des sediments et sur les
    parametres d'interstratification des illites-smectites. Bull.
    Miner., 104, 47-55.
MILLOT,G. 1964. Geologie des Argiles. Masson et Cie, Paris.
MITSUI, K. 1975. Diagnetic alteration of some minerals in
    argillaceous sediments in western Hokkaido, Japan. Sci.Rep.
    Tohoku Univ., 3rd ser., 13, 13-65.
MIZUTANI, S.1970. Silica minerals in the early stages of
    diagenesis. Sedimentology, 15, 419-436.

VAN MOORT, J.C. 1971. A comparative study of the diagenetic
    alteration of clay minerals in Mesozoic shales from Papua,
    New Guinea, and in Tertiary shales from Louisiana, U.S.A.
    Clays Clay Miner., 19, 1-20.
MUFFLER, L.J.P. & WHITE, D.E. 1969. Active metamorphism of
    Upper Cenezoic sediments in the Salton Sea Geothermal field
    and the Salton Trough, Southeastern California. Bull.geol.Soc.
    Am., 80, 157-182.
NICOT, E. 1981. Les phyllosilicates des terrains pre-cambriens
    du Nord-Ouest du Montana, U.S.A. Bull.Miner., (in press).
PERRY, E.A. & HOWER, J. 1970. Burial diagenesis in Gulf Coast
    pelitic sediments. Clays Clay Miner., 18, 165-178.
PERRY, E.A. & HOWER, J. 1972. Late stage dehydratation in deeply
    buried pelitic sediments. Bull.Am.Ass.Petrol.Geol., 56, 2013-2021.
REYNOLDS, R.C. & HOWER, J. 1970. The nature of interlayering in
    mixed-layer illite-montmorillonites. Clays Clay Miner., 18,
    25-36.
ROSS, C.S. & HENDRICKS, S.B. 1945. Minerals of the montmorillonite
    group. Prof.Pap. U.S.geol.Surv., 205,
SCHELLMANN, W. 1969. Die Bildungsbedingungen sedimentarer Chamosit
    und Hamatit-Eisenerze am Beispiel der Lagerstatte Echte.
    Neues Jb.Miner.Abh., 111, 1-31.
SCHOEN, R. 1964. Clay Minerals of the Silurian Clinton ironstones,
    New York State. J.sedim.Petrol., 34, 855-863.
SCHOEN, R. & WHITE, D.E. 1965. Hydrothermal alteration in GS-3
    and GS-4 drill holes, main Terrace, Steamboat Springs, Nevada.
    Econ.Geol., 60, 1411-1421.
SRODON, J. 1980. Precise identification of illite/smectite
    interstratification by X-ray powder diffraction. Clays Clay
    Miner., 28, 401-411.
VELDE, B. 1965. Experimental determination of muscovite polymorph
    stabilities. Am. Miner., 50, 436-449.
VELDE, B. 1968. The effect of chemical reduction on the stability
    of pyrophyllite and kaolinite in pelitic rocks. J.sedim.Petrol.,
    38, 13-16.
VELDE, B. 1969. The compositional join muscovite-pyrophyllite at
    moderate temperatures and pressures. Bull.Soc.fr.Miner.
    Cristallogr., 92, 360-368.
VELDE, B. 1971. A discussion of the stability and natural
    occurrence of margarite. Mineralogy Mag., 295, 317-323.
VELDE, B. 1973. Phase equilibria studies in the system
    MgO-Al O -SiO -H O: Chlorites and associated minerals.
    Mineralogy Mag., 39, 297-312.
VELDE, B. 1976. The chemical evolution of glauconite pellets as
    seen by microprobe determinations. Mineralogy Mag., 40, 733-
    760.
VELDE, B. 1977a. Clays and Clay Minerals in Natural and Synthetic
    Systems. Elsevier, Amsterdam.
VELDE, B. 1977b. A proposed phase diagram for illite, expanding
    chlorite, corrensite and illite-montmorillonite mixed-layered
    minerals. Clays Clay Miner., 25.

VELDE, B. & BYSTROM-BRUZEWITZ, A.M. 1972. Transformation of
    natural clay minerals at elevated temperatures and pressures.
    Geol. For Stockh.Forh., 94, 450-458.
VELDE, B. & ODIN, G.S. 1975. Further information related to the
    origin of glauconite. Clays Clay Miner., 23, 276-381.
VELDE, B. RAOULT, J.F. & LEIKINE, M. 1974. Metamorphosed
    berthierine pellets in Mid-Cretaceous rocks from North-Eastern
    Algeria. J. sedim. Petrol., 44, 1275-1280.
VELDE, B. & RUMBLE, D. 1977. Alumina content of chlorite in
    muscovite-bearing assemblages. Carnegie Inst.Washington
    Yearbook, 76, 421-423.
WEAVER, C.E. 1959. The clay petrology of sediments. Clays Clay
    Miner., 6, 154-187.
WEAVER, C.E. & BECK, K.C. 1971. Clay water diagenesis during
    burial: how mud becomes gneiss. Spec.Pap.geol.Soc.Am., 134.
WINKLER, H.G.F. 1964. Das P-T feld der Daigenese und
    niedrigtemperierten Metamorphose auf Grund von Mineral-
    reaktionen. Beitr. Miner. Petrogr., 10, 70-93.

FACIES CONTROLS ON SANDSTONE DIAGENESIS

Hans Fuchtbauer
Geological Institute of Ruhr-University
Postfach 2148
D-463 BOCHUM
GERMANY

INTRODUCTION

Before discussing the possible influences of depositional
environment on diagenesis, one general and two more specific
problems have to be considered:

(a)   which reliable criteria exist for the time sequence
       of diagenetic events?
(b)   how is the diagenesis influenced by the original
       grain size?
(c)   which relationships exist between diagenesis and
       original mineralogical composition of the sand-
       stones?

Only if all of these points are taken into account can the
early diagenetic cements and alterations be selected in order
to investigate the influence of the depostional environment.
To my knowledge, only the following criteria relate to depositional
controls:

(a)   haematite seams point to an arid continental
       environment.
(b)   early anhydrite (gypsum) and calcite cements may
       indicate dune and wadi deposits, respectively.
       Even early halite cement occurs in dune or sabkha
       sandstones.
(c)   siderite is, in general, non-marine and is frequently
       associated with carbonaceous sequences.
(d)   zeolites (e.g. analcime), if not associated with

269

A. Parker and B. W. Sellwood (eds.), Sediment Diagenesis, 269–288.
© *1983 by D. Reidel Publishing Company.*

        volcaniclastics, and authigenic tourmaline point
        to slightly evaporitic environments (e.g. alkaline
        lakes).
(e)   as to authigenic feldspars, albite can (but need
        not necessarily) indicate a marine to slightly
        evaporitic environment, and potassium feldspar
        indicates a continental environment, if both are
        present in the same basin.
(f)   early chlorite cement and chamosite ooids may
        suggest marine sandstones, and sudoite non-marine
        sandstones.
(g)   glauconite cement (and grains), frequently
        associated with phosphorite, are indications of
        marine deposition (or lakes with increased salinity).

These criteria will be demonstrated and discussed in natural
occurrences.

Diagenesis depends on the interaction between pore fluids
and rock framework.   Only if the pore fluids are very similar
to the fluids in which the sediments were deposited can relationships between environment and diagenesis be expected.   This
means that only early diagenetic processes may reflect the
depositional environment.   If this is so, it is important to
define reliable criteria for early diagenetic processes and of
time sequences of cementation in general.

CRITERIA FOR RECOGNISING TIME SEQUENCES IN CEMENTATION

In Fig. 1a, the plagioclase overgrowth clearly predates the
quartz overgrowth.   Most people will also agree that cement
near the pore walls or in pore throats formed earlier than in
the pore centre.   This is certainly true for systems with only
one cement (e.g. calcite filling the pore from the walls to the
centre).   It may also be true for a pore lined with quartz
overgrowths and sealed with carbonate or anhydrite (Fig. 1b),
especially if the latter is "aggressive" towards the quartz.
(Fig. 1c).   But is it also correct for a pore with quartz overgrowths and kaolinite booklets in the centre?   The latter did
not fill the whole pore space so that there was place for solutions to precipitate quartz which pushed the kaolinite crystals
aside.   Only if we find kaolinite inclusions in the quartz
overgrowth are we sure that kaolinite grew earlier than quartz
in this particular pore.   But this does not preclude the
possibility that both minerals precipitated in the same period
of time, or that quartz growth began even earlier than kaolinite
formation but continued after the end of the latter.

These are the many different interpretations which all apply

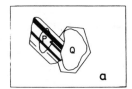 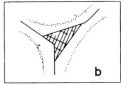 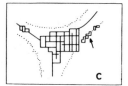

Fig. 1. Fabric relations in cements. Examples from
the Buntsandstein (Fuchtbauer, 1967b, 1974a).
(a) Overgrowth of plagioclase (P) is older
than the one of quartz (Q). (b) Three
quartz grains with overgrowths. Remaining
pore space filled with anhydrite. (c) Like
(b), but anhydrite replaced quartz marginally
as well as between detrital grain and overgrowth.

to the lucky case that more than one cement is present in one
pore. This leaves us uncomfortable with such fabric interpretations, and we feel we should look for more reliable indications
of cement successions. Six possible criteria, which also have
their limitations, are discussed below.

(1) Quartz Overgrowth, where occurring over a long
period of time, serves as a time scale for accompanying cement minerals. If, for instance, in rocks
with continuous silicification, calcite cement was
precipitated earlier than anhydrite cement, the
quartz overgrowths will be smaller in pores sealed
with calcite than in those sealed with anhydrite.

(2) The Minus-Cement Porosity concept (Fig. 2a, b) of
Rosenfeld (1949). Minus-cement porosity (i.e. the
volume percentage of porosity plus cement or the
porosity after removing the cement) portrays the
fabric at the beginning of cementation. The concept is that before cementation occurs, porosity
decreases continuously by mechanical rearrangement
of the grains supported by surficial solution at
grain contacts. For these reasons, the cement in
Fig. 2a was filled-in earlier than in Fig. 2b. One
can apply this concept only to the first cement
stabilizing the framework, and where at least two
minerals occur as a first cement in different pores
of the same sandstone. As an example consider the
Dogger beta sandstones in northwestern Germany.
According to Drong (1965), calcitic cementation

fixed the original porosity of 40% (= minus-cement porosity) in places, whereas in other samples of the same occurrence the minus-cement-porosity was only 28%. It was assumed that the latter sandstones were cemented later. This was confirmed by unstable heavy minerals (e.g. kyanite, staurolite), which were preserved only in the sample with 40% minus-cement-porosity, but were removed by intrastratal solution in the other samples.

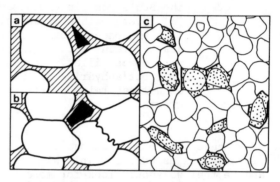

Fig. 2.   Minus-cement-porosity (a, b) and contact strength (c) (Fuchtbauer, 1974a, b). The minus-cement-porosity including cement (hatched) and porosity (dark) is 33 Vol-% in (a) and 16 Vol-% in (b), indicating an earlier cementation of (a) compared with (b). The contact strength (see text) is 1.2 for grain contacts sealed by albite cement (stippled) and 1.6 for contacts sealed by quartz cement (white). The conclusion is that albite cementation was prior to quartz cementation.

(3)   <u>The Contact Strength</u> concept (Fig. 2c; (Fuchtbauer, 1967b). It is also based on the assumption that the porosity decreases with increasing time and depth of subsidence. At the same time, the grain contacts become closer (Taylor, 1950). Again this is a measure to compare different cementing minerals in the same rock, provided each mineral occurs in places as the first cement. The contact strength can be quantified by the formula $(a + 2b)/(a + b)$, in which a is the number of point contacts and b the number of long contacts including straight, concave-convex, and sutured grain contacts. Originally, the latter contacts have been counted separately, but the differences are in many cases not

significant and are only similated by overgrowth
fabric, as shown by cathodoluminescence.   Fig. 2c
demonstrates an example from the Bunter sandstone
(Lower Triassic).   The contact strength of grains
separated by albite cement (dotted) is 1.2 (10 point
contacts, 3 long contacts, i.e., (10 + 3.2)/(10 + 3)
= 1.2), whereas the contact strength of grains
separated by quartz cement (white) is 1.6 (Fuchtbauer,
1967a, b, and 1974a, p. 135).   The conclusion is that
albite cementation occurred prior to quartz cementation.   This method should only be used for well-
sorted, cemented sandstones.   It is not applicable
to investigations of compaction in sandstones with
clay matrix (Nickel, in Zimmerle, 1976, 233).

(4)   Oxygen isotope analyses can be compared for different
authigenic silicates with the enclosing pore fluids
to show equilibrium at present or for a presumed
earlier period of diagenesis.   This may help to
unravel diagenesis (Land and Dutton, 1978).

(5)   A few authigenic minerals indicate merely by their
shape an early diagenetic origin.   This applies to
chamosite ooids and chamosite cement (Horn, 1965),
glauconite, and analcime flasers in siltstones which
point to an early formation, at least for the precursor mineral.

(6)   Last but not least, the depth (and temperature) at
which diagenetic features first appear in drilled
sections, correlates with the time sequence of
these features, provided the sandstones in the
sequence are comparable to each other in their
primary composition, and the area of the boring did
only subside and was never uplifted (which might
cause reversals in diagenesis) (e.g. Scherp, 1963;
Boles and Franks, 1979).   This is because temperature and age, the main factors in diagenesis, increase
downhole.   Replacements which also increase with
depth may obliterate this picture.

GRAIN SIZE INFLUENCES

As grain size is the most important facies attribute and
therefore also a depositional criterion, we approach the main
theme of this paper when discussing the influence of sand grain
size on diagenesis.   According to Mellon's important paper of
1964, which seems to be forgotten by many sandstone petrographers,
quartz cementation occurred preferentially in fine-grained layers.

Subsequent calcite cement filled the coarse sand layers, in a
Cretaceous non-marine sandstone.   This seems to be an important
model (e.g. Hawkins, 1978) and may also explain uncemented coarse
layers in quartz-cemented fine sandstones reported by Colter and
Ebbern (1978).

Tillman and Almon (1979) found quartz cement prevalent in
non-reworked sandstones, whereas calcite cement was more important
in reworked sandstones, which seem to be coarser than the non-
reworked ones.   Quartz and albite cement in the fine-grained
sandstones and anhydrite cement in the coarse-grained sandstones
are reported by Fuchtbauer (1976b);   but if enough silica was
supplied, even coarse sandstones became cemented by quartz.

The following combinations are frequent: (a) clay minerals,
quartz, and feldspar in fine-grained sandstone, (b) calcite,
ankerite, and anhydrite in coarse-grained sandstones.

A possible explanation is based on four lines of evidence:

(1) If the time sequence of these cement minerals could
     be determined, (a) was precipitated earlier than
     (b), in general.

(2) The group (a)-minerals are pertinent to the sand-
     stones and therefore can form overgrowths on detrital
     grains, whereas group (b)-minerals are generally not
     detrital components of these sandstones;   nucleation
     therefore cannot occur.

(3) Since the specific surface (cm^2/cm^3 solid material)
     is much higher in fine- than in coarse-grained sand-
     stones, overgrowths of a given thickness will reduce
     the porosity and especially the permeability much
     more in fine- than in coarse-grained sandstones.

(4) Diffusion rather than advection driven by compaction
     is responsible for the symmetrical decrease in
     porosity as shown in Fig. 3 (left) and increase in
     quartz overgrowth (Fig. 3, right).   In both cases,
     silica is the prevailing cement.   The source was
     evidently outside the sandstones, i.e. in the adja-
     cent shales and siltstones.

Based on these facts and findings, it is suggested that
silica migrated by diffusion into the sandstones and increased
the differences in permeability between fine- and coarse-grained
sandstones.   Because the silica was dissolved in the adjacent
shales and siltstones (see below), the travel-distances were
small compared with such ions which were not pertinent to the

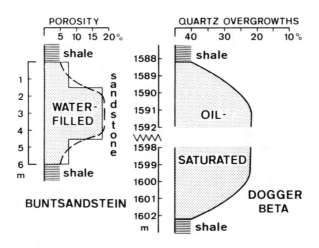

Fig. 3.   Sandstones cemented mainly by supplies from
          adjacent shales (Fuchtbauer, 1974c). Left:
          Porosity decrease towards over- and underlying
          shales (72 samples from different borings in
          N Germany.   Maximum depth of burial approx.
          2500m).   Right: In spite of oil saturation,
          silica cementation was able to go on, to a
          certain degree, by diffusion through the
          connate water film.   In water-filled sand-
          stones of the same area, 70% quartz overgrowths
          are common.   Since the porosity (25-30%) was
          not much affected by the differences in dia-
          genesis of water- and oil-saturated sandstones,
          the more sensitive number of quartz overgrowths
          was used.   On dry mounts the following grain
          numbers were counted: a = grains without over-
          growths, b = with overgrowth covering less than
          half, and c = covering more than half of the
          grain surface.   The abscissa is $(0.5\ b + c)$
          · $100/(a + b + c)$.

sequence (e.g. sulphates and carbonates).   When the latter sub-
stances arrived, only the coarser sands were permeable enough to
let them pass and/or precipitate.

Irrespective of whether this model is correct or not, the
grain size should always be considered in diagenesis studies.   For
instance, out of 15 papers on diagenesis published in vol. 135
(1978) of the Jl. geol. Soc. Lond., only five provide information
on grain size.

INFLUENCES OF THE ORIGINAL COMPOSITION

Clearly the composition of the detrital minerals of a sandstone must have a bearing upon diagenesis.   This will be discussed below for different groups of unstable minerals and grains.

(1) Hornblende, augite, olivine, biotite
These minerals are disintegrated early in diagenesis
and during weathering.   Under humid conditions they
are mainly dissolved, but under arid conditions
haematite and a mixed layer mineral illite-smectite
with ferric iron are developed (Walker, 1976, and
unpublished work by the author), which form coatings
around the sand grains.   The iron is originally precipitated as amorphous $Fe_5 HO_8$. $4H_2O$ which
alters by ageing under weakly acid conditions and
relatively high temperature to haematite (Schwertmann
and Fischer, 1974).   Part of the haematite may be
formed during diagenetic replacement of the above
mixed-layer mineral by illite.   Biotite resists
diagenesis but is unstable during weathering.

(2) Argillaceous sandstones
Clay minerals in fine-grained sandstones are
generally a primary matrix.   In coarse sandstones,
they can also form (a) a secondary matrix infiltrated
from the surface and forming thin pellicles parallel
to grain surfaces (Walker, 1976; Galloway, 1979,
233) or (b) authigenic pore fillings which are
whisker-like (illite), irregular (smectite), booklets floating in the pores (kaolinite), or platelets
oriented perpendicular to grain surfaces (chlorite,
chamosite, illite).   The tangential illite skins
(a) are thought to be diagenetic precipitates, by
Colter and Ebbern (1978), but I would prefer either
Walker's interpretation or a mild recrystallization
of infiltrated or of primary matrix during late
diagenesis.

The significance of complete clay mineral coatings
is that they prevent the quartz grains from being
enlarged by silica cement (Horn, 1965; Tillman and
Almon, 1979).   Such sandstones retain porosity
and may contain oil.

(3) Feldspathic sandstones
Feldspar grains in sandstones can be replaced by
clay minerals during weathering or diagenesis under
neutral to acid conditions.   Hawkins (1978) described
feldspars replaced by kaolinite in coarse-grained

and by illite in fine-grained late Carboniferous
sandstones. He suggested that in the latter "this
leaching process never extended much beyond the
illitization stage because porosity reduction restricted fluid movement, but in coarse-grained sandstone, where porosity retention was maintained,
leaching continued, resulting in transformation of
illite to well-crystallized kaolinite 'booklets' and
'rouleau'". The fine-grained sandstones were a
rather closed system compared with the coarse-grained
ones. The illitic replacements could also be
alteration products of smectites or mixed layer
minerals. These minerals are still preserved in
the Tertiary Molasse and are interpreted as alteration products of feldspars (Fuchtbauer, 1964, p.268).
In general, illitization of feldspars and fibrous
or palisade-like cement rims of illite occur only
at considerable depth of burial (Nagtegaal, 1978;
Sommer, 1978, using K/Ar dating). Feldspars are
stable in basic solutions of high alkali and silica
concentration, e.g. in evaporitic basins (Nagtegaal,
1978). In such rocks authigenic feldspars can be
found. In sandstones they occur only as overgrowths of detrital feldspars and are therefore
missing in quartzose sandstones.

(4) Lithic sandstones
These sandstones include varieties (a) with
relatively stable quartzitic rock fragments, (b)
with less stable carbonate rock fragments, (c) with
mechanically and chemically unstable volcanic rock
fragments, and (d) with mechanically unstable phyllitic rock fragments.

(a) Sandstones with quartzitic rock fragments behave
nearly like quartzose sandstones during diagenesis.
(b) Sandstones with carbonate rock fragments or
skeletons are generally characterized by an early
carbonate cement (Fuchtbauer, 1964).
(c) Sandstones with volcanic rock fragments which
generally are also rich in plagioclase show a
peculiar variety of diagenetic alterations. Most
characteristically the porosity of those sandstones
is reduced by a combination of squeezing and alteration of volcanic rock fragments (Hawkins, 1978).
This leads to an increase of pseudomatrix which
drastically reduces the porosity ("greywackisation",
e.g. Marfil and de la Pena, 1980). Several clay
minerals, zeolites, and microquartz are typical
authigenic minerals in such sandstone which frequently

occur near converging plate margins (Barrows, 1980;
Coombs, 1954; Davies et al., 1979 with discussion of
phase relations; Galloway, 1979; Hayes, 1978;
Obradovic, 1980; Stanley and Benson, 1979; Surdam
and Boles, 1979, for chemical reactions).   However,
they occur also in intraplate areas, e.g. in the
Rotliegendes of N Germany and the North Sea, in
which volcanic detritals are presumably the source
material of illite cement (Hancock, 1978) and microquartz.
(d) Sandstones with phyllitic rock fragments are
similar to those discussed above because such
fragments are also squeezed into the pores during
subsidence.   However, they are chemically stable
(Nagtegaal, 1978, p.104).   The Upper Carboniferous
sandstones of northwestern Germany belong occasionally to this group.

DIAGENETIC CRITERIA FOR THE DEPOSITIONAL ENVIRONMENT

Influences of the depositional environment will concentrate
on the early diagenetic processes.   Criteria to identify an
early diagenetic origin have been discussed already, have
fluences of special rock constituents.   In sandstones composed
of "normal" constituents (i.e. quartz with small percentages of
rock fragments, feldspars, mica, and matrix) the following
succession of cement minerals is common (Fig. 4, and Fuchtbauer,
1979):

(1) aluminium silicates (clay minerals, feldspar)
(2) quartz
(3) carbonates and sulphates
(4) chlorides

This sequence corresponds more or less to the order of
increased solubility and therefore reflects the ionic strength
of the interstitial water increasing with time and depth of
burial.   It implies also a tendency of increasing pH with time.
Though there are certainly many exceptions to this sequence, it
may be worthwhile to record the exceptions and to investigate
the conditions which are responsible for them.   In the first
stage (1), special conditions occur: according to Curtis (1978)
only the most reactive phases will participate in diagenetic
processes.   "Probably most significant in this context are the
extremely fine-grained and amorphous constituents derived from
soils: intimate mixtures of hydrated aluminium and iron oxides
together with degraded organic matter".   The aluminium and iron
compounds are able to precipitate silica even out of undersaturated solutions to form aluminium silicates (Harder, 1974).

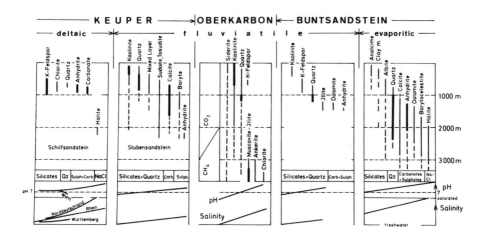

Fig. 4.   Sequences of cementation in sandstones of
          different depositional environment and age
          (Upper Triassic Keuper, Heling 1963 and 1963;
          Upper Carboniferous, Scherp 1963; Lower
          Triassic, Buntsandstein, Fuchtbauer 1967b,
          1974a, b). Thick lines = strong cementation;
          interrupted lines = uncertain; Ph and salinity
          schematically.

These compounds as well as other early-diagenetic minerals may
reflect the depositional environment.

    Those environments which eventually can be distinguished are
listed below.

        (a)   Continental environments - arid
              Red colour is a very sensitive criterion of haematite
              (Schwertmann, pers.comm.).   Sandstones stained by
              haematite indicate a more or less arid continental
              environment (Folk, 1976).   According to Walker
              (1976), fluvial or aeolian sands obtain coatings of
              clay and haematite by mechanical infiltration, as
              shown by geopetal fabric (Kessler, 1978).

    In the Rotliegendes, wadi and dune sandstones are character-
ized by calcite and (locally) by gypsum/anhydrite cement,
respectively (Kessler, 1978; Glennie et al., 1978).   Tourmaline
overgrowths are reported from the German Rotliegendes by Hancock
(1978).   Authigenic tourmaline is frequent in the Zechstein
anhydrites (and dolomites) but occurs also in normal marine
limestones.

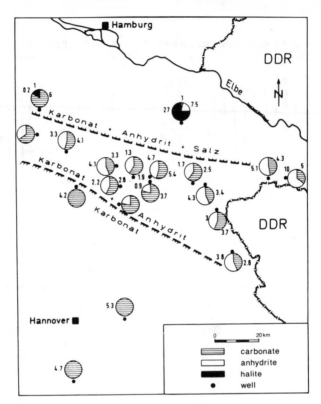

Fig. 5.   Cement minerals in Rotliegendes sandstones,
          N Germany (Drong, 1979). Sectors indicate
          relative amounts of cement minerals. Figures
          = cement percentages related to the volume of
          solid rock. Karbonat = calcite. Bohrung
          = boring. The cement distribution indicates
          an influence of the depositional environment.

        The regional distribution of cement minerals can also
indicate an influence of the depositional environment (Figs. 5
and 6b).   Fig. 5, from the Rotliegendes (lower Permian) shows
three diagenetic facies belts, from south to north, i.e. towards
the interior of the basin, (1) calcite cement, (2) calcite and
anhydrite cement, and (3) halite and anhydrite cement.   (1) and
(2) are presumably wadi and dune sandstones, (3) is possibly a
sabkha environment.   Whereas the volume of cement (1) and (2) is
only 4-15% of the rock, it is 35.5% for (3) indicating an early
cementation (minus-cement porosity concept).

        Fig. 6B from the Buntsandstein ("Bunter"; Lower Triassic)

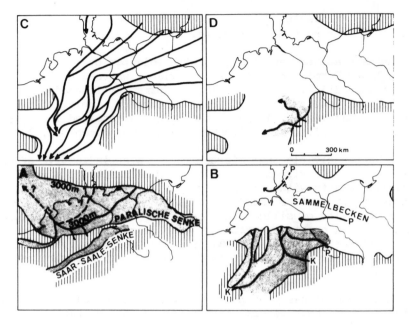

Fig. 6. Palaeogeographic sketches of (A) Upper
         Carboniferous, (B) Buntsandstein, (C) Schilf-
         sandstein, (D) Stubensandstein (s. Fig. 4;
         from Fuchtbauer, 1974b). Hatched = no sedi-
         mentation; shaded = fluvial sedimentation;
         white = deltaic (C) and brackish to evaporitic
         potassium feldspar; P = rich in plagioclase
         and feldspar.

shows a river system (shaded) oriented to the North, into a
large brackish to marine alkaline lagoon, the "Sammelbecken".
The detrital minerals are similar in both areas though the river
sands, especially in the southernmost area, contain only potassium
feldspar (K in Fig. 6B), whereas an admixture of plagioclase
appears in the north. The diagenesis, however, is different as
shown in Fig. 4. Whereas analcime, albite, and locally chlorite
are the first cements in the lagoon, potassium feldspar appears
in the fluvial sandstones. Where albite is frequent, analcime
is missing, and vice versa. Analcime is one of the few zeolites
which does not necessarily derive from volcanic material. It
formed in the slightly evaporitic lagoon presumably from a
zeolitic precursor and is possibly replaced by albite. Authigenic feldspars occur only as overgrowths. On detrital plagioclase and potassium feldspar grains albite overgrowths occur
exclusively in the "lagoon" and potassium feldspar overgrowths in
the fluvial sandstones (Fuchtbauer, 1967a, b). According to

Waugh (1978), potassium feldspar overgrowths occur also in the
fluvial Bunter sandstone of England.

The same author reported potassium feldspar overgrowths from
aeolian Lower Permian sandstones in England, whereas Hancock
(1978) found albite overgrowths in the German Rotliegendes.   The
environmental interpretation is open.   Careful mapping of the
diagenetic features may in many cases provide detailed information on the depositional environment (Jansa and Fischbuch, 1974).

It is conceivable that microquartz is also indicative of an
arid-continental environment.   Microquartz is fibrous microcrystalline quartz with strong undulosity.   It is interpreted
as an ageing product of amorphous silica and therefore requires
a supersaturation of silica about ten times that for crystallization of ordered quartz overgrowths.   As discussed, however, under
4(c), microquartz is also typical of sandstones with volcanic
rock fragments or with opaline tests even of marine environment
(Sears, 1980).

Silica cement has many possible sources which can be listed
in the order of their general importance as follows (Fuchtbauer,
1979, p.1134):

(1)   dissolution of quartz grains in associated siltstones.
      This has been proven microscopically by comparing
      the shape of such grains in siltstones with grains
      of the same size in sandstones.   As shown in Fig.
      7, the quartz grains in siltstones are partially
      dissolved on their upper and lower surfaces.   A
      calculation indicates that about 10% of the silt-
      stone can be dissolved and can contribute to the
      cementation of sandstone.

(2)   replacement of feldspar and quartz by clay minerals,
      carbonate or anhydrite.

(3)   "pressure" solution and stylolites in sandstones.

(4)   The alteration of smectite to illite in shales
      delivers silica which can amount to about 1% of the
      shale.

(5)   dissolution of volcanic glass, volcanic rock frag-
      ments, and opaline tests.

(6)   it is often considered that huge volumes of compact-
      ion water are necessary and able to provide the
      silica required for the cementation of sandstones
      (Land and Dutton, 1978).   In the present writer's

opinion, diffusion (see Fig. 3) is also an important
vehicle of silica transfer from a microenvironment
of higher concentration to a microenvironment of
lower concentration.

Only the mechanism (5) is relevant to early diagenesis.

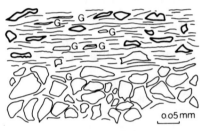

Fig. 7.   Top: Fine-grained siltstone with quartz grains
          (marked by thick lines) which are dissolved on
          their upper and lower surfaces.   Bottom:
          Coarse-grained siltstone with no visible
          quartz dissolution.   Lines = clay, G = mica.
          Upper Triassic, N Germany, 2600 m depth
          (Fuchtbauer, 1979).

(b)   Continental environments - humid in part
      As river water is a solution with very low ion
      concentration, we cannot expect typical minerals to
      form in fluvial sandstones, during early diagenesis.
      However, if early chlorites were formed, which are
      characterized by merely pore-filling but not replace-
      ment structures, we would expect Mg-bearing
      chlorites only in marine environments.   This is
      shown in Fig. 4, Keuper, and Fig. 6C and 6D.
      Whereas "normal" chlorite occurs in the deltaic
      Schilfsandstein, only sudoite (Al-chlorite) and
      tosudite (regular interlayering of sudoite and
      montmorillonite) are formed in the fluvial
      Stubensandstein.   The non-marine origin of the
      sudoite-type chlorite is particularly striking in
      the aeolian Permian sandstone of Cornberg/Hessen,
      where the variety "cookeite" occurs which contains
      about 1 percent lithium (Flehmig and Menschel,
      1971).   Such a peculiar enrichment is only con-
      ceivable in a non-marine environment which in this
      case was arid.   Humid environments favour the
      formation of kaolinite, which is common in lateritic
      soils.   The cations of the source minerals (e.g.
      potassium feldspar) are removed by an effective
      drainage system, whereas in swamps the removal of
      cations is due to the low pH.   Patches of dia-

genetic kaolinite occur adjacent to plant remains
also in marine sandstones.

Tuff layers can be good environmental indicators.
Burger and Stadler (1971) were able to trace volcanic tuff layers ("tonsteins") of the Upper
Carboniferous from the coal seams, where they are
kaolinitic, into the adjacent shales, where they
are mixed layers.   The interpretation as volcanic
tuff layers is based on zircon and quartz varieties
(Fuchtbauer and Riedel, 1979).

In general, however, it is not advisable to use
even high kaolinite contents in sandstones as a
criterion of non-marine deposition, inasmuch as
more than one generation of kaolinite precipitation
can occur (Zimmerle, 1963).

Very characteristic of non-marine sandstones and
siltstones, especially in carbonaceous sequences,
however, is fine-grained, early-diagenetic siderite.
Only occasionally it forms in marine sediments,
even early in diagenesis, provided that sulphate
ions in the pore fluid were sufficiently reduced
by desulphurizing bacteria to avoid precipitation
of ferroan sulphides.

(c) Marine environments
Glauconite consists of sand-sized green particles
but it can also occur as cement (Scholle, 1979).
It forms early in diagenesis as a smectitic clay
mineral which later changes into an illite-like
mineral.   It is generally considered indicative
of a marine environment and is frequently associated with phosphorite.   According to experiments
conducted by Harder (1974) a K-rich solution
similar to seawater is needed.   He obtained glauconite in the transition zone between oxidizing
and reducing conditions.   In nature, glauconite
grains are formed in oxidizing environment but
probably with reducing conditions in their interior.
Richter (pers. comm.) found glauconite grains in
lacustrine Keuper sandstones.   Small numbers of
glauconite grains can also occur in fluvial sandstones due to reworking of marine sediments.

7 A chamosite ooids and berthierine pellets (Velde,
1977) which are Fe-rich and Fe-poor, respectively
(Troger, 1967) occur in marine sandstones as well.
Rohrlich et al. (1969) found chamositic faecal

pellets with 38.1% Fe O   and a 14 $\overset{o}{A}$ line which
faded upon heating in Recent sediments of an inlet
(Loch Etive, Scotland).   Von Engelhardt et al.
(1955, p.575) described Mg-chamosite ooids
(n_Z 1.60,   = 0.003-0.007, misprinted in the
publication) disseminated in Jurassic and Cretaceous
sandstones of W Germany.   Chamosite at present is
found especially off river deltas (Giresse and Odin,
1973).   The conditions under which they form are
more reducing than for glauconite (Velde, 1977).
The Fe-content of the chamosite ooids is presumably
correlated with their abundance.   If sparsely
disseminated, they are Mg-chamosites, while they
are concentrated in chamositic iron ores.   The
Mg-chamosites are certainly more typical for
normal-marine environments.

REFERENCES

BARROWS, K.J. 1980. Zeolitization of Miocene volcaniclastic rocks,
    southern Desatoya Mountains, Nevada. Bull.geol.Soc.Am., 91,
    199-210.
BOLES, J.R.G., and FRANKS, S.G. 1979. Clay diagenesis in Wilcox
    sandstones of southwest Texas: Implications of smectite dia-
    genesis on sandstone cementation. J.sedim.Petrol., 49, 55-70.
BURGER, K. and STADLER, G., 1971. Petrographische Ausbildungs-
    form, Fazies, Chemismus und Genese des Kaolin-Kohlentonsteins
    Zollverein 8 in Ruhrkarbon. Forschungsber. Land NRW, Nr.
    2126, 266p.
COLTER, V.S. and EBBERN, J., 1978. The petrography and reser-
    voir properties of some Triassic sandstones of the Northern
    Irish Sea Basin. J.geol.Soc.Lond., 135, 57-62.
COOMBS, D.S. 1954. The nature and alteration of some Triassic
    sediments from Southland, New Zealand. Trans. Soc.N.Z., 82,
    65-109.
CURTIS, C.D. 1978. Possible links between sandstone diagenesis
    and depth-related geochemical reactions occurring in enclosing
    mudstones. J.geol.Soc.Lond., 135, 107-117.
DAVIS, D.K., ALMON, W.R. BONIS, S.B. and HUNTER, B.E. 1979. De-
    position and diagenesis of Tertiary-Holocene volcaniclastics,
    Guatemala. In: Scholle, P.A. and Schluger, P.R. (Eds.):
    Aspects of Diagenesis. Spec. Publ. Soc.Econ.Paleont.Miner.,
    26, 281-306.
DRONG, H.J. 1965. Die Schwerminerale des Dogger beta und ihre
    diagenetischen Veranderungen. Vortr., Dt.Miner.Ges., Hannover.
DRONG, H.L. 1979. Diagenetische Veranderungen in den Rotliegend
    Sandsteinen im NW-Deutschen Becken. Geol.Rdsch., 68,
    1172-1183.

FLEHMIG, W., and MENSCHEL, G. 1971. Cookeit als Lithiumtrager
in Sandsteinen des unteren Zechsteins von Nordhessen.
Fortschr. Mineral., 49, Beiheft 1, 96-97.

FOLK, R.L. 1976. Reddening of desert sandstones: Simpson Desert,
N.T., Australia. J.sedim.Petrol., 46, 604-615.

FUCHTBAUER, H. 1964. Sedimentpetrographische Untersuchungen in
der alteren Molasse nordlich der Alpen. Eclog.geol.Helv., 57,
157-298.

FUCHTBAUER, H. 1967a. Influence of different types of diagenesis
on sandstone porosity. Proc. Seventh World Petrol. Congr.,
Mexico, 2, 353-369.

FUCHTBAUER, H. 1967b. Der Einflu des Ablagerungsmilieus auf
die Sandsteindiagenese im mittleren Buntsandstein. Sediment.
Geol., 1, 159-179.

FUCHTBAUER, H. 1974a. Sediments and Sedimentary Rocks 1. E.
Schweizerbart, Stuttgart and J. Wiley & Sons, New York.
Halsted Press Div., 464pp.

FUCHTBAUER, H. 1974b. Zur Diagenese fluviatiler Sandsteine.
Geol. Rdsch., 63, 904-925.

FUCHTBAUER, H. 1974c. Some problems of diagenesis in sandstones.
Bull.Cent.Rech. Pau, SNPA. 8, 391-403.

FUCHTBAUER, H. 1979. Die Sandsteindiagenese im Spiegel der
neueren Literatur. Geol.Rdsch., 68, 1125-1151.

FUCHTBAUER, H., and RIEDEL, D. 1979. Zirkon-und Quarzvarietaten
in Kaolin-Kohlentonsteinen der Bochumer Schichten (Westfal A)
des Ruhrkarbons. Gluckauf-Forschungshefte, 40, 130-132, Essen.

GALLOWAY, W.E. 1979. Diagenetic control of reservoir quality in
arc-derived sandstones: Implications for petroleum exploration.
In: Scholle, P.A. & Schluger, P.R. (Eds.): Aspects of Diagenesis.
Spec. Publ. Soc.Econ.Paleont. Miner., 26, 251-262.

GIRESSE, P., and ODIN, G.S. 1973. Nature mineralogique et origine
des glauconies du plateau continental du Gabon et du Congo.
Sedimentology, 20, 457-488.

GLENNIE, K.W., MUDD, G.C. and NAGTEGAAL, P.J.C. 1978. Depositional environment and diagenesis of Permian Rotliegendes sandstones in Leman Bank and Sole Pit areas of the UK southern
North Sea. J.geol.Soc.Lond., 135, 25-34.

HANCOCK, N.J. 1978. Possible causes of Rotliegend sandstone diagenesis in northern West Germany. J.geol.Soc.Lond., 135, 35-40.

HARDER, H. 1974. Illite mineral synthesis at surface temperatures.
Chem.Geol., 14, 241-253.

HARDER, H. 1978. Synthese von Glaukoniten unter naturlichen
Bedingungen. Naturwisse., 65, 59.

HAWKINS, P.J. 1978. Relationship between diagenesis, porosity
reduction, and oil emplacement in late Carboniferous sandstone
reservoirs, Bothamsall Oilfield, E Midlands. J.geol.Soc.Lond.,
135, 7-24.

HAYES, J.B. 1978. Sandstone diagenesis - recent advances and un-
solved problems. In: AAPG: Clastic diagenesis - Its relation to
hydrocarbon reservoirs quality and hydrocarbon entrapment School,
Boulder, Colorado, June 5-9, 1978.

HELING, D. 1963. Zur Petrographie des Stubensandsteins. Disser-
tation Univ. Tubingen, 56pp.
HELING, D. 1965. Zur Petrographie des Schilfsandsteins. Beitr.
Mineral. Petrogr., 11, 272-296.
HORN, D. 1965. Diagenese und Porositat des Dogger-beta-Haupt-
sandsteins in den Olfeldern Plon-Ost und Preetz. Erdol &
Kohle. 18, 249-255.
JANSA, L.F., and FISCHBUCH, 1974. Evolution of a Middle and Upper
Devonian sequence from a clastic coastal plain-deltaic complex
into overlying carbonate reef complexes and banks. Sturgeion-
Mitsue Area, Alberta. Bull.geol.Surv.Can., 234, 105pp.
KESSLER, L.G. 1978. Diagenetic sequences in ancient sandstones
deposited under desert climatic conditions. J.geol.Soc.Lond.,
135, 41-49.
LAND, L.S., and DUTTON, S.P. 1978. Cementation of a Pennsylvanian
deltaic sandstone: Isotopic data. J.sedim.Petrol., 48, 1167-
1176; with discussions on "Cementation of sandstones" by K.
Bjørlykke (J.sedim.petrol., 49, 1358-1359) and Land & Dutton
(J.sedim.Petrol., 49, 1359-1361).
MARFIL, R., and de la Pena, J.A. 1980. Diagenetic aspects of
Permian sandstones in Central Spain: Implication for gray-
wacke genesis. Internat.Assoc.Sedimentologists 1st Europ. Mtg.,
Bochum, Abstr., 158-160.
MELLON, G.B. 1964. Discriminatory analysis of calcite - and
silicate - cemented phases of the Mountain Park sandstone.
J.Geol., 72, 786-809.
NAGTEGAAL, P.J.C. 1978. Sandstone-framework instability as a
function of burial diagenesis. J.geol.Soc.Lond., 135, 101-105.
OBRADOVIC, J. 1980. Diagenetic processes in some pyroclastic
rocks. Internat.Assoc. Sedimentologists 1st Europ.Mtg.,
Bochum, Abstr., 260-262.
ROHRLICH, V., PRICE, N.B. and CALVERT, S.E. 1969. Chamosite in
the Recent sediments of Loch Etive, Scotland. J.sedim.Petrol.,
39, 624-631.
ROSENFELD, MA. 1949. Some aspects of porosity and cementation.
Prod.month., 13, 39-42.
SCHERP, A. 1963. Die Petrographie der palaozoischen Sandsteine
in der Bohrung Munsterland 1 und ihre Diagenese in Abhangig-
keit von der Teufe. Fortschr. Geol. Rheinl. Westf., 11, 251-
282, Krefeld.
SCHOLLE, P.A. 1979. A color illustrated guide to constituents,
textures, cements and porosities of sandstones and associated
rocks. Mem. Am.Assoc.Petrol.Geol., 28, 201pp.
SCHWERTMANN, U., and FISCHER, W.R. 1974. Natural "amorphous"
ferric hydroxide. Geoderma, 10, 237-247.
SEARS, S.O. 1980. Porcellanous cement in Upper Miocene sandstone
reservoir rocks, offshore California: Origin and effect on
fluid flow properties. Geol. Soc.Amer.Abstr. & Progr., 12, 7,
519.

SOMMER, F. 1978. Diagenesis of Jurassic sandstones in the Viking
    Graben. J.geol.Soc.Lond., <u>135</u>, 63-67.
STANLEY, K.O. and BENSON, L.V. 1979. Early diagenesis of High
    Plains Tertiary vitric and arcosic sandstone, Wyoming and
    Nebraska. In: Scholle, P.A. & Schluger, P.R. (Eds.): Aspects
    of Diagenesis. Spec.Publ. Soc.Econ.Paleont. Mineral., <u>26</u>, 401-
    423.
SURDAM, R.C., and BOLES, J.R. 1979. Diagenesis of volcanic sand-
    stones. In: Scholle, P.A. & Schluger, P.R. (Eds.): Aspects of
    Diagenesis. Spec.Publ. Soc.Econ.Paleont.Mineral., <u>26</u>, 227-242.
TAYLOR, J.M. 1950. Pore space reduction in sandstones. Bull.Am.
    Assoc.Petrol.Geol., <u>34</u>, 701-716.
TILLMAN, R.W., and ALMON, W.R. 1979. Diagenesis of Frontier
    formation offshore bar sandstones, Spearhead Ranch field,
    Wyoming. In: Scholle, P.A. & Schluger, P.R. (Eds.): Aspects of
    Diagenesis. Spec.Publ.Soc.Econ.Paleont. Mineral., <u>26</u>, 337-378.
TROGER, W.E. 1967. Optische Bestimmung der gesteinsbildenden
    Minerale. Teil 2 Textband. - Schweizerbart, Stuttgart, 822pp.
VELDE, B. 1977. Clays and Clay Minerals in Natural and Synthetic
    Systems. Elsevier, Amsterdam, Oxford, New York. 218pp.
VON ENGELHARDT, W., FUCHTBAUER, H. and GOLDSCHMIDT, H. 1955.
    Einige Ergebnisse der quantitativen Rontgenanalyse feinkorniger
    Sedimente. Geol. Rdsch., <u>43</u>, 572-577.
WALKER, T.R. 1976. Diagenetic origin of continental red beds.
    In: Falke, H. (Ed.): The Continental Permian in Central, West
    and South Europe. NATO Adv. Study Inst. Ser.C. Math.Phys.
    Sci., Reidel, Dordrecht, Holland, 240-282.
WAUGH, B. 1978. Authigenic K-feldspar in British Permo-Triassic
    sandstones. J.geol.Soc.Lond., <u>135</u>, 51-56.
ZIMMERLE, W. 1963. Zur Petrographie und Diagenese des Dogger-
    beta-Hauptsandsteins im Erdolfeld Plon-Ost. Erdol & Kohle,
    16, 9-16.
ZIMMERLE, W. 1963. Petrographische Beschreibung und Deutung der
    erbohrten Schichten. In "Die Tiefbohrung Saar", Geol. Jahrb.,
    A, <u>27</u>, 91-305.

DEPOSITIONAL MODELS FOR CARBONATE ROCKS

Noel P. James
Department of Geology
Memorial University of Newfoundland
St. John's, Newfoundland
Canada A1B 3X5

INTRODUCTION

Intensive study of modern carbonate sediments and a substantial library of fossil examples have brought us to the point
where it is now possible to formulate actual facies models for
carbonate deposits which may serve as frameworks or guides for
future observations and, most important in terms of hydrocarbon
exploration, predictors in new geological situations.

This chapter is an outline of the components which make up
carbonate sedimentary rocks and the formulation of a series of
depositional models for several of the more common environments of
accumulation.   The main points presented in this article were
first published as a series of papers on facies models in
Geoscience Canada (James, 1977a, b, 1978; McIlreath and James,
1978) and the following is a synthesis and updating of these
contributions.

Carbonate versus Terrigenous Clastic Sediments

Terrigenous clastic sediments are made primarily by the
disintegration of parent rock, are transported to the environment
of deposition and once there the patterns of texture and fabric
are impressed upon them by the hydraulic regimen.   The signature
of terrigenous clastic facies is thus largely in sedimentary
structures and grain size variations.

Carbonate sediments are born, not made.   They are created in
or close to the environment of deposition.   Thus, in addition to
the purely physical sedimentary parameters used in the analysis of
289

A. Parker and B. W. Sellwood (eds.), Sediment Diagenesis, 289–348.
© 1983 by D. Reidel Publishing Company.

CARBONATE SEDIMENTS	TERRIGENOUS CLASTIC SEDIMENTS
The majority of sediments occur in shallow, tropical environments	Climate is no constraint, sediments occur worldwide and at all depths
The majority of sediments are marine	Sediments are both terrestrial and marine
The grain size of sediments generally reflects the size of organism skeletons and calcified hard parts	The grain size of sediments reflects the hydraulic energy in the environments
The presence of lime mud often indicates the prolific growth of organisms whose calcified portions are mud size crystallites	The presence of mud indicates settling out from suspension
Shallow water lime sand bodies result primarily from localized physiochemical or biological fixation of carbonate	Shallow water sand bodies result from the interaction of currents and waves
Localized buildups of sediments without accompanying change in hydraulic regimen alter the character of surrounding sedimentary environments	Changes in the sedimentary environments are generally brought about by widespread changes in the hydraulic regimen
Sediments are commonly cemented on the sea floor	Sediments remain unconsolidated in the environment of deposition and on the sea floor
Periodic exposures of sediments during deposition results in intensive diagenesis, especially cementation and recrystallization	Periodic exposure of sediments during deposition leaves deposits relatively unaffected
The signature of different sedimentary facies is obliterated during low-grade metamorphism	The signature of sedimentary facies survives low-grade metamorphism

Figure 1. Differences between terrigenous clastic and carbonate sediments

non-carbonate sediments, the composition of the sedimentary
particles themselves, either precipitated from seawater (e.g.
ooids) or formed by organisms (e.g. corals, brachiopods) is
equally important in characterizing the depositional environment.

The main differences in the style of accumulation between
carbonate and terrigenous clastic sediments are outlined in
Figure 1.

Carbonate Accumulations

Since carbonate secreting plants and animals are most
abundant in shallow (0-10 m), warm, tropical seawater, and the
direct precipitation of CaCO from seawater is easiest in similar
settings, the rate of carbonate sediment accumulation in shallow
water is orders of magnitude greater than in deep water.   In
addition, the distribution of benthic invertebrates in directly
related to energy distribution and nutrient supply.   These two
factors together result in the formation of a series of zoned
biological communities which quickly transform the shallow sea
floor into separate depositional environments.   Because accumulation in shallow water is so much greater than in surrounding deep
environments, as sea level rises, a carbonate platform, retaining
the basic facies zonation, evolves (Figure 2).   For a detailed
discussion of this facies pattern the reader is referred to the
excellent synthesis by Wilson (1975).

On any carbonate platform, there are basically three styles
of sediment accumulation:
    In-place production: The bulk of carbonate sediments are
produced in this way (the subtidal carbonate factory), by the
growth or precipitation of CaCO in one locality, which even
though reworked somewhat, remains more or less stationary.
Typical of such accumulations are: (1) open shelf muddy carbonates;
(2) ooid or skeletal sand shoals; and (3) reefs.
    Fallout: The calcareous zooplankton and phytoplankton which
live in the upper part of the water column, upon death rain down
upon all environments.   They are generally not significant on
the shallow platform, but are important in basinal and deep sea
settings.
    Resedimentation: Sediments produced on the shelf are often
transported either wholly or in part and redeposited elsewhere,
and in this way resemble many terrigenous clastic sediments.
These processes dominate deposition on: (1) muddy tidal flats;
and (2) platform margins.

COMPONENTS OF CARBONATE ROCKS

Particles

The sediments on a carbonate platform come from may different

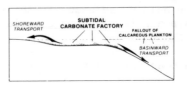

CARBONATE PLATFORM

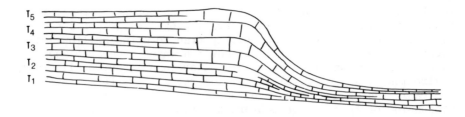

Figure 2. A sketch of the growth of a carbonate platform with an inset illustrating the main
zones of carbonate accumulation.

sources.

Skeletal: In spite of the wide spectrum of benthic and
planktic plants and animals which precipitate CaCO , there are
only two ways in which these hard parts are arranged: (1) as
whole rigid skeletons (foraminifers, snails, corals); and (2) as
numerous individual segments held together in life by organic
material (trilobites, crinoids, green algae) and upon death the
skeleton disintegrates into a series of calcareous particles.

Precipitated: CaCO is precipitated from seawater either
directly or indirectly in association with biological elements.
The crystals may be aragonite or Mg-calcite.   They may be
individual crystallites in seawater (?), cements within or
between sediment grains or accretionary rinds on nucleii of
various sorts.   The most common particles are ooids but various
skeletal and non-skeletal grains may be cemented to form larger
particles such as grapestone lumps.

Processed: Fine-grained sediments, because of their small

grain size are often reducing just below the sediment/water interface.   The adsorbed organic material is used by burrowing
invertebrates and as mud-sized particles pass through these
organisms they are aggregated into pellets.   These pellets may or
may not be preserved.

Eroded: In addition to carbonate-secreting plants and animals
there are a host of invertebrates that have adapted to an
endolithic way of life.   These boring organisms, penetrating both
skeletons and lithified limestone, produce a substantial amount
of fine-grained sediment (worms - mud; bivalves - mud; sponges -
fine sand and silt).   Rasping by echinoids and fish also creates
sand and silt sized grains.

Clastic: Sediments on the platform are commonly hardened
early on the sea floor, by cementation or desiccation. These
limestones are often broken up and eroded during storms to form
limestone clasts.

For a recent outline of the microscopic attributes of all
these particles the reader is referred to excellent books by
Majweske (1969), Horowitz and Potter (1971), Bathurst (1975), and
Scholle (1978).

The Sedimentary Aspect of Skeletal Grains

Even though carbonate-producing plants and animals have
changed markedly through geological time, if they are viewed as
sediment producing organisms rather than biological entities, then
fossils do have living equivalents in the modern ocean - even
though they may not even be in the same phyla.   The most important
carbonate-producing and binding organisms and their fossil
equivalents are outlined in Figure 3.

Classification

Several classifications of carbonates have been introduced
over the last 50 years (see particularly Ham, 1962) but those in
most common use today were synthesized by Folk (1962) and Dunham
(1962).   The more general of these by Dunham is used throughout
the following and is outlined in Figure 4.

Carbonate Mud and Carbonate Sand

Carbonate mud: In spite of a long history of investigation
it is still not clear how much carbonate mud is produced by
organisms and how much is precipitated directly (Bathurst, 1975).

Mud-sized particles are shed by both plants and animals upon
their post-mortem disintegration.   Volumetrically, the most
important mud-producers in the modern ocean are calcareous green
algae.   Mud is also formed by the breakdown of delicate worm

CARBONATE BUILDUPS ARE LIKE SHAKESPEARE; THE
PLAYS GO ON — ONLY THE ACTORS CHANGE
(R. N. GINSBURG)

ROLE	MODERN ORGANISM	ANCIENT COUNTERPART
LARGE REEF BUILDERS	CORALS	CORALS, STROMATOPOROIDS, RUDISTS, BRYOZOA, ARCHAEOCYATHIDS
LARGE WHOLE SKELETONS (SAND + GRAVEL)	BIVALVES, GASTROPODS FORAMINIFERS	BRACHIOPODS, BIVALVES, CEPHALOPODS, TRILOBITES, FORAMINIFERS, ALGAE
SKELETONS THAT DISINTEGRATE (SAND)	CODIACEAN ALGAE (HALIMEDA)	PELMATOZOANS, SPONGES ALGAE
SKELETONS THAT DISINTEGRATE (MUD)	CODIACEAN ALGAE	CODIACEAN ALGAE
ENCRUSTERS	ALGAE, FORAMINIFERS	ALGAE, FORAMINIFERS
BINDERS	BLUE-GREEN ALGAE	BLUE-GREEN ALGAE

Figure 3. The sedimentary aspect of modern carbonate-
producing and binding organisms and their
counterparts in the fossil record.

tubes, foraminifera tests and coralline algae skeletons which live
primarily as epibionts on sea grasses. Although large and more
robust skeletons are rarely reduced to mud-sized grains by
abrasion, mastication of weakly calcified large skeletons by
browsing organisms is important in the production of mud-sized
particles. Finally, the silt and clay size particles generated
by boring organisms, especially in reef environments, locally
contribute significant amounts to the mud fraction. Several
studies (Stockan et al., 1967; Neumann and Land, 1975) confirm
that algae alone produce more carbonate mud than can be accounted
for by the Holocene record.

The orgin of carbonate mud by direct precipitation is more
equivocal. Most evidence to date is circumstantial. In the
Bahamas, and other areas, clouds of fine-grained sediment in the
seawater (whitings) have often been cited as evidence of
spontaneous precipitation, but they may well be due to the stirring

CLASSIFICATION ACCORDING TO DEPOSITIONAL TEXTURE

DEPOSITIONAL TEXTURE				Original components were bound together during deposition... as shown by intergrown skeletal matter, lamination contrary to gravity, or sediment-floored cavities that are roofed over by organic or questionably organic matter and are too large to be interstices.
Original components not bound together during deposition				
Contains mud (particles of clay and fine silt size)		Grain supported	Lacks mud and is grain supported	
Mud supported				
Less than 10 per cent grains	More than 10 per cent grains			
Mudstone	Wackestone	Packstone	Grainstone	Boundstone

Figure 4. The classification of carbonate rocks
proposed by Dunham (1962).

up of sediment by feeding fish.   It appears, however, that in
areas slightly elevated salinity (Persian Gulf) to very saline
water bodies (Dead Sea) aragonite mud is being formed by direct
precipitation.   Perhaps a significant observation in this regard
is that early lithification of sediment by precipitation of
aragonite and Mg-calcite is a common, though not ubiquitous,
phenomenon.

Carbonate Sand: Carbonate sands in the modern ocean, like
carbonate muds, are predominantly due to the breakdown of
calcified algae, especially the form Halimeda.   In the Paleozoic
pelmatozoans appear to be the dominant producers of skeletal sand.
Other accessory components which are volumetrically less important
but often critical in paleoenvironmental interpretation are
foraminifers, bivalves, gastropods, brachiopods, trilobites,
echinoderms, coralline algae and fragments of corals, stromatoporoids, bryozoans, and archaeocyathans.

Ooids are ubiquitous particles throughout the geological
record.   There appears to be a general correlation between the
nature of the ooid cortex and the environment of formation:
concentric, formed in open ocean, agitated settings of normal
salinity: radial, formed in open ocean to restricted settings that
are relatively quiet and may have slightly elevated salinity:
asymmetric, located in restricted and tranquil environments.   Most
ooids appear to accumulate at the shoreline or in the banks at the
platform margin.   Early lithification into sand sheets and hardgrounds is a common phenomenon in these settings.

The last group of grains that contribute significantly to
carbonate sand are grouped under the term peloids.   The most
common are those produced as faecal pellets which are held together by organic mucus.   They are preserved if cemented early,
otherwise they are destroyed by compaction.   As soft pellets
they can be moved and deposited as if they were sand grains, so
banks of mud composed of "soft sand" may accumulate, but little is
left in the record of their genesis (R.N.Ginsburg, pers.comm.).
Another group of peloids is produced by the alteration of skeletal
particles to mud by the boring action of cyanobacteria and precipitation of microcrystalline cement in the vacated holes.   Regardless of their orginin these grains are usually generated in the
quiet environments of the platform interior.

Carbonate Sediments in Geological History

When thinking in general terms of carbonate sedimentation
different organisms can be treated in a similar fashion.   When
discussing the attributes of specific accumulations, however, the
evolutionary record of $CaCO_3$ -producing plants and animals must
be taken into account.   As pointed out by Lowenstam (1963) and
more recently by Wilkinson (1979), the major organism groups, and
so the original mineralogy of sediment, is markedly different at
different times in geological history (Figure 5).

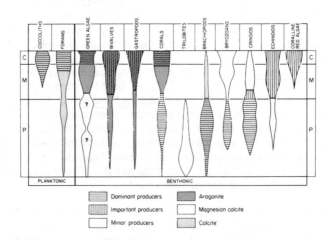

Figure 5.   The approximate diversity, abundance and
            composition of principal groups of calcareous
            marine organisms (after Wilkinson, 1979).

On the basis of current knowledge it appears that Paleozoic

benthic faunas were dominated for the most part by metazoans
which secreted either calcite or Mg-calcite skeletons (bryozoans,
brachiopods, corals, stalked echinoderms and locally trilobites).
In contrast Cenozoic biotas were mostly organisms that secreted
aragonite (green algae, scleractinian corals, molluscs) and to a
much lesser extent by groups that precipitated Mg-calcite
(coralline algae, echinoids).   The record of Mesozoic forms is
intermediate although aragonitic skeletons are locally very
abundant (e.g. rudists).   There is also compelling evidence that
Mg-calcite or perhaps even calcite ooids may have been much more
common in the Paleozoic than they are today, when most are
aragonitic.

SHOALING-UPWARD SEQUENCES

     Perhaps the most easily recognizable and at the same time
most widespread and potentially economically important depositional
system on the carbonate platform is that developed when sediments
build to sea level and so are periodically exposed as a series of
tidal flats.   A quantum jump in our understanding of these
deposits occurred when modern carbonate tidal flats were examined
in detail, notably by Robert Ginsburg and his colleagues in
Florida and the Bahamas about 20 years ago.   It was quickly
realised that there were a host of sedimentary structures and
textures on these flats that would allow a much more precise
definition of environments of deposition than was possible before:
these findings were quickly applied to fossil sequences (Fischer,
in Merriam, 1964; Laporte, 1967; Roehl, 1967).   This application
in turn generated two different lines of investigation: (1) description of other areas of modern tidal flat deposition, in
particular the southern shores of the Persian Gulf where evaporites
are common and Shark Bay, Western Australia, where a great variety
of modern stromatolites are forming; and (2) documentation of
different styles of tidal flat deposits in the geological record.

     In the following section the main attributes of modern tidal
flats are considered, and then using this framework together with
the many well documented fossil examples, various depositional
models are outlined.

Modern Carbonate Tidal Flats

     The main elements of a modern carbonate tidal flat system
as exemplified by the narrow shelf and embayments of Shark Bay,
Western Australia, the southern coast of the Persian Gulf and the
wide platform of the Bahama Banks are shown in Figure 6.   A
characteristic of most modern examples is that they occur in
protected locations: protected that is from the open ocean waves
and swells, yet still affected by tides and severe storms.   This

unique setting is commonly afforded by the presence of a semi-
protective barrier composed of lime sand shoals, locally
associated with reefs and/or islands.   The barrier commonly is
dissected by tidal channels through which flow high velocity
tidal currents.   A shallow muddy lagoon lies in the less of this
barrier.   The lagoon may be enormous as in the case of the
Bahamas, relatively narrow and elongate as in the Persian Gulf,
or very small as in the pocket embayment of Shark Bay.   In such
an arrangement, tidal flats are present as: (1) small areas atop
and on the lee side of the emergent sand shoals of the barrier,
and (2) large flats along the shoreline of the shallow lagoon
(Figure 6).   Thus tidal flats occur in association with two
separate carbonate accumulations, high energy sand bodies and
low energy lime muds.   A third type of association which is less
common in modern situations is the association with reefs,
especially the interior of large reef complexes.

Intertidal Environments

     On wide, gradually sloping tidal flats, the intertidal zone
can be the familiar gradual transition from land to sea or a
complex area of many subenvironments.   At one end of the spectrum
the flats have few, very shallow, short tidal creeks (Figure 7).
At the other end of the spectrum the flats are dissected by many
tidal creeks flanked by levees.   Slight depressions between the
creeks are occupied by tidal ponds (which fill and partially
empty during each rise and fall of the tide), and the whole
complex is fronted by small beach ridges or erosional steps
(Figure 7).   Perhaps in this case it would be better to refer to
the whole zone as the "pond and creek belt" because some of the
areas are dry most of the time (levees and beaches) whereas
others are continuously submerged (ponds and creeks).   The
complications have led Ginsburg and Hardie (in Ginsburg, 1975)
to despair of conventional terms and instead to relate different
zones to the percent of time that they are exposed rather than to
their position.

     On some tidal flats, in the Bahamas for example, where there
are many tidal creeks and noticeable relief between levee and
tidal pond (about 1 m), the true intertidal zone which lies
between the two may comprise only 60 to 70 per cent of the intertidal environment.   In other areas such as the Persian Gulf,
where there are fewer creeks and less relief, almost the whole
flat is truly intertidal.   The important point to grasp is that
numerous environments may exist in very close proximity, not only
perpendicular to the shoreline but parallel to it as well, so
that in the geological record rapid, local lithological variations
are to be expected, both vertically and laterally, rather than a
smooth succession of progressively shallower environments.

MODERN TIDAL FLAT COMPLEX

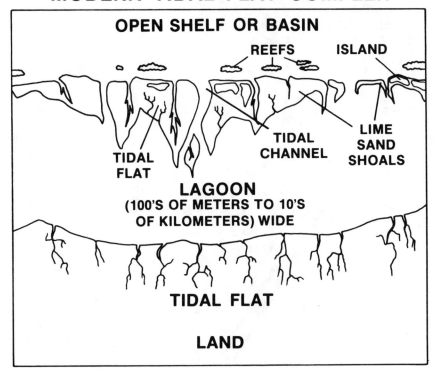

Figure 6. Plan view of the geometry of a modern tidal
flat complex. Note that tidal flats can be
present both adjacent to the land or in the
lee of lime sand shoals. (From James, 1977b,
permission of Geological Association of
Canada).

The tidal flat wedge is built up of fine-grained sediments
brought onto the flats from the adjacent offshore marine zone,
by storms rather than by daily tides. Large storms such as
hurricanes which flood the flat with sheets of water white with

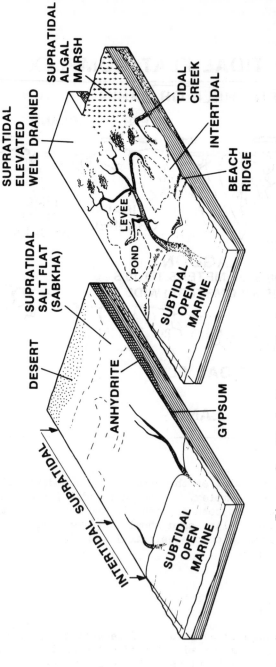

Figure 7. Block diagram showing the major morphological
           elements of a tidal flat, left – a hyper-
           saline tidal flat with few channels and
           bordering a very arid desert (similar to the
           modern Persian Gulf), right – a normal marine
           tidal flat with many channels and ponds and
           bordering an elevated well-drained area of
           low swamp algal marsh in a humid climate
           (similar to the modern Bahamas). (From
           James, 1977b, permission of Geological
           Association of Canada).

suspended sediment, are particularly effective.   Shinn, Lloyd
and Ginsburg (1969) have suggested that the tidal flat is a river
delta turned wrong-side out with the sea as the "river" supplying
sediment to the channelled flats as the "delta".

Sediments of the intertidal zone are characterized by three
distinctive features, not found elsewhere: (1) algal mats; (2)
irregular to even laminations (cryptalgal laminites), with
fenestral porosity; and (3) desiccation features.

The algal mats, gelatinous to leathery sheets of blue-green
algae growing on top of the sediment surface, are widely regarded
as the signature of intertidal deposition; they may occur
throughout the intertidal zone but their precise distribution is
controlled by climate and the presence or absence of other
organisms.   The upper limit is controlled by climate; in arid
areas they cannot grow above the high intertidal into the supratidal zone, whereas in areas of high rainfall where the supratidal zone is moist or flooded for days at a time, mats are
prolific.   The lower limit is more variable and appears to be
controlled by the presence of gastropods that eat algae.   In
areas of normal salinity, mats are prevented from developing
below the middle intertidal zone because they are browsed by
gastropods; in areas of hypersalinity (lethal for many species
of gastropods), mats grow down into the subtidal zone.   In
addition algal mats will colonize only a temporarily or
permanently stable bottom, and will not grow on shifting sand.

Although the algal mats may themselves vanish with time,
evidence of their presence during deposition remains because of
the peculiar pores that they help to create, generally referred
to as "laminoid fenestrae".   These are irregular, elongate to
mostly sub-horizontal sheet-like cavities (loferites or birds-eye
of some workers) with no obvious support and much larger than
can be explained by grain packing.   They are simply due to the
fact that the mats are covered with sediment and eventually rot
away as they are buried, leaving voids as well as holes due to
entrapped gas, and shrinkage.

Another structure recording the presence of blue-green algal
mats are the finely laminated carbonates ranging from stratiform
and slightly crenulated to the familiar arched domes of stromatolites.   These have been called cryptalgal (hidden, algal)
laminations by Aitken (1967) in reference to the fact that the
influences of algae in the rock-forming process is more commonly
inferred than observed.   On modern tidal flats many subenvironments have distinctive types of laminations (Hardie and Ginsburg,
1977).

This intertidal environment in turn displays a distinct

zonation.

Lower intertidal zone: Much of the subtidal character remains evident in sediments from this part of the environment, and the deposits are commonly well burrowed and bioturbated. In hypersaline areas, however, the surface of the sediment is veneered with a thick algal mat frequently broken into desiccation polygons. Beneath the mat carbonate grains are blackened due to reducing conditions and altered by boring algae to peloids of lime mud.

Tidal ponds and the creeks that drain them on hypersaline tidal flats support the most prolific growth of algal mats anywhere on the flat. The algal mat flourishes in water depths greater than those in the immediate offshore area because of relatively elevated salinities in the ponds. On tidal flats where the salinity is closer to normal, marine tidal ponds are populated by a restricted but prolific faunas of foraminifers and gastropods and the gastropods prevent the growth of algal mats. Similarly, if tidal creeks are common in such areas, the channels are devoid of mats but do contain concentrations of the pond fauna, which may accumulate as bars of skeletal lime sand. As the channels migrate these skeletal sands commonly form a basal lag deposit.

Middle and upper intertidal zone: Sediments here are commonly light-grey to light-brown (oxidizing conditions), have good fenestral porosity (the variable growth of algal mats), are graded (episodic storm deposition) and are broken into desiccation polygons (prolonged exposure). There is generally good growth of algal mats throughout. In the lower parts thick leathery mats are separated into desiccation polygons a few centimetres to a metre in diameter with cracks filled by lime mud. In the central parts, thinner, leathery mats have surfaces that are puffed up into blisters and convoluted into crenulated forms and in the upper parts shrivelled, crinkled and split mats are found. Bedding generally is irregular especially in the upper zones with mats alternating with graded storm layers.

In some settings sediment in the upper intertidal zone dries out to form chips of lime mud while in others the sediment below the mats is lithified to a depth of as much as 10 cm.

Although sediments commonly are laminated throughout the intertidal environment, they are also riddled with small-scale tubules produced by insects and worms, as well as larger tubes produced by crabs and other crustaceans. Sediments also may be penetrated by the prolific shallow roots systems of salt-tolerant plants.

Supratidal environment: In all situations (including channel levees) this area is characterized by long periods of exposure. This is reflected by the lithification of storm-deposited sediments

in the form of surface crusts, several centimetres thick, and
which in turn are fractured into irregular polygons.   These
polygons may be pushed up by the force of crystallization (or by
plant roots) to form "teepees", or dislodged completely to form
pavements of flat-pebble breccia.   Clasts are commonly cemented
on modern tidal flats by cryptocrystalline aragonite or calcite,
and characteristically contain considerable (25 to 50%) fine
crystalline dolomite.

     If the creek levees in the intertidal zone have built up
above normal high tide level, they consist of hard, finely to
very finely laminated sediment, extremely regular, and composed
of alternating layers of sediment and thin algal mats with
excellent fenestral porosity.

     The landward parts of the supratidal zone may grade into
various terrestrial environments, the end members of which are:
(1) areas of elevated, pre-existing bedrock and no sedimentation
in which the surface of the rock is characterized by intensive
subaerial diagenesis, and the development of caliche (calcrete
crusts); (2) areas of contemporaneous sedimentation which grade
between: a) low-lying environments in regions of high rainfall,
occupied by algal marshes; b) low-lying environments in arid,
desert regions, characterized by evaporite formation; and c)
well drained zones, often slightly elevated and with little
deposition.

     Algal marshes, flooded by fresh water during the rainy
season are an ideal environment for the growth of algal mats
and these mats are periodically buried by layers of sediment
swept in during particularly intense storms; thus, the preserved
record is one of thick algal mats alternating with storm layers.
With progressive aridity the supratidal zone dries out.   If the
chlorinity of the groundwater remains constantly above 39%,
cementation, particularly by aragonite, is common.   Cementation
is most common if there is minor but consistent input of fresh
water from inland to dilute the hypersaline groundwaters somewhat.   If the chlorinity of groundwaters remains constantly
above 65% then authigenic evaporites precipitate within the
sediment below ground level.   In this setting (called a supratidal sabkha, or salt flat in the Middle East) dolomitization is
also common in the subsurface, saline brine pools occur at the
surface and terrigenous wind-blown sand occurs in the sediment.

     In relatively well-drained zones the supratidal environment
is a deflation surface, occasionally cut by the upper reaches of
tidal creeks, sometimes damp from rising capillary waters and
covered by a thin film of algal mat.   Scoured and rippled sediment is common and clasts are sometimes encrusted with algae to
form oncolites.

The Depositional Model

    The ideal carbonate shallowing-upward sequence comprises
four units, illustrated in Figure 8.   The basal unit, which is
generally thin, records the initial transgression over pre-
existing deposits and so is commonly a high energy deposit.   The
bulk of the sequence, which may be of diverse lithologies, consists
of normal marine carbonate, as discussed below.   The upper part
of the sequence consists of two units: the intertidal unit within
the normal range of tides;   the other a supratidal unit, deposited
in the area covered only by abnormal, windblown or storm tides.
Each of these units exhibits the characteristic criterial of subaerial exposure.

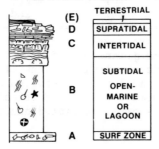

Figure 8. Five divisions of the shallowing-upward model
          for carbonates: A - lithoclast rich lime
          conglomerate or sand, B - fossiliferous lime-
          stone, C - stromatolitic, mud-cracked cryptalgal
          limestone or dolomite, D - well laminated dolo-
          mite or limestone, flat-pebble breccia, E -
          shale or calcrete, bracketed to emphasize that
          the unit is often missing - see text.   Symbols
          used throughout are from Ginsburg (1975).
          From James, 1977b (permission Geological
          Association of Canada).

    The thread that binds all such sequences together is the
presence of the distinctive intertidal unit, which, once
recognized, allows one to interpret the surrounding lithologies
in some kind of logical sequence.

    First-order variation on the basic model revolves around
the two main types of intertidal environment: (1) quiet, low-
energy situations, commonly referred to as tidal flats; and (2)
agitated, high-energy situations, or quite simply, beaches.

Second-order variation involves the kind of subtidal units below
and of supratidal units above; the subtidal reflects the type
of marine environment adjacent to the tidal flat and the supratidal reflects the adjacent terrestrial environment, in particular the climate (Figure 9).

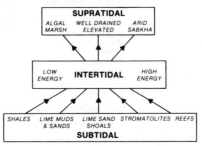

Figure 9. A flow diagram indicating the various
possible environmental transitions present
in a carbonate shallowing-upward sequence.
(From James, 1977b, permission of
Geological Association of Canada).

Muddy sequences: These sequences developed either by progradation of the wide continental tidal flat or by shoaling of
the lime sand bodies that formed the barrier offshore (Figure
10). The climate in the region of deposition was generally too
wet or the groundwater table too low or diluted by fresh water
to permit precipitation of evaporites.

The muddy sequences, those in which skeletal lime muds or
muddy lime sands are the main subtidal unit, are well developed
today in well-drained areas of Shark Bay where salinities are
too high to permit browsing of the algal mats by gastropods, and
in the tidal creek and pond belt of the Bahamas. These sequences
are generally regarded as the classic tidal flat sequences. The
basal unit, if present, records the initial incursion of the sea
onto land and as such is commonly coarse-grained, composed of
clasts, etc., all diagnostic of surf-zone deposition. The subtidal unit is characteristically a bioturbated lime wackestone
to packstone with a normal and diverse marine fauna, commonly
containing stromatolites in deposits older than middle Paleozoic.
In such deposits the characteristic tidal flat features such as
desiccation polygons, well-laminated sediments and fenestrae will
occur at the base of the intertidal zone. In deposits younger
than middle Paleozoic, the prolific browsing and burrowing
activity in the lower intertidal zone (unless the water mass was

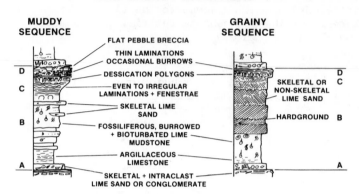

Figure 10. Two hypothetical sequences with a low
            energy tidal flat unit developed on a low
            energy subtidal unit (left) and a high
            energy lime sand unit (right). (From James,
            1977b, permission of Geological Association
            of Canada).

hypersaline) have homogenized the sediment, so that the signature
of intertidal deposition is recorded only within the mid and
upper intertidal sediments.

    If the tidal flat was extensively chanelled, the migration
of channels back and forth may also have destroyed some of the
subtidal character, forming instead a partial fining-upward
sequence (much like that of a river), with a basal skeletal lime
sand.

    Where fenestrae are present they show a zonation horizontal
to laminated in the lower intertidal environments (smooth mat),
irregular and in some cases vertical in the middle and upper
intertidal environments (pustular, shrivelled and crinkled mats).
Desiccation polygons are most common near the top, apparently
coincident with cementation. The supratidal zone is characterized by very evenly laminated deposits or flat pebble breccias.

    Readers interested in the finer details of such sequences
are referred to studies by Laporte (1967) and Fischer (in
Merriam, 1964), the latter outlining and documenting a similar
facies sequence but in reverse order, forming a deepening-upward
sequence.

    An interesting variation on this model has recently been

proposed by Hardie and Ginsburg (1977) as a result of their study
of the Bahama tidal flats. Here the sedimentary record during
the Holocene transgression is not one of continued seaward progradation of the shoreline and stacking of progressively shallower
facies. Rather the accumulation is vertical behind a protecting
barrier beach ridge. They envisage tidal flat accretion as
beginning when the offshore bar emerges above mean high water to
become a barrier. The complex series of environments in the
pond and channel belt will with time become choked with sediment
and turn into one vast supratidal coastal plain. The resultant
stratigraphic sequence would be a supratidal laminate cap over
unlayered, bioturbated subtidal and intertidal pelleted mud.
In such a situation the intertidal portion of the succession will
be difficult indeed to discern.

Grainy sequence: The same style of shoaling sequence also
may be present offshore from the low energy tidal flat on the
lime-sand shoals. Here low energy tidal flats developed in the
lee of the leading edge of the shoal once beach ridges were
developed or currents had swept sand together to form islands.
This will be reflected in the sequence as a sudden change from
obvious high energy intertidal deposition. The subtidal unit is
generally well-sorted, oolitic, pelletoidal or skeletal lime
sand (pelmatozoans are particularly common in the Palaeozoic),
with a few containing oncolites. Bedding is characteristically
planar, with herringbone cross-laminations, large at the base
and becoming smaller upwards, and individual bedding planes are
commonly covered with small-scale ripples. Early cementation
is characteristic, and so deposits contain many intraclasts of
cemented lime sand, and bored surfaces. Once the shoals, or
parts of the shoals are inactive they may be burrowed and much of
the original cross-bedding may be destroyed.

The intertidal to supratidal units are similar to those
described above, but are generally relatively thin. If the
shoal is exposed for a long time caliche and soil profiles
commonly develop, reflected by brown irregular laminations,
breccias, and thin shale zones.

Stromatolite and reef sequences: One common variation in the
model is the development of shoaling-upward sequences in association with abundant stromatolites (especially in the lower
Paleozoic/Precambrian, and with reefs in the Phanerozoic in
general.

In Shark Bay, Western Australia, where all environments are
hypersaline and so stromatolites abound, the interrelationship
between stromatolie morphology and environment has only recently
been documented (Hoffman, 1976). In the intertidal zone,
columnar to club-shaped forms up to one metre high are found
rimming headlands. In relatively high energy, exposed environments the relief of the columns is proportional to the intensity

of wave action.   These grade laterally away from the headlands
to the lower energy bights where the stromatolites are more
prolate and elongate, oriented normal to the shoreline.   In tidal
pools digitate columnar structures abound.

      These growth forms are the result of active sediment move-
ment: algal mats only grow on stabilized substrate, thus
columns are nucleated upon pieces of rock, etc.: growth is localized there and does not occur on the surrounding shifting sands.
Early lithification of the numerous superimposed layers of mat
and sediment turns the structures into resistant limestone.
Moving sand continuously scours the bases of the stromatolites.
The mounds of pillars are largest in subtidal or lower intertidal
environments and decrease in synoptic relief upwards, finally
merging with stratiform mats in upper intertidal zones, above the
zone of active sediment movement.

      The resulting model sequence, summarized in Figure 11, is
integrated from the Shark Bay example and the summary sequence of

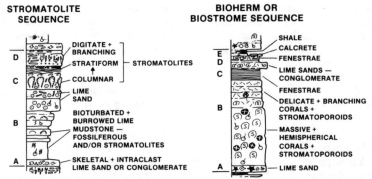

Figure 11. Two hypothetical sequences with a low
           energy intertidal unit developed in
           conjunction with stromatolites (left) and on
           top of a skeletal metazoan bioherm or
           biostrome (right). (From James, 1977b,
           permission of Geological Association of
           Canada).

200 or more shoaling sequences present in the Rocknest Formation
of middle Precambrian age near Great Slave Lake (Hoffman, 1976).
In the intertidal zone deposits reflect higher energy than normal,
indicating a more exposed shoreline.   These sediments underlie

and surround the domal to columnar stromatolites, which in turn
grade up into more stratiform stromatolites, and finally into
very evenly bedded structures.   The supratidal unit of this
sequence is characterized by both desiccation polygons and flat-
pebble breccias as well as occurrences of delicate branching
stromatolites, forming in supratidal ponds.   Care should be taken
in delineating this sequence because stromatolites that are similar to those in the intertidal zone also occur in the subtidal
(Playford and Cockbain, 1976).

     Shallowing-upward sequences are also common as the last stage
of sedimentation in large bioherms, as numerous successions within
the large back-reef or lagoonal areas of reef complexes, and as
"caps" on widespread biostromes (Figure 11).   In this type of
unit itself, generally as a transition from large massive hemispherical colonial metazoans of the reef facies, to the more
delicate, stick-like forms that are common in the shallow protected locations.   These stick-like skeletons may be swept
together on beaches at the edge of the tidal flat.   As a result,
the intertidal unit commonly contains a conglomerate within it, or
at the base.   The upper part of the sequence is otherwise similar
to the others described.   For a more detailed description of
"reefy" sequences see the two studies by Havard and Oldershaw
(1976) and Read (1973).

     Recognition of shoaling-upward sequences within biostromal
or reef complexes has profound implications in terms of reservoir
potential.   Each period of exposure probably results in at least
some meteoric diagenesis of the underlying carbonate and so a wide
variety of dissolution and cementation phenomena.   In addition,
what appears to be a more or less homogeneous carbonate rock body
is, in terms of porosity and permeability distribution, now a
stratified unit, with the top of each shoaling-upward cycle being
a permeability barrier.
     Carbonate-evaporite sequences: The other major variation on
the model is at the opposite end of the environmental spectrum,
in the supratidal zone, in this case emergent in a very arid
environment and flushed by hypersaline groundwaters.   The hypersalinity of the groundwaters and attendant high evaporation results
in the formation of authigenic evaporites which in turn raises
the Mg^{++}/Ca^{++} ratio of the groundwaters and induces dolitiza-
tion of the sediment.   The processes occur within the sediment,
above the water table in the supratidal zone.   If the water
compositions are barely within the field of gypsum precipitation,
and there are fluctuations due to brackish flow of groundwater
from the mainland, evaporites will occur in the form of isolated
masses or crystals in the upper part of the sequence.   If the
groundwater compositions are continuously well within the field
of gypsum precipitation, growth of evaporite minerals takes place
as a mush of gypsum crystals in the intertidal zone or as layers

of anhydrite nodules, as complex masses with a characteristic
chickenwire texture, and as layers contorted into enterolithic
(intestine-like) shapes (Figure 12).   The important point, which
is often ignored, is the growth of the evaporites within the sediment, as a diagenetic overprint on depositional facies of various
environments.   As evaporite growth is porphyroblastic, the best
sediment commonly is displaced to intercrystalline areas and
earlier fabrics are destroyed.   Accompanying dolomitization
commonly is intense, with sediments of the intertidal and much
of the subtidal zone affected.

These sequences are probably amongst the best potential
reservoir lithologies because dolomitization creates good porous
and permeable carbonate while the interbedded evaporites are
excellent seals.

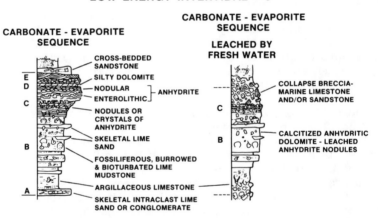

LOW ENERGY INTERTIDAL − 3

Figure 12.   Two hypothetical sequences with a low-
energy intertidal unit and a supratidal
unit developed under arid conditions;   on
the right the evaporites have been dissolved
by percolating fresh waters.   (From James,
1977b, permission of Geological Association
of Canada).

Evaporites, however, are very soluble when exposed to
percolating meteoric waters of low salinity and have a tendency
to vanish from the record.   Dissolution of the evaporites affects
the sequence in several ways, but the most important is the
formation of collapsed breccias.   This collapse occurs when the
evaporites dissolve leaving no support for the overlying sediments,
which subside into the void created by evaporite removal.   Thus
the top of the sequence is a breccia of marine limestone from the

overlying sequence with a mixture of terrigenous sand if a
terrigenous facies capped the original sequence (Figure 12).
Isolated anhydrite crystals in lower parts of the sequence may be
leached out, forming vugs which may be subsequently filled with
quartz or chalcedony (usually length-slow). The dolomite, at
least in the upper part, is commonly altered to calcite, in the
reverse of the dolomitization process (so-called "dedolomitization").

Sequences with a High Energy Intertidal Unit: In contrast to
the low-energy intertidal (the tidal flat) the higher energy beach
zone is not commonly recognized in the rock record. This may be
partly because it resembles many subtidal grainstone deposits and
so is not obviously distinctive. Also, it is relatively narrow
compared to the tidal flat, and has a lower preservation potential,
and finally, the beach deposits lack the distinctive sedimentary
features of the tidal flat. These very reasons illustrate the
value of the concept of a shoaling-upward sequence as a guide.
Once the potential for such a sequence is recognised in the
geological record, then one can concentrate on the search for
subtle features that characterize beach deposition, which otherwise might go unnoticed.

The beach is characterized by two zones: (1) the lower foreshore, that zone unusually below the zone of wave swash; and (2)
the upper foreshore, the zone of wave swash. Sediments of the
lower foreshore are coarse-grained, poorly sorted, have a matrix
of lime mud (if it is available), and are characterized by small
and large-scale festoon cross-bedding, oriented parallel to the
shoreline and generally attributed to longshore drift. The upper
foreshore comprises thick-bedded, internally laminated, very well-
sorted lime sands and gravels in planar cross-bedded accretionary
beds that dip gently seaward (generally less than 15 degrees).
Sediments in the upper foreshore zone may have many open-space
structures, the equivalent of the fenestrae of muddy intertidal
sediment, called keystone vugs (Dunham, 1969) or microcaverns
(Purser, 1973). These are due to gas escape and in the geological
record are partly to completely filled with cement.

As on the tidal flat, periodic exposure of beach deposits
leads to cementation and partial subaerial diagenesis. The
textures thus created are difficult to recognize in the field but
are important keys to recognizing the beach environment. The two
most important of these diagenetic phenomena are beachrock and
calcrete.

Beachrock is composed of seaward-dipping beds of lime sand
and gravel that are generally cross-laminated and occur in the
lower intertidal to middle intertidal environment. It is formed
by the precipitation of carbonate cement out of seawater or mixed
seawater and rainwater. The beds of limestone may be up to one
metre thick, are commonly jointed at right angles to the beach

and are encrusted and/or bored by numerous intertidal organisms.
Lithification disappears seaward and rarely extends higher than
the intertidal zone.   The partly-cemented beds may be broken up
and redeposited as conglomerates, made up of cemented sand clasts.
In the upper parts of the intertidal zone cementation takes place
in intergranular voids partly filled with air;   the cements, as a
result, are often stalactitic (more extensively developed on the
undersides of grains).

       If exposed for long periods of time and if located in an
environment where there is at least periodic rainfall, the lime
sands will begin to undergo subaerial diagenesis (see Bathurst,
1975 and Chapter   , this volume for a discussion of subaerial
diagenesis.   In addition the upper metre or so of such sub-
aerially exposed deposits develop calcrete or caliche horizons
which have many features that closely resemble those produced by
laminar to laterally-linked stromatolites and oncolites.

       The supratidal unit in these sequences may be any of the
one described above, although calcrete (caliche) is very common.
Beaches may act as small barriers protecting supratidal ponds
and flats so that the cap in such sequences will be in thin beds
of lime mud (often dolomitized) with all of the associated supratidal features.   One variation not found elsewhere occurs where
the high energy surf zone of the overlying sequence erodes the
top of the sequence down to the cemented portions, resulting in
truncation layers or hardgrounds that separate sequences.   The
lower two units of this type of sequence are similar to those
described in the preceding sections on sequence with low-energy
intertidal units (Figure 13).   In this sequence, however,
characteristic subtidal carbonates grade upward into coarse-
grained lime sands with all the characteristics of the lower and
upper foreshore described above (Figure 10).   The supratidal unit,
in the form of a thin shale (soil) may be present, but more
commonly the supratidal is represented not by a deposit but by
intensive diagenesis of the upper unit (cementation, dissolution,
calcrete formation and microkarst).   This is in many ways similar to the diagenetic overprint of other facies by supratidal
evaporite formation.

REEFS

       A reef, rising above the sea floor, is an entity of its
own making - a sedimentary system within itself.   The numerous,
large calcium carbonate-secreting organisms stand upon the remnants
of their ancestors and are surrounded and often buried by the
skeletal remains of the many small organisms that once lived on,
beneath and between them.

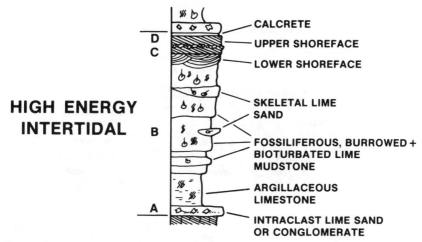

Figure 13. A hypothetical sequence with a high-energy
            intertidal unit: a beach, developed, in
            this case, adjacent to a low energy sub-
            tidal environment. (From James, 1977b,
            permission Geological Association of
            Canada).

Because they are built by organisms, fossil reefs are
storehouses of paleontological information and modern reefs are
natural laboratories for the study of benthic marine ecology.
This, together with the fact that fossil reefs buried in the
subsurface contain a disproportionately large amount of our oil
and gas reserves compared to other types of sedimentary deposits,
has resulted in their being studied in detail by paleontologists
and sedimentologists, perhaps more intensely than any other
single sedimentary deposit, yet from two very different viewpoints.

Dynamics of Reef Sedimentation

The present stage of any thriving reef is a delicate balance
between the upward growth of large skeletal metazoans, the continuing destruction of these same organisms by a host of rasping,
boring and grazing organisms, and the prolific sediment production
by rapidly growing, short-lived, attached calcareous benthos
(Figure 14)

The large skeletal metazoans (e.g. corals) generally remain
in place after death, except where they are so weakened by bio-
eroders that they are toppled by storms.   The irregular shape
and growth habit of these reef-builders results in the formation
of roofed-over cavities inside the reef that may be inhabited by
smaller, attached calcareous benthos, and may be partly to

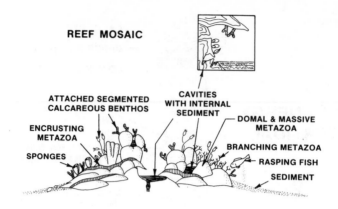

Figure 14. A sketch illustrating the different aspects
of the organism/sediment mosaic that is a
reef. (From James, 1978, permission of
Geological Association of Canada).

completely filled with fine-grained "internal" sediment. Encrusting organisms grow over dead surfaces and aid in stabilizing the
structure. Branching reef-builders frequently remain in place
but just as commonly are fragmented into sticks and rods by
storms to form skeletal conglomerates around the reef.

Most reef sediment is produced by the post-morterm disintegration of organisms that are segemented (crinoids, calcareous
green algae) or non-segmented (bivalves, brachiopods, foraminifers,
etc.) and that grow in the many nooks and crannies between the
larger skeletal metazoa. The remainder of the sediment is produced by the various taxa that erode the reef; boring organisms
(worms, sponges, bivalves) produce lime mud: rasping organisms
that graze the surface of the reef (echinoids, fish) produce
copious quantities of lime sand and silt. These sediments are
deposited around the reefs as an apron of sediment and also filter
into the growth cavities to form internal sediment, which is
characteristically geopetal.

The fabric of reef limestones is in turn affected by two
early diagenetic processes, early cementation and bioerosion.
Early cement in the form of Mg-calcite (as isopachous rinds of
spar and as micrite) and/or aragonite (as needles and large
spherulites) has been found to be particularly common in ocean-
facing reefs. This precipitation has the deleterious effect of
filling many growth cavities and so limestone platform margin
reefs, unaffected by later diagenesis, are poor reservoir rocks.
Bioerosion while just as common in lagoon reefs, with the occurrence of early lithification to create ever-new substrates for

infestation can, with time, often alter the depositional fabric of
the reef limestone.   This process involves repeated boring by
sponges, bivalves and other endoliths, death of the organisms,
infill of the cavity by sediment, lithification of that sediment
and then reboring by another generation of endoliths.

Diversity Amongst Reef-building Metazoans

      Very diverse faunas, in terms of both growth form and taxa,
occur when a community is well established and conditions for
growth are optimum, i.e., nutrients are in good supply, chemical
and physical stresses are low.   In such optimum environments the
division of biomass amongst various species is due mainly to
complex biological controls (Stoddart, 1969; Ginsburg and James,
1974; Frost et al. 1977).

      In contrast, low diversity environments commonly fall into
three general categories: (1) unpredictable environments; (2) new
environments (faunas moving into a new environment); and (3)
severe environments (high chemical and physical stress).   Among
the factors most likely to stress modern and fossil reef-building
communities are: a) temperature and salinity flucutations - most
modern and likely most ancient reef-builders grow or grew best in
tropical seawater of normal salinity; b) intense waves and swell
- the skeletons of most reef-builders will be broken or toppled
by strong wave surge; c) low light penetration - in modern reef-
building organisms rapid calcification takes place because
symbiots, which are light dependent, take over some of the bodily
functions of the host; d) rapid sedimentation - all reef-builders
are sedimentary filter-feeders or micropredators and water
abundantly charged with fine-grained sediments would clog the
feeding apparatus.

The Growth Form of Reef-building Metazoans

      The relationship between organisms shape and environment is
one of the oldest and most controversial topics in biology and
palaeobiology.   In terms of reef-building metazoans, however,
many observations of the interrelationship between organisms and
surrounding sediments from the rock record combined with studies
of modern coral distribution of tropical reefs allow us to make
some generalizations about form and environment that are very
useful in reef facies analysis (Figure 15).

Description of Reef Limestones

      Many different classifications have been proposed for the
resulting reef carbonates but the most descriptive and widely
accepted is a modification of Dunham's (1970) classification of
lime sand and mudrocks proposed by Embry and Klovan (1971) at the

GROWTH FORM AND ENVIRONMENT OF REEF BUILDING SKELETAL METAZOA			
GROWTH FORM		ENVIRONMENT	
		Wave Energy	Sedimentation
	Delicate, branching	low	high
	Thin, delicate, plate-like	low	low
	Globular, bulbous, columnar	moderate	high
	Robust, dendroid, branching	mod-high	moderate
	Hemispherical, domal irregular, massive	mod-high	low
	Encrusting	intense	low
	Tabular	moderate	low

Figure 15. The growth form of reef-building metazoans and the types of environments in which they kost commonly occur.

University of Calgary (Figure 16).   They recognize two kinds of reef limestone, allochthonous and autochthonous. The alloch-thonous limestones are the same as the finer grained sediments,

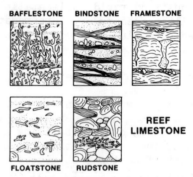

Figure 16. An interpretative sketch of the different types of reef limestone recognized by Embry and Klovan (1971). (From James, 1978, permission of Geological Association of Canada).

but with two categories added to encompass large particles.   If more than 10 per cent of the particles in the rock are larger than 2 mm and they are matrix supported it is a Floatstone;   if the rock is clast supported it is a <u>Rudstone</u>. The autochthonous limestones are more interpretative; <u>Framestones</u> contain in place, massive fossils that formed the supporting framework; <u>Bindstones</u> contain in place, tabular or lamellar fossils that encrusted or bound the sediment together during deposition; <u>Bafflestones</u> contain in place, stalked fossils that trapped sediment by baffling.

Reef Facies

    While it is a simple thing to recognize a modern reef or bank while swimming or flying over it, in the fossil most exposures are in vertical section, reef-building metazoans have changed much with geological time and diagenesis has often obliterated the skeletons themselves. The problem of fossil reef nomenclature has pervaded the literature on carbonate rocks for over 80 years (Nelson et al., 1962; Heckel, 1974). The term <u>bioherm</u> is a useful one for lens-like bodies of organic origin that are embedded in rocks of different lithology (Figure 17).

REEF FACIES

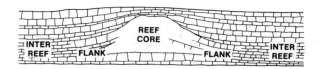

        Figure 17. A sketch of a typical bioherm illustrating
                    the major facies. (From James, 1978,
                    permission of Geological Association of
                    Canada).

Another similar term is <u>buildup</u> (Wilson, 1967) which is defined as a body of locally formed (laterally restricted) carbonate sediment which can be demonstrated to possess topographic relief, again having no reference to internal composition. Dunham (1970) has suggested separating the different types of carbonate buildups in the rock record. The first of these he called <u>stratigraphic reefs</u> or thick, laterally restricted masses of pure or largely pure carbonate rock (Figure 18). Implicit in this concept is that it is objective and refers only to the geometry of the buildup. On the other hand, an <u>ecologic reef</u> is a rigid wave-resistant, topographic structure produced by actively building and sediment binding organisms. Study of many modern reefs, however, indicates that many are not organically bound but rather sites of pervasive early lithification.

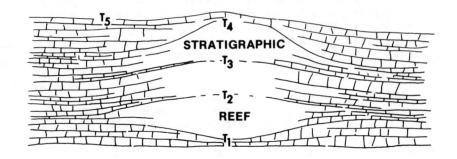

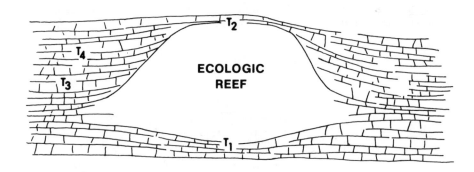

Figure 18. A sketch illustrating the difference between
a stratigraphic reef which is a thick
laterally restricted mass of carbonate rock
often composed of several superimposed
bioherms which individually had little
relief above the surrounding sea floor and
an <u>ecologic reef</u> which is a rigid wave
resistance topographic structure generally
formed during one specific interval of time.

Reef Structure

    Buildups may occur on a carbonate platform in a variety
of guises and in many settings.   In some instances such as modern
Pacific atolls or fossil Devonian structures the whole platform,
often 10's of km in size and dominated by reef growth, is
described as a reef complex.   At the other end of the spectrum

a small but complex community of organisms in Ordovician strata
and properly called a bioherm may be the same size as one coral
head of a Jurassic reef.   Buildups themselves may occur attached
to rocky shorelines, as individual patch reefs (of all shapes
and sizes) as a complex barrier system along the platform margin
or as mounds on the seaward slope.

Regardless of size, location or associated facies the key
to understanding the reef proper is unravelling the complex
series of lithologies that comprise the reef core (James, 1982).

It has long been recognized that there is an ecological
succession in many Palaeozoic reefs (Lowenstam, 1959), i.e. the
replacement of one community of reef-building organisms by another
as the reef grew.   A recent synthesis by Walker and Alberstadt
(1975) of reefs ranging in age from Early Ordovician to Late
Cretaceous suggests that a similar community succession is present
in reefs throughout the Palaeozoic and Mesozoic.   Application of
this concept to Oligocene reefs (Frost, 1977) which are dominated
by scleractinian corals (the reef-builders in today's oceans)
allows us now to equate ancient reef community succession with
observations on modern reef communities with some measure of confidence.   In most cases four separate stages of reef growth can
be recognized, and these stages along with the types of limestone,
relative diversity of organisms and growth form reef-builders in
each, are summarized in Figure 19.

STAGES OF REEF GROWTH

STAGE	TYPE OF LIMESTONE	SPECIES DIVERSITY	SHAPE OF REEF BUILDERS
DOMINATION	bindstone to framestone	low to moderate	Laminate encrusting
DIVERSIFICATION	framestone (bindstone) mudstone to wackestone matrix	high	domal massive lamellar branching encrusting
COLONIZATION	bafflestone to floatstone (bindstone) with a mud stone to wackestone matrix	low	branching lamellar encrusting
STABILIZATION	grainstone to rudstone (packstone to wackestone)	low	skeletal debris

Figure 19.   A sketch of the four divisions of the reef-
            core facies with a tabulation of the most
            common types of limestone, relative species
            diversity and shape of reef-builders found in
            each stage.(From James, 1980, permission of
            Geological Association, Canada).

Pioneer (stabilization) Stage

This first stage is most commonly a series of shoals or other accumulations of skeletal lime sand composed of pelmatozoan or echinoderm debris in the Palaeozoic and Mesozoic, and plates of calcareous green algae in the Cenozoic. The surfaces of these sediment piles are colonized by Algae (calcareous green), plants (sea grasses) and/or animals (pelmatozoans) that send down roots or holdfasts to bind and stabilize the substrate. Once stabilized, scattering branching algae, bryozoans, corals, soft sponges and other metazoans begin to grow between the stabilizers.

Colonization Stage

This unit is relatively thin when compared to the reef structure as a whole, and reflects the initial colonization by reef-building metazoans. The rock is generally characterized by few species, sometimes massive or lamellar forms but more commonly thickets of branching forms, often monospecific. In Cenozoic reefs the one characteristic common to all corals in this stage of reef growth is that they are able to get rid of sediment and clearn their polyps, and so are able to grow in areas of high sedimentation. The branching growth form creates many small subenvironments or niches in which numerous other attached and encrusting organisms can live - forming the first stage of the reef ecosystem. Stromatatactis (cavity filling of laminated fibrous calcite and sediment) is common in rocks representing this stage.

Diversification Stage

This stage usually provides the bulk of the reef mass and is the point at which most pronounced upward-building towards sea level occurs and easily definable, lateral facies develop. The number of major reef-building taxa is usually more than doubled, and the greatest variety in growth habit is encountered. With this increase in form and diversity of framework and binding taxa, comes increased nestling space, i.e. surfaces, cavities, nooks and crannies, leading to an increase in diversity of debris-producing organisms.

Domination (climax) Stage

The change to this stage of reef growth is commonly abrupt. The most common lithology is a limestone dominated by only a few taxa with only one growth habit, generally encrusted to laminated. Most reefs show the effect of surf at this stage, in the form beds of rudstone.

The reason for this ecologic succession is at present a

topic of much debate. Some workers feel that the control is
extrinsic and reflects a progressive replacement of deep-water
communities by shallow water ones as the reef grows to sea level
and into more turbulent water - yet there is often abundant
evidence that the first two stages are developed in shallow water.
Other workers feel that the control is intrinsic and reflects a
natural succession as the organisms gradually alter the sub-
stratum and change the energy flow pathways as the community
develops - yet there is abundant evidence of increasing water
turbulence as the structure grows.

Superimposed Reefs

Reef structures in the rock record are often impressive
because of their size, not only laterally but vertically. Careful examination of stratigraphically thick reefs, however, often
reveals that they are not a single structure, but a series of
superimposed or stacked reefs that grew on top of one another in
more or less the same place. Individual episodes of reef growth
are commonly separated by periods of exposure, reflected in the
rock by intensive diagenesis, calcrete horizons, or shales
(palaeosols). When the ocean floods one of these surfaces that
has been exposed, reef growth begins at the diversification stage
because there is already a hard, often elevated, substrate
present.

Reefs Through Time

The foregoing model is predicated on the assumption that
a full spectrum of reef-building organisms are present, as we
see in tropical oceans today, but such was not the case for much
of the Phanerozoic. The critical element that is often missing
and without which the four stages of development in the reef
core cannot occur, is the presence of skeletal metazoa that
secrete large robust, branching, hemispherical or tabular
skeletons. Without them the reef cannot exist in the zone of
constant turbulence, usually wave induced, because smaller and
more delicate forms would be broken and swept away (unless submarine cementation is very rapid, pervasive and near-surface).
This zone of turbulence is the optimum area for growth and
diversity because sediment is constantly removed, water is clear
and nutrients are constantly swept past the sessil organisms.
Such large skeletal metazoa were, however, present only at certain
times during the Phanerozoic (Figure 20) and each period has its
own specialized group of frame-builders: a) Middle and Upper
Ordovician - bryozoa, stromatoporoids, tabulate corals; b)
Silurian and Devonian - stromatoporoids, tabulate corals; c)
Late Triassic - corals, stromatoporoids; d) Late Jurassic -
corals, stromatoporoids; e) Upper Cretaceous - rudist bivalves;
f) Oligocene, Miocene (?), Plio-Pleistocene - scleractinian corals

(Laporte, 1974).

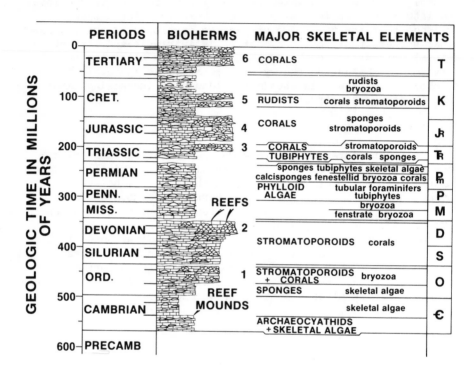

Figure 20. An idealized stratigraphic column representing the Phanerozoic and illustrating times
when there were only reef mounds and times
when there were both reefs and reef mounds.

What then of the rest of the Phanerozoic record - were there
no reefs? While there were certainly periods when no reefs at
all formed, these periods were generally short and represent
either climatic/tectonic crises or the complete lack of any reef
builders, even small ones (e.g. Middle and Upper Cambrian).
During most of the Phanerozoic there were structures that some
workers call reefs, some call mounds, some call banks: they lack
many of the characteristics we ascribe to reefs, yet were clearly
rich in skeletal organisms and had relied above the sea floor.
The origin of these structures, which I have called reef mounds,
has probably caused more discussion than any other topic in the
literature on reefs (Heckel, 1974). When viewed against the
backdrop of the general reef facies model, however, I think of
them as half-reefs or incomplete reefs because they represent
only stages one and two of the model. These structures did not

develop the other upper two stages either because the environment
was not conducive to the growth of large skeletal metazoa or
because there larger metazoa simply did not exist at the time
when the structure formed.

Reef mounds are, as the name suggests, flat lenses to steep
conical piles with slopes up to 40 degrees consisting of poorly
sorted bioclastic lime mud with minor amounts of organic bound-
stone. With this composition they clearly formed quiet water
environments and from the rock record appear to occur in three
preferred locations: 1) arranged just downslope on gently-dipping
platform margins; 2) in deep basins; and 3) spread widely in
tranquil shelf lagoons or wide shelf areas. When viewed in
section, reef mounds display a similar facies sequence in each
case (Wilson, 1975 (Figure 21).

REEF MOUND

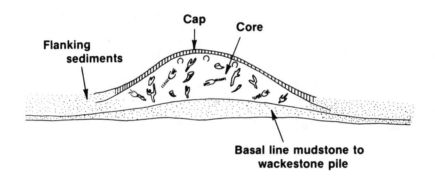

Figure 21. Cross-section through a hypothetical reef
            mound illustrating the geometry of the
            different facies. (From James, 1980,
            permission of Geological Association of
            Canada).

These mounds are usually formed by a basal bioclastic lime
mudstone to wackestone pile and an upper core of lime mudstone
to bafflestone. The core consists of delicate to dendroid forms
with upright growth habits in a lime mudstone matrix. The limestone is frequently brecciated, suggesting partial early lithification, dewatering and slumping, and contains stromatactis.
Each geological age has its own special fauna that forms this
stage: a) Lower Cambrian - archaeocyathids; b) Middle to Lower
Ordovician - sponges and algae; c) Middle Ordovician, Late
Ordovician, Silurian, Early Carboniferous (Mississippian) -

bryozoa; d) Late Carboniferous (Pennsylvanian) and Early Permian
- platy algae; e) Late Triassic - large fasciculate dendroid corals;
f) Jurassic - lithistid sponges; g) Cretaceous - rudist bivalves.
The mounds are usually capped by a thin layer of encrusting or
lamellar forms, occasionally domal or hemispherical forms, or
winnowed lime sands.

The mounds are surrounded by massive, commonly well-bedded
accumulations of archaeocyathids, pelmatozoan, fenestrate bryozoan,
small rudist, dendroid coral, stromatoporoids, branching red algae
or tabular foraminfer debris and chunks of wholly to partly lithified lime mudstone. Volumetrically these flank beds may be
greater than the core itself and almost bury it.

Although the origin of most of these mounds can clearly be
related to some combination of baffling and encrusting by
organisms, localized prolific production of carbonate sediment,
and possible shaping by currents and storms, those found in rocks
of Mississippian age (Tournaisia-Visean) are particularly puzzling.
Commonly called Waulsortian and mounds (from the name of a village
in Belgium) these structures are just as large as most reef
mounds and have sides as steep but possess no major large organisms,
only crinoids and bryozoa as tiny fragments which make up no more
than 20 percent of the rock - the rest is lime mud.

In summary (Figure 20), there are times when the model is
inapplicable because there are no reefs at all, there are times
when only reef mounds form, and there are times when both reef
mounds and reefs occur, but in different environments.

Zoned Platform Margin Reefs

Once a reef has reached the colonization stage, and
especially the diversification stage, the structure is frequently
high enough above the surrounding sea floor to affect water
circulation and thus to alter sedimentation patterns. At this
point not only are the surrounding sedimentary environments
altered by the reef itself develops a zonation of different
organism/sediment associations, because its margins reach from
shallow to deep water.

Modern reefs are best developed and most successful on the
windward sides of shelves, islands and platforms where wind and
swell are consistent and onshore. The asymmetry of many ancient
reefs and distribution of sediment facies suggests that this was
so in the past as well. The reason for the preferential development of reefs on the windward side is by no means established but
sedimentation is likely the most important. Shallow water reef-
building species characteristically produce abundant fine sediment, yet the major reef-builders, because they are filter feeders

and micropredators, are intolerant of the sediment.   The open
ocean, windward locations are the only ones in which fine sediment
is continuously swept away.

Growth of reefs into the zone of onshore waves and swell
forms a natural breakwater and so creates a relatively quiet
environment in the lee of the reef.   Frequently, this restriction
significantly changes water circulation on the shelf, platform or
lagoon behind the reef.   In such a marginal location, the
symmetrical reef facies model comprising a reef-core facies
surrounded on all sides by reef-flank facies is no longer dis-
cernable.   Instead facies are more asymmetrically distributed
with the reef-core facies flanked on the windward side by the
fore-reef facies and on the leeward side by the platform facies
(often called the back-reef facies).

Reef Core Facies

The massive bedded limestones of the reef core commonly
illustrate several different lithologies which develop in one of
the following four zones (Figure 22).

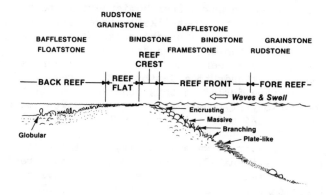

Figure 22.   Cross-section through a hypothetical, zoned
marginal reef illustrating the different reef
zones, spectrum of different limestones produced in each zone and environment of different reef-building forms. (From James,
1980, permission of Geological Association
of Canada).

Reef crest zone: This is the highest part of the reef at any
stage in its growth, and if in shallow water, it is that part of
the reef top that receives most of the wind and wave energy. The
composition of the reef crest depends upon the degree of wind
strength and swell.   In areas where wind and swell are intense
only those organisms that can encrust, generally in sheet-like
forms, are able to survive.   When wave and swell intensity are
only moderate to strong, encrusting forms still dominate but are
commonly also bladed or possess short, stubby branches.   In
localities where wave energy is moderate, hemispherical to
massive forms occur with scattered clumps of branching reef-
builders, although the community is still of low diversity.   The
lithologies formed in these three cases would range from bindstones to framestones.

Reef front zone: This zone extends from the surf zone to an
indeterminate depth, commonly less than 100 metres, where the
zone of abundant skeletal growth grades into sediments of the
fore-reef zone.   Direct analogy between modern reefs, especially
Caribbean reefs, and ancient reefs is difficult because today the
sea floor from the surf zone to a depth of 12 metres or so is
commonly dominated by the robust branching form Acropora palmata
a species which developed only recently, in the late Pleistocene.
Such branching forms are rarely found in ancient reefs and instead
the most abundant forms are massive, laminar to hemispherical
skeletons, forming framestones and sometimes bindstones.

The main part of this zone supports a diverse fauna with
reef-builders ranging in shape from hemispherical to branching
to columnar to dendroid to sheet-like.   Accessory organisms and
various niche dwellers such as brachiopods, bivalves, coralline
algae, crinoids and green segmented calcareous algae (Halimeda),
are common.   On modern reefs where the reef-builders are corals
this zone commonly extends to a depth of 30 metres or so.   The
most common rock type formed in this zone would still be frame-
stone but the variety of growth forms also leads to the formation
of many bindstones and bafflestones as well.

Below 30 metres or so wave intensity is almost non-existent
and light is very attenuated.   The response of many reef-building
metazoans is to increase their surface area, by having only a
small basal attachment and a large, but delicate, plate-like shape.
Rock types from this zone look like bindstones, but binding plays
no role in the formation of these rocks and perhaps another term
is needed.

The deepest zone of growth of coral and green calcareous
algae on modern coral reefs is around 70 metres.   The lower limit
may depend upon many factors, perhaps one of the most important
being sedimentation, especially in shale basins which border many
reefs, so that this lower limit should be used with caution in the

interpretation of fossil reefs.

Sediments on the reef front are of two types: (1) internal
sediments within the reef structure, generally lime mud giving
the rocks a lime mudstone to wackestone matrix; (2) coarse sands
and gravels in channels running seaward between the reefs.   These
latter deposits have rarely been recognised in ancient reefs.

As a result of numerous observations on modern reefs it
appears that most of the sediment generated on the upper part of
the reef front and on the reef crest is transported episodically
by storms up and over the top and accumulates in the lee of the
reef crest.   Sediments on the intermediate and lower regions of
the reef front, however, are transported down to the fore-reef
zone.   Shallow-water material is contributed to the fore-reef
zone only when it is channelled by way of passes through the
reef.

Reef flat zone: The reef flat varies from a pavement of
cemented, large skeletal debris with scattered rubble and
coralline algae nodules in areas of intense waves and swell, to
shoals of well-washed lime sand in areas of moderate wave energy.
Sand shoals may also be present in the lee of the reef pavement.
Vagaries of wave refraction may sweep the sands into cays and
islands.   These obstructions in turn creat small protected
environments very near the reef crest.   Water over this zone is
shallow (only a few metres deep at most) and scattered clumps of
reef-building metazoans are common.   The resulting rock types
range from clean skeletal lime grainstones to rudstones.

Back-reef zone: In the lee of the reef flat conditions are
relatively tranquil and much of the mud formed on the reef front
comes out of suspension.   This, coupled with the prolific growth
of mud and sand-producing bottom fauna such as crinoids,
calcareous green algae, brachiopods, ostracods, etc., commonly
results in mud-rich lithologies.   The two most common growth
habits of reef-builders in these environments are stubby, dendroid
forms, often bushy and knobby, and/or large globular forms that
extend above the substrate to withstand both frequent agitation
and quiet muddy periods.

The rock types characteristic of this environment are bafflestones or floatstones to occasional framestones with a skeletal
wackestone to packstone matrix.   In some reefs there are beds
of nothing but disarticulated branches in lime mud (e.g.
Amphipora limestone of the Upper Devonian), but there is little
evidence of much transport.

Fore-Reef Facies

This facies consists of thin to thick and massively bedded
skeletal lime grainstones to lime packstones which are composed

of whole or fragmented skeletal debris, blocks of reef limestone
and skeletons of reef-builders, and which grade basinward into
shales or lime muds.   In contrast to the reef facies these beds
are rarely dolomitized.

Platform Facies

    The most abundant limestones are thin-bedded, skeletal-rich,
often bioturbated lime wackestones to packstones.   Evaporites
are commonly interbedded with carbonates if the reef has severely
restricted water circulation.

Reefs and Carbonate Platform Geometry

    Reef-rimmed platforms (Figure 23): At times in geological
history, when a complete spectrum of reef-builders is present,
reefs are an integral part of carbonate platforms and characteristically form a rim along the seaward or leading edge of the

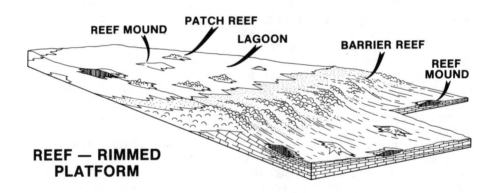

Figure 23.   A sketch illustrating the disposition of
             facies on a carbonate platform at times
             when a complete spectrum of reef-building
             organisms was present.

structure.   The edge of the shelf or platform is occupied by a
marginal reef.   The reef is well-zoned.   The front is steep and
wave action intense, but zonation is weak if the front slopes
gradually seaward and the seas are relatively quiet.   The linear
reef is cut by passes through which platform sediments are
funnelled basinward.

Patch reefs on the platform in the lee of the barrier reef range from circular to elliptical to irregular in plan and are sometimes large enough to enclose a lagoon themselves.

Reef mounds occur on the inner shallow parts of the platform, in areas of normal salinity but turbid water. Reef mounds also occur at depth, in front of the barrier reef down on the reef front or fore reef.

Patch reefs or reef mounds commonly form a widespread lithofacies compared to the barrier reef.   The stratigraphic thickness of these reefs is dependent upon the rate of subsidence; if subsidence rate is high, reefs may be spectacular in their thickness.
Shoal-rimmed platforsm (Figure 24): At times in geological history when large skeletal metazoans are absent, however, the

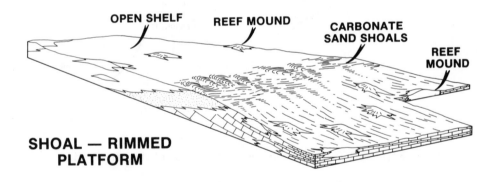

Figure 24. A sketch illustrating the disposition of
            facies on a carbonate platform at times when
            only small delicate skeletal metazoans
            prevailed and most bioherms were reef mounds.

small and delicate organisms cannot live in the turbulent shallow water at the shelf edge.   As a result, the margin of the shelf or platform is normally a complex of oolitic or skeletal (generally crinoidal) sand shoals and islands (Figure 24).   The only reef structures are of reef mounds which occur below the zone of active waves down on the seaward slopes of the shelf or platform and if conditions are relatively tranquil behind the barrier, on the shelf itself.

CARBONATE PLATFORM MARGINS

The marginal zone of carbonate platforms, extending from
the shallow rim down the seaward-dipping slope to the basin,
although spanning a wide range of depths and encompassing a
complex series of depositional environments, is interconnected as
a single depositional system (Figure 25).

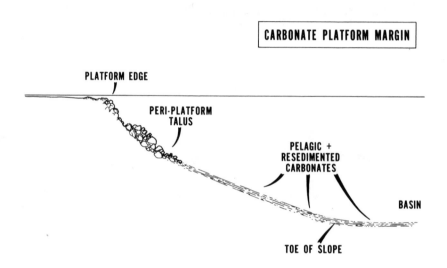

Figure 25.   A sketch of a typical platform margin.

The deposits are characterized by alternations of delicately
laminated mudstones, carbonate sands and gravity-flow conglomerates
or breccias in varying proportions.   These sediments are often
the only remaining clues as to the nature and composition of a
now dolomitized or tectonically obliterated platform margin.
Furthermore, the very presence of this debris is an excellent
indicator of a nearby carbonate platform or reef complex and this
principle has been successfully used to locate reefs in the subsurface.   The lime sands of these deposits, where intercalated
with organic-rich basinal sediments, can be reservoirs.

In the past most of our understanding of platform margin
sediments has come from the rock record, the modern deep sea floor
could not be studied with the same facility as shallow carbonate

environments.   In the last 5 to 10 years, however, the use of
high resolution seismic together with small research submersibles,
especially by W.Schlager and R.N.Ginsburg (University of Miami)
and A.C.Neumann (University of North Carolina), have provided
important insights into the processes and products of platform
margin sedimentation.   This article first outlines the principles
of platform margin sedimentation and then integrates this data,
from the modern and the fossil record, into a sequence of depositional models.   Many of the principles discussed and much of
the material presented has been published previously in an article
written together with Ian McIlreath (McIlreath and James, 1978).

THE SEDIMENTS

    Deposits on the platform margin (Figure 26) originate in
one of three ways: (1) from reef and associated skeletal sands
or from oolitic and peloidal sands formed at the shallow margin
rim; (2) as a pelagic rain of planktic mud and skeletal remains;
and (3) as resedimented shallow-water and slope sediments deposited
from gravity flows in deep water.

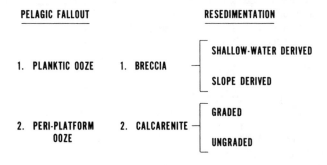

CARBONATE SLOPE SEDIMENTS

PELAGIC FALLOUT		RESEDIMENTATION	
1. PLANKTIC OOZE	1. BRECCIA	SHALLOW-WATER DERIVED	
		SLOPE DERIVED	
2. PERI-PLATFORM OOZE	2. CALCARENITE	GRADED	
		UNGRADED	

        Figure 26.   The different types of carbonate sediments
                    which occur on a carbonate platform margin.

The Shallow Rim

    Since much of the sediment in deep water settings along the
margins of carbonate platforms is derived, the style of sedimentation is largely tied to the nature of the source, the shallow
marginal rim.   This rim is either a series of reefs or lime sand

shoals and in some instances these facies grade laterally into one another.

Carbonate sand bodies: These accumulations have been most studied in the Bahamas (Hine et al., 1981). The nature, geometry and transport direction of these sands appears to reflect their setting, i.e. whether these sand bodies are on the windward or leeward sides of the bank, or stated another way, whether it is a high energy or low energy setting. In high energy situations sediment is generally transported onto the bank and so results in a prograding wedge of sands thinning towards the platform interior. In leeward or low energy situations sands are swept off the bank into deep water, often in large quantities.

Local, less obvious factors can, however, dramatically alter this pattern. If, for example, islands are present just behind the margin then even in high-energy situations, sediment accumulates in front of these islands and is moved seaward to the deep by gravity processes or storm return flows. In other instances, if tidal currents are dominant then the overall effects of energy are masked and as flood generally dominates over ebb, sand movement is predominantly bankward.

Reefs: The nature of reefs at the platform margin has been outlined in the previous section.

Peri-platform Talus

Directly seaward of the shallow water reefs, or lime-sand shoals that form a platform margin, there is commonly a debris apron of limestone blocks skeletons of reef building metazoa, lime sands and muds.   These accumulations are the result of rock-fall and sand-streams from shallow water and are common along the seaward margins of modern reef complexes (James and Ginsburg, 1979; Land and Moore, 1977).   The blocks themselves may be multi-generation in composition because the reefs, and shoals and other deposits at the platform margin are characteristically susceptible to early lithification, either by submarine cementation, or if there are slight fluctuations in sea level, by complex subaerial diagenesis.   In addition, parts of the talus wedge are commonly cemented on the sea floor.   The lithified portions of these lime-stones become hard and brittle, and so far particularly susceptible to fracturing and fragmentation.

Large passes through a reef also act as conduits, funnelling back-reef sediments into this zone so that, along strike, areas of chaotic breccia may alternate with fans of lime sand.   The latter sediment is also commonly cemented, forming numerous hard-grounds.

Examination of sediment dispersal seaward of the platform in areas with low to intermediate slopes (up to 30 degrees) indicates

that this talus does not travel any significant distance away
from the margin by day-to-day processes.

Pelagic Fallout

   Planktic Carbonates: These sediments, deposited in the open
sea are derived from the skeletons of planktonic micro-organisms
which inhabit the overlying water column.   Such deposits include
ooze and chalk and consist primarily of the remains of various
nannofossil groups, especially coccoliths, the tests of planktonic
and sometimes benthic foraminifers.   Macrofossils such as
pteropods, pelecypods, echinoderms and, in older units, ammonites
are present as accessory components.   An excellent summary of
such deposits can be found in Hsu and Jenkyns (1974), Scholle
(1077) and Cook and Enos (1977).

   True pelagic carbonates are apparently not known from the
early Paleozoic (Figure 27) and are first recognized from rocks

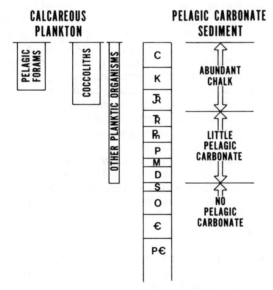

Figure 27.   The geological range of calcified micro-
             organisms which are important in the
             formation of pelagic carbonate sediment.

of Upper Silurian age (Tucker, 1974).   Planktonic foraminifers
and coccoliths appear to have evolved in the Jurassic and during
post-Jurassic time pelagic carbonate has increased to the point
that in the last 100 Ma it comprises about 2/3 of world-wide

carbonate deposition.   It is important, in terms of their future
diagenetic history, to note that both of the major components of
planktic carbonates, planktic foraminifers and coccoliths, are
low magnesium calcite.

Most chalks accumulate at a rate of between one and 30 m/my.
The sedimentary structures and colours depend upon the degree of
circulation and oxygenation.   Dark colours and preserved laminations reflect stagnation;   lighter colours, more burrows and
fewer preserved sedimentary structures reflect stronger bottom
circulation.

The water depth of pelagic carbonate deposition ranges from
less than 100 m to greater than 4500 m.   The limiting factors
for such accumulations are the relative rates of sedimentation of
carbonates versus non-carbonate components, physical erosion and
chemical dissolution.   Chemical dissolution is particularly
important in carbonate slope facies because the environment
passes, with depth through several important increasing pressure
and decreasing temperature boundaries.   Aragonite components,
such as pteropods, may be selectively removed by dissolution in
water as shallow as 500 m (the aragonite compensation depth) while
calcite components are completely dissolved at the carbonate compensation depth, between 4000 and 5000 m in today's oceans.   Much
less is known about the removal or recrystallization of Mg-calcite.
This progressive removal by dissolution results in a residual
sediment composed largely of siliceous skeletons, red hemipelagic
clays and wind-blown silt.   Dissolution also takes place in the
oxygen minimum layer, that zone just below the thermocline in the
modern ocean where, due to the increased metabolism of aerobic
organisms and the lack of oxygen replacement, the oxygen level
is often reduced to less than 0.2 ml/l.   The higher levels of
CO associated with the oxygen deficiency lead to an increase in
the $CaCO$ solubility.   Where this zone impinges on the sea floor
sedimentary carbonate is removed and the resulting sediment is
enriched in organic matter, contains more opaline silica and
where the oxygen values are too low even for burrowing organisms,
the sediment is dark and laminated.

In some areas of the modern ocean the production of siliceous
plankton (silicioflagellates, diatoms and radiolaria) exceeds that
of calcareous nanno- and microplankton.   During the Paleozoic,
when pelagic carbonate was reduced or absent, siliceous sediment
was much more widespread in deep-water areas.
Peri-platform ooze: Sediments that make up the fine-grained
pelagic component of most slope deposits come not only from the
water column but from the adjacent platform as well (Wilson, 1969).
While contributions at any one time from the water column is
more or less constant, that portion derived from the platform is
episodic.   Most often storms stir up the wide, shallow mud-floored

areas of the shelf and the milk-white water streams out across the
shelf margin to settle in deep water.   A less voluminous but more
regular transfer process exists at such near-vertical shelf-to-
deep-oceanic-basic transitions as St. Croix, Virgin Islands where
warm sediment-rich shelf waters "float" over the cooler basinal
waters by tidal exchange.   These fine-grained, shallow-water
derived slope sediments have been called "peri-platform ooze" by
Schlager and James (1978) because they occur as an apron around
the platform and because they are significantly different in
their composition and mineralogy (significant amounts of aragonite
and high magnesium calcite) from the wholly pelagic sediments of
the open sea.

In the Precambrian and Paleozoic most pelagic slope carbonates
may well have been almost wholly peri-platform ooze.

The resultant slope deposits are monotonous, uniform dark
grey, fine-grained lime mudstones, generally thin-bedded with
flat planar contacts and internal micro-lamination.   Mudstone
beds are often separated by partings into very thin beds of
similar mudstone or beds of shale, forming characteristic
"rhythmites" or "ribbon limestones".   The original depositional
textures and fabrics are often modified by sedimentary boudinage,
while differential compaction and/or cementation frequently trans-
forms the evenly-bedded sediments into a nodular limestone.   The
irregular nodules may, in some cases, be so packed together to
form a jig-saw puzzle resembling an in situ breccia.

Resedimented Carbonates

Lime Breccias: These deposits, which have been called debris
flows, submarine mass flows, mass breccia flows, breccia and mega-
breccia beds, rudite sheets, or clistostromes (in the non-tectonic
sense) are certainly the most impressive parts of the slope
sequence.   They originate in two very different areas, high up
on the slope in shallow water or lower down on the slope proper.
Breccias derived from the shallow marginal rim: These
breccias are generally exposed in discontinuous to laterally
extensive sheets, channels with lenticular cross sections or
irregular masses.   They stand out as resistant masses of light-
coloured carbonate against a background of dark-coloured, well-
bedded limestone and shale.   They are characterized by blocks of
all sizes and shapes, but often equi-dimensional and somewhat
rounded.   Some of the blocks are so enormous that they have been
mistaken for bioherms (see Mountjoy et al. 1972).   One exceptional
clast in the Cow Head Group (Cambro-Ordovician) at Lower Head,
Newfoundland is 0.2 km x 50 m in size, with surrounding blocks
often 30 x 15 m in dimension (Kindle and Whittington, 1958).   The
breccias commonly have a matrix of lime mud, lime sand or
argillaceous lime mud.

The deposits are bedded, with a planar to undulating basal contact accentuated by differential compaction and an irregular to hummocky upper contact. The nature of the bedding contacts often cannot be determined accurately because the bedding planes are styolitic, and so any original bedding-plane features are often destroyed. Davies (1977) made the interesting observation that the common occurrence of crinoids, bryozoa and ammonoids at the upper surface of Permo-Pennsylvanian deposits on Ellesmere Island may represent an indigenous fauna inhabiting the "reef-like" upper surface of the deposit.

The polymict nature of the clasts reflects the complexity of the source area; the platform margin consisting of partly lithified reefs and/or lime-sand shoals, down slope (yet still shallow reef mounds, or peri-platform talus. Well-sorted and well-bedded lime sands which can be differentially submarine cemented, individual colonies of reef builders, multigeneration reef rock, limestones with subaerial karst features, tidal flat lithologies and even cemented talus that has been refractured to give breccia clasts within breccia, are all to be expected.

The fabrics of such coarse clastic deposits range from mainly chaotic through imbricated to horizontal and wave-form. They are rarely graded or even reverse graded. They range from clast-supported to most commonly matrix-supported, with the matrix ranging from shale to argillaceous lime mud, lime mudstone and occasionally lime sand. As Hopkins (1977) points out, however, what is often taken to be lime mud in outcrop turns out to be peloid lime sand in thin section, so that sand-sized matrix may be more common than supposed.

The exact mechanism by which these sediments are transported are not yet clear. Submarine debris flows (Hampton, 1972) are sediment gravity flows in which granular solids such as boulders, pebbles and sand are more or less "floated" during transport by the yield strength of the matrix composed of interstitial fluid and fine sediment. Buoyancy of the fluid matrix also contributes to the support. Since not all such deposits have a clay mineral matrix the transport mechanism is thought to be a combination of debris flow and grain flow (Middleton and Hampton, 1973).

Breccias Derived from the Slope: The evenly-bedded calcilutites or lime muds of the slope facies are often prone to downslope creep. Individual beds can be seen to neck or wedge out, or whole intervals will move downslope within a series of slump folds. Dislocation and movement of large masses of slope material downslope leads to the formation of breccias of submarine glide masses composed of numerous tabular clasts of slope lime-stone that have been bent or fractured, that are poorly-sorted and that exhibit random to subparallel orientations, often resembling shallow-water "flat pebble conglomerates". Enormous blocks of

bedded slope sediments, often internally folded, are caught up in
the breccias.

The source of these breccias is thought to be the large
"intraformational truncation surfaces" or "cut-and-fill structures"
(Wilson, 1969) which are sharp concave-up discontinuity surfaces
that truncate underlying beds and are overlain by a downslope
thickening wedge of sediment with an angular relationship on the
truncated beds. In these deposits, reduction of shear stress
occurs by displacement of coherent masses along discrete shear
planes and not usually by deformation within the mass as occurs
in slumps.

The tabular clasts of slope material clearly indicate that
the slope sediments were partly consolidated very early, probably
by submarine cementation. Cementation may have been similar to
that in shall-water with lithified and unlithified layers reflecting times of slow and rapid sedimentation respectively. If
deposition and neomorphism took place below the thermocline,
dissolution of aragonite and possible precipitation of calcite
may have caused the same effect in layers of different original
composition. Alternatively, if the lime mudstone is interlaminated with shale, cementation of the carbonate may have taken
place while the shale remained soft.

Deposits of the two end members, one originating high on
the slope and the other down on the proper slope are sometimes
found intermixed in extensive breccia masses. Such deposits
are similar to those termed marl-flaser breccia by Schlager and
Schlager (1973), characterized by a chaotic fabric of plastically
deformed, dark grey argillaceous lime-mudstone lithoclasts separating irregular lenses of subangular limestone and other lithoclasts,
with the deformed marls forming the flaser fabric. These are
thought to be shallow-water derived breccia flows that incorporated
lime mudstone clasts from the floor of the slope environment as
they moved basinward and they may grade downslope, as do many
other breccias, into turbidites.

Graded Calcarenites

A large proportion of any slope sequence is commonly size-
graded beds of clastic textured limestone, mainly of sand size,
interpreted to be the carbonate equivalent of siliciclastic
turbidites. The deposits have also been called allodapic limestones (Meischner, 1964). Such sediments are well-bedded and
characteristically have sharp planar bases that can be coplanar
with, or locally scour and truncate, underlying slope beds.
Sole marks and load structures are usually absent although in
some cases they may be obliterated because of stylolitization
and solution along bedding contacts. Calcareous turbidites can

exhibit all five of the typical ABCDE divisions of the Bouma
sequence but most commonly it is the A, and sometimes the B and C
divisions that characterize the deposits.   The particles in the
basal parts of division A are often cobble size and larger and the
more common grain types are lithoclasts, skeletal debris and ooids,
the petrology of which indicates a shallow water origin.

The most obvious sources for these units are the unstable
accumulations of lime sand and gravel that build up near the
platform margin and are occasionally set into motion.   It is also
possible that they are the distal parts of carbonate debris flows.
Davies (1977) has suggested a third origin, the indigenous fauna,
especially pelmatozoans, that live on the slopes and produce
abundant skeletal material that may be easily remobilized.

Post-Paleozoic graded calcarenites derived from sediments
further down the slope profile can be virtually indistinguishable
compositionally from pelagic limestone.   These calcarenites are
generally rich in pelagic components such as coccoliths and
foraminifers but may also contain lesser amounts of pteropods,
sponge spicules, radiolarians, and coarser-grained skeletal
debris (especially pelmatozoans).   The sediments are size-sorted
and may be mixed with clastic terrigenous or volcaniclastic sediment if they have travelled great distances.   Although the sedimentary structures such as horizontal laminations, convolutions,
occasional channels, flute and groove casts and trace fossils
may be present, the A and B divisions of the Bouma sequence are
commonly missing and they generally start with C or D divisions.
Non-graded Calcarenites:Massive to cross-bedded and ripple-
marked calcarenites are an enigmatic type of deposit found in
many slope sequences.   These deposits are fine- to coarse-grained
wackestones to grainstones with occasional large clasts or fossils.
Individual beds have sharp bases and vary in geometry from
lenticular to irregular masses.   The fabric may be random or
grains may be aligned parallel to the paleoslope.

The grains in these deposits are variable, ranging in
composition from shallow-water derived particles to pelagic grains.

These deposits may result from one of three depositional
mechanisms, liquified flow, grain flow, or reworking of pre-
existing sediments by bottom current.   Perhaps the massive deposits
having an apparent lack of sedimentary structures are nothing more
than the product of downslope mass movement of well-sorted lime
sands produced at a rapid rate near the platform margin.

Sedimentary structures in the cross-bedded deposits indicate
some sort of bottom currents, often running parallel to the slope
(contour currents).   Well-sorted, rippled lime sands, sometimes
with large scale bed forms, and composed of ooid sand occur in

the deeper parts of the slope around the margins of the Tongue of the Ocean, Bahamas and are also common on the slopes along the western part of the Bahama Banks (Mullins and Neumann, 1977) where currents flow along and parallel to the slope at speeds of 50 cm/sec and more (although such currents are rapid and not characteristic of today's oceans). These currents may rework pre-existing pelagic slope deposits, leaving only the larger foraminifers and pteropods together with lithoclasts of cemented pelagics to form a deep-water grainstone. They may also winnow the upper parts of turbidites, removing the finer layers and leaving a sequence composed only of shallow-water clasts, and divisions A and B of the Bouma sequence, capped by a cross-bedded lime sand.

Such clean, well-sorted sands are commonly sites of submarine cementation and hardground formation. In such areas precipitation of cement may lead to displacive expansion of grain-to-grain distance, resulting in fracturing and the formation of in situ breccias.

THE DEPOSITIONAL SYSTEM

Even though there is a continuous rain of pelagic sediment on the slope, by and large the bulk of slope sediment come from the platform, as resedimented gravity flows or peri-platform ooze. Thus, the nature of slope deposition closely reflects events on the platform, or more specifically whether the platform is close to sea level, drowned or exposed (Figure 28). If the platform surface is close to sea level then there is a constant source of carbonate sediment being produced and so slope as well as adjacent basin deposits are correspondingly abundant. If, however, sea level drops the platform is exposed to meteoric diagenesis and the sediments are turned to limestone, thus shutting off the supply of sediment and leading to a sharp reduction in slope deposition. Starved deep water sedimentation can also reflect rapid sea level rise, to the extent that carbonate production is outpaced and a tranquil deep water shelf with no shallow rim is formed and the deposits are primarily muddy skeletal carbonates. In discussing the topic of sea level fluctuation and slope deposition, Schlager and Ginsburg (1981) make the important point that sedimentation on carbonate slopes is opposite to that in deep water adjacent to a siliclastic shelf, where lower sea level increases erosion and delivery of terrigeneous material to the deep.

Carbonate slope sedimentation also differs from that adjacent to terrigeneous shelves in another important way. Carbonate sand is delivered all along the platform margin and because no dendritic drainage systems that would funnel sediment

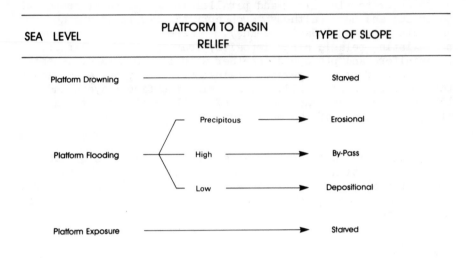

SEA LEVEL	PLATFORM TO BASIN RELIEF	TYPE OF SLOPE
Platform Drowning	⟶	Starved
Platform Flooding	Precipitous ⟶	Erosional
	High ⟶	By-Pass
	Low ⟶	Depositional
Platform Exposure	⟶	Starved

Figure 28. The type of carbonate margin developed in
response to various combinations of sea
level and relief between platform and
basin.

into canyons are developed, then deep sea fans are absent.
Instead the platform, whether reefs or sand shoals, acts as a
line source, creating a continuous belt of overlapping turbidites
and gravity-flow deposits at the toe-of-slope (Schlager and
Chernak, 1979).

If the carbonate platform is near sea level then the style
of slope and adjacent basin sedimentation is dependent upon the
relief between platform and basin. Three types of carbonate
slopes are recognised in this instance, depositional, by-pass
and erosional.

Depositional (Figure 29): These slopes, developed when the
relief between platform and basin is relatively low, are characterized by a gentle incline that decreases gradually to merge
with the flat basin floor. Most deposits originate on the
upper slope or at the shallow rim. Because the slope is so
gentle turbidites and debris sheets are deposited on the lower
slope as well as in the basin.

If the shallow rim is reef dominated, then the full spectrum

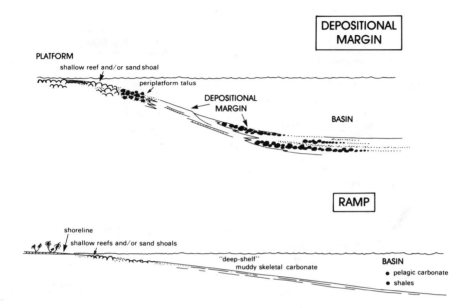

Figure 29. A sketch illustrating the morphology of
and sediments on a depositional margin and
a carbonate ramp.

of allochthonous deposits is to be expected.   On the other hand,
if the shallow platform edge is a series of sand shoals then
turbidites and grain flows are the predominant transport
mechanisms, with conglomerates and breccias rare, although hardgrounds and incipient brecciation may be common.

An important style of depositional margin, which differs
from most of the others described in this section, has been
called a carbonate ramp, by Ahr (1973).   In this case there is
no striking break in slope or rim, instead the facies patterns
are apt to be wide irregular belts with the highest energy zone
relatively close to the shore (Figure 29).   Ginsburg and James
(1974) have called these open shelves and good modern examples
are present on the Yucatan Shelf, West Florida Shelf and in the
Persian Gulf.   On ramp style margin there is little or no
resedimentation.

By-pass (Figure 30): If relief between platform and basin
is appreciable then resedimented carbonates, originating high on
the slope, by-pass most of the slope proper and are deposited at
the toe-of-slope and adjacent basin.   Relief may range from an
actual cliff to a platform that stands high above the surrounding
basin.

This style of margin is seen on a small scale today around

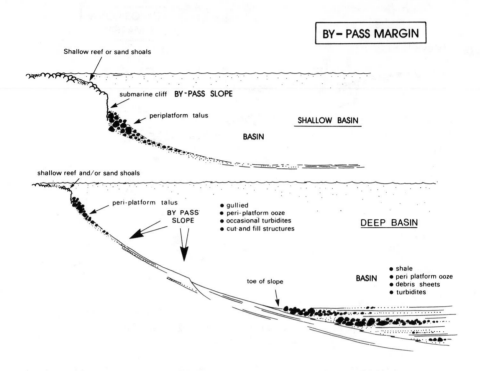

Figure 30. A sketch illustrating the morphology of
           and sediments on by-pass margins fronting
           shallow and deep basins.

some modern carbonate platforms.   Quaternary fluctuations in sea
level have resulted in the formation of an escarpment 50 to 100 m
high, which is crowned by the modern living reef and footed by a
wedge of peri-platform talus (James and Ginsburg, 1979).   Corals,
limestone blocks and skeletal sand from the reef are transported
over the cliff by rock fall and sand streams to accumulate as a
wedge of talus.   This pile of debris becomes progressively more
fine-grained and thinner as it grades into basinal sediment.

    On a larger scale, the wedge of peri-platform talus may be
fronted by a long, basinward-dipping incline.   Flows, which
are triggered on the upper slope by oversteepening of peri-
platform talus, collapse of the reef or by slumping of previously
deposited sediments, are seen to erode gullies on modern slopes
(Schlager, Hooke and James, 1976).   The slope itself is a trap
for peri-platform ooze and the record of deposition is one of
bioturbated mud, cut and fill structures, hardgrounds and

occasional sand and boulder beds.

Erosional: This type of margin, with virtually no cement slope deposition and predominantly erosion of the platform margin is seen today along the margins of some of the Bahama Banks.

Summary: The foregoing article paints with a broad brush the spectrum of different depositional systems to be expected at carbonate platform margins.   Individual cases, besides being controlled by local geographic setting and the interplay between eustasy and tectonics are also controlled to a greater degree by geological history.   Because slope deposition is very much dependent upon the nature of the shallow rim the presence or absence of shallow-water metazoans capable of building reefs is a critical factor in the equation.   Such organisms are only present at specific times in geological history (see previous section on reefs).   In addition the pelagic calcareous zoo-plankton and phytoplankton are insignificant in the early Paleozoic, minor in the middle and late Paleozoic and prolific in Mesozoic and Cenozoic.   Thus fine-grained slope deposits are almost entirely peri-platform ooze in the Precambrian and Early Paleozoic while they are perhaps one-half peri-platform ooze and one-half true pelagic carbonate in the Mesozoic and Cenozoic. Consequently shale is an important component of such settings in geological history but much less abundant in the recent past.

Geometry of the adjacent deep water settings and paleocean-ography also play an important part in the resultant facies patterns.   If the platform fronts an open ocean or basin then sedimentation falls in to the previously discussed patterns.   If, however, the platform or platforms are adjacent to narrow seaways or troughs with strong currents then the slope and basin sediments may be reworked by axial currents and early lithification to form hardgrounds, synsedimentary clasts or even "lithoherms" (Neumann et al. 1977) may be pervasive.

Finally, because a carbonate platform is largely generated in place it is a continually developing system.   With time the nature of the margin will change, depending largely upon the source of the sediments and relative sea level.   It may well evolve as the platform grows from depositional to by-pass to possibly erosional in nature.

ACKNOWLEDGEMENTS

A portion of this paper was published originally in Geoscience Canada and reprinted in Geoscience Canada Reprint Series No. 1 Facies Models, edited by R.G.Walker.   I am grateful to the Geological Association of Canada for permission to reproduce portions of the text and some of the diagrams here.   Much of the material in this article is directly or indirectly the

result of ongoing research funded by the National Science and
Engineering Council of Canada.

REFERENCES

AHR, W.M. 1973. The carbonate ramp: an alternative to the shelf
model. Trans.Gulf Coast Ass. geol.Soc., 23, 221-225.

AITKEN, J.D. 1967. Classification and environmental significance
of cryptalgal limestones and dolomites, with illustrations from
the Cambrian and Ordovician of southwestern Alberta. J. sedim.
Petrol., 37, 1163-1178.

BATHURST, R.G.C. 1975. Carbonate sediments and their diagenesis.
Elsevier, Amsterdam, 2nd edition, 658p.

COOK, H.E., and ENOS, P. Eds. 1977. Deep-water carbonate
environments: Soc. econ.Paleont.Miner., (Spec.Publ.) 25, 336p.

DAVIES, G.R. 1977. Turbidities, debris sheets and truncation
structures in upper Paleozoic deep water carbonates of the
Sverdrup Basin, Arctic Archipelago. In: Deep-water carbonate
environments.(Ed. by H.E.Cook and P.Enos).Soc.econ. Paleont.
Miner. Spec.Publ.), 25, 221-249.

DUNHAM, R.J. 1962. Classification of carbonate rocks according
to depositional texture: In:Classification of carbonate rocks
(Ed. by W.E.Ham), pp.108-121, Mem.Am.Ass.Petrol.Geol., 1, Tulsa.

DUNHAM, R.J. 1969. Vadose pisolite in the Capitan Reef (Permian),
New Mexico and Texas. In: Depositional environments in carbonate
rocks: a symposium. (Ed. by G.M.Friedman). Spec.Publ.Soc.econ.
Paleont.Miner., 14, 182-191.

DUNHAM, R.J. 1970. Stratigraphic reefs versus ecologic reefs.
Bull.Am.Ass.Petrol.Geol., 54, 1931-1932.

EMBRY, A.F., and KLOVAN, J.E. 1971. A Late Devonian reef tract on
northeastern Banks Island, N.W.T. Bull.Can.Petrol.Geol., 19,
730-781.

FISCHER, A.G. 1964. The Lofer Cyclothems of the Alpine Triassic.
In: Symposium on cyclic sedimentation. (Ed. by D.F.Merriam),
107-149, Bull.geol.Surv., Kansas, 169.

FOLK, R.L. 1962. Spectral subdivision of limestone types. In:
Classification of carbonate rocks. (Ed. by W.E.Ham), pp. 62-84,
Mem.Am.Ass.Petrol.Geol., 1, Tulsa.

FROST, S.H. 1977. Ecologic controls of Caribbean and Mediterranean
Oligocene reef coral communities: In: Proceedings of Third
International Coral Reef Symposium. (Ed. by D.L.Taylor), Miami,
Florida, 367-375.

FROST, S.H., WISS, M.P. and SAUNDERS, J.B. 1977. Reefs and
related carbonates - ecology and sedimentology. Am.Ass.Petrol.
Geol., In: Studies in Geology, 4, 421 p.

GINSBURG, R.N. (Ed.) 1975. Tidal Deposits: a casebook of
Recent examples and fossil counterparts. Springer-Verlag,
Berlin, 428 p.

GINSBURG, R.N., and JAMES, N.P. 1974. Spectrum of Holocene reef-
building communities in the western Atlantic. In: Principles
of Benthic Community Analysis (notes for a short course). (Ed.
by A.M.Ziegler et al.). Univ. of Miami, Fisher Island Station,
7, 1-7.22.
GINSBURG, R.N., and JAMES, N.P. 1974. Holocene carbonate sediments
of continental shelves. In: The Geology of Continental Margins.
(Ed.by C.A.Burke and C.L.Drake), Springer-Verlag, New York,
137-155.
HAM, W.E. (Ed.) 1962. Classification of carbonate rocks.
Mem.No.1, 1279p, Am.Ass.Petrol.Geol.,
HAMPTON, M.A. 1972. The role of subaqueous debris flow in
generating turbidity currents. J.sedim.Petrol., 42, 775-793.
HAVARD, C., and OLDERSHAW, A. 1976. Early diagenesis in back-
reef sedimentary cycles. Snipe Lake, reef complex, Alberta.
Bull.Can.Petrol.Geol., 24, 27-70.
HECKEL, P.H. 1074. Carbonate buildups in the geologic record:
a review. In: Reefs in Time and Space. (Ed.by L.F.Laporte).
Spec.Publ., Soc.econ.Paleont.Miner., 18, 90-155.
HINE, A.C., WILBER, R.J. and NEUMANN, A.C. 1981. Carbonate sand
bodies along contrasting shallow bank margins facing open sea-
ways in Northern Bahamas. Bull.Am.Ass.Petrol.Geol., 65, 261-291.
HOFFMAN, P.1976. Stromatolite morphogenesis in Shark Bay, Western
Australia. In: Stromatolites. (Ed.by M.R.Walter), Elsevier,
Amsterdam, 261-273.
HOFFMAN, P. 1976. Environmental diversity of middle Precambrian
stromatolites. In: Stromatolites. (Ed. by M.R.Walter), Elsevier,
Amsterdam, 599-613.
HOPKINS, J.C. 1977. Production of fore-shore breccia by
differential submarine cementation and downslope displacement of
carbonate sands, Miette and Ancient Wall buildups, Devonian,
Canada. In: Deep-water carbonate environments. (Ed. by H.C.
Cook and P.Enos). Spec.Publ., Soc.Econ.Paleont.Miner., 25,
155-170.
HOROWITZ, A.S., and POTTER, P. 1971. Introductory Petrography
of Fossils. Springer-Verlag, New York, 302p.
HSU, K.J., and JENKYNS, H.C. (Eds.), 1974. Pelagic sediments: on
land and under the sea. Spec.Publ., int.Ass.Sediment., 1, 448p.
JAMES, N.P. 1977a. Facies Models, 7 - Introduction to Carbonate
Facies Models. Geoscience Canada, 4, 123-126.
JAMES, N.P. 1977b. Facies Models, 8 - Shallowing upward sequences
in Carbonates. Geoscience Canada, 4, 126-137.
JAMES, N.P. 1978. Facies Models, 10 - Reefs. Geoscience Canada,
5, 12-26.
JAMES, N.P. 1982. Reefs. In: Carbonate Depositional Environments.
(Ed. by Scholle, P., Moore, C and Bebout, D.). Mem.Am.Ass.Petrol.
Geol., (in press).
JAMES, N.P. and GINSBURG, R.N. 1979. The seaward margin of
Belize Barrier and Atoll Reefs. Spec. Publ., int. Ass.Sediment.,
3, 206p.

KINDLE, C.H., and WHITTINGTON, H.B. 1958. Stratigraphy of the Cow Head Region, Western Newfoundland. Bull. geol.Soc.Am., 69, 315-342.

LAND, L.S., and MOORE, C.H. 1977. Deep forereef and upper island slope, North Jamaica. In: Reefs and related carbonates: ecology and sedimentology. (Ed. by S.H.Frost, M.P.Weiss and J.B. Saunders. Am.Ass.Petrol.Geol., Studies in Geology, 4, 53-67.

LAPORTE, L. 1967. Carbonate deposition near mean sea-level and resultant facies mosaic: Manilius Formation (Lower Devonian) of New York State. Bull.Am.Ass.Petrol.Geol., 51, 73-101.

LAPORTE, L.F. (Ed.), 1974. Reefs in Time and Space. Spec.Publ., Soc.econ.Paleont.Miner., 18, 256p.

LOWENSTAM, H.A. 1959. Niagaran reefs in the Great Lakes area. J.Geol., 58, 430-487.

LOWENSTAM, H.A. 1963. Biologic problems relating to the composition and diagenesis of sediments. In: The Earth Sciences - Problems and Progress in Current Research. (Ed. by T.W. Donnelly). Univ.Chicago Press, 137-195.

McILREATH, I.A., and JAMES, N.P. 1978. Facies Models, 13 - Carbonate Slopes. Geoscience Canada, 5, 189-199.

MAJEWSKE, O.P. 1969. Recognition of invertebrate fossil fragments in rocks and thin sections. Brill, Leiden, 101p.

MEISCHNER, K.D. 1964. Allodapische Kalke, Turbidite in riff-nahen Sedimentations-Becken. In: Turbidites. (Ed.by A.H.Bouma and A.Brouwer), Elsevier, Amsterdam, 156-191.

MERRIAM, D.E. (Ed.), 1964. Symposium on Cyclic Sedimentation. Bull.geol.Surv., Kansas, 169, 636p.

MOUNTJOY, E.W., COOK, H.E. PRAY, L.C. and McDaniel, P.N. 1972. Allochthonous carbonate debris flows - worldwide indicators of reef complexes, banks or shelf margins. Proc.24th Internatl. Geol.Congress, Sect.6, 172-189.

MIDDLETON, G.V. and HAMPTON, M.A. 1973. Sediment gravity flows; mechanics of flow and deposition. In: Turbidites and deep-water sedimentation. Soc.econ.Paleont.Miner., Pacific Section, Short Course, Anaheim, 1-38.

MULLINS, H., and NEUMANN, A.C. 1979. Carbonate slopes along open seas and seaways in the Northern Bahamas. In: Geology of Continental Slopes. (Ed. by O.Pilkey and R.Doyle), Spec.Publ., Soc.econ. Paleont.Miner., 27, 165-193.

NELSON, H.F., BROWN, C.W. and BRINEMAN, J.H. 1962. Skeletal limestone classification. In: Classification of carbonate rocks: a symposium. Mem.Am.Ass.Petrol.Geol., 1, 224-253.

NEUMANN, A.C., and LAND, L.S. 1975. Lime mud deposition and calcareous algae in the Bigh of Abaco, Bahamas: a budget. J.sedim.Petrol., 45, 763-786.

NEUMANN, A.C., KOFOED, J.W. and KELLER, G.H. 1977. Lithoherms in the Straits of Florida. Geology, 5, 4-10.

PLAYFORD, P.E., and COCKBAIN, A.E. 1969. Algal stromatolites: deepwater forms in the Devonian of Western Australia, Science, 165, 1008-1010.

PURSER, B.H. (Ed.), 1973. The Persian Gulf. Springer-Verlag,
Berlin, 471p.
READ, J.F. 1973. Carbonate cycles, Pillara Formation (Devonian)
Channing Basin, Western Australia. Bull.Can.Petrol.Geol., 21,
38-51.
ROEHL, P.O. 1967. Stony Mountain (Ordovician) and interlake
Silurian Facies analogs of recent low-energy marine and
subaerial carbonates, Bahamas. Bull.Amer.Assoc.Petrol.Geol.,
51, 1979-2032.
SCHLAGER, W., HOOKE, R, and JAMES, N.P. 1976. Episodic erosion
and deposition in the Tongue of the Ocean, Bahamas. Bull.geol.
Soc.Am., 87, 1115-1118.
SCHLAGER, W., and CHERNAK, A. 1979. Sediment facies of platform-
basin transition, Tongue of the Ocean, Bahamas. In: Geology of
Continental Slopes. (Ed. by O. Pilley and R. Doyle), Spec.
Publ., Soc.econ.Paleont.Miner., 27, 193-208.
SCHLAGER, W., and GINSBURG, R.N. 1981. Bahama carbonate platforms
- the deep and the past. Marine Geology (in press).
SCHOLLE, P.A. 1977. Deposition, diagenesis and hydrocarbon
potential of "deeper-water" limestones. Am.Ass.Petrol.Geol.,
Continuing Education Course Notes. Ser. 7, 25p.
SCHOLLE, P.A. 1978. A color illustrated guide to carbonate rock
constituents, textures, cements and particles. Mem.Am.Ass.
Petrol.Geol., 27, 241p.
SCHLAGER, W., and JAMES, N.P. 1978. Low-magnesian calcite
limestones forming at the deep-sea floor, Tongue of the Ocean,
Bahamas. Sedimentology, 25, 675-702.
SCHLAGER, W., and SCHLAGER, M. 1973. Clastic sediments associated
with radiolarites (Tauglboden-Schichten, Upper Jurassic, eastern
Alps). Sedimentology, 20, 65-89.
SHINN, E.A., LLOYD, R.M. and GINSBURG, R.N. 1969. Anatomy of a
modern carbonate tidal flat, Andros Island, Bahamas. J.sedim.
Petrol., 39, 1202-1228.
STOCKMAN, K.W., GINSBURG, R.N. and SHINN, E.A. 1967. The
production of lime mud by algae in south Florida. J.sedim.
Petrol., 37, 633-648.
STODDART, D.R. 1969. Ecology and morphology of recent coral
reefs. Biol.Rev., 44, 433-498.
TUCKER, M.E. 1974. Sedimentology of Palaeozoic pelagic limestones: the Devonian Griotte (Southern France) and Cephalopod-
enkalk (Germany). In: Pelagic Sediments: on land and under the
sea. (Ed. by J.Hsu and H.C. Jenkyns), Spec. Publ. int.Ass.
Sediment. 1, 71-92.
WALKER, K.R., and ALBERSTADT, L.P. 1975. Ecological succession
as an aspect of structure in fossil communities. Paleobiol.,
1, 238-257.
WILKINSON, B.H. 1979. Biomineralization, paleoceanography and
the evolution of calcareous marine organisms. Geology, 7,
524-527.

WILSON, J.L. 1969. Microfacies and sedimentary structures in
"deeper-water" lime mudstones. <u>In</u>: Depositional environments
in carbonate rocks. (Ed. by G.M. Friedman). Spec.Publ.Soc.
econ.Paleont.Miner., <u>14</u>, 4-19.
Wilson, J.L. 1975. Carbonate Facies in Geologic History.
Springer-Verlag, Berlin, 471p.

EARLY DIAGENESIS OF CARBONATE SEDIMENTS

R.G.C. Bathurst
Department of Geology,
University of Liverpool,
P.O. Box 147, Liverpool L69 3BX, U.K.

INTRODUCTION

The realm of early diagenesis will be regarded here as
synonymous with eogenetic environment of Choquette and Pray (1970).
In this the various diagenetic processes act during that stage
between accumulation of the sediment on the Earth's surface and
its burial below the depth of significant influence by processes
related to that surface.   As most limestones are marine in
origin, attention will be concentrated on this group of marine
carbonate sediments.

The path followed in the diagenetic evolution of such
sediments depends first of all on their compositions and the
environments of deposition.   A sand composed of Mg-calcite (high-
magnesian calcite) ooids on an active oolite bank will experience
chemical changes that differ from those undergone by an aragonite
needle-mud in an evaporitic sabkha.   Matters of crucial concern
are mineralogy, water chemistry, the flow of pore-water and the
porosity and permeability.   Shallow marine sediments are
dominated by tests of aragonite (molluscs, green algae, corals)
and Mg-calcite (echinoderms, forams, trilobites).   Brachiopod
shells are calcite (low-magnesian calcite) and so are coccoliths,
as in deep-sea oozes and in the Cretaceous chalk of shallower
origin (see Chapter 6).   During diagenesis sediments may
encounter any of four main types of pore-water, namely: sea-water,
meteoric fresh-water, water squeezed from compacting shales and
other formation waters with diverse histories that move upward
and outward through the compacting sediments of a subsiding basin.
The movements of these waters will be controlled by the pattern of

349

A. Parker and B. W. Sellwood (eds.), Sediment Diagenesis, 349–377.
© *1983 by D. Reidel Publishing Company.*

permeabilities.   Precipitation and dissolution will commonly be
closely related and will influence, and be influenced by, the
porosity.   All changes are likely to be affected by rate and
depth of burial, by the relief of load caused by erosion, and by
related changes in pressure and temperature.

DIAGENESIS ON THE HOLOCENE SEA FLOOR

In modern (Holocene) carbonate sands, in such regions as the
Great Bahama Bank, the Trucial Coast embayment, the Yucatan shelf,
in tropical or subtropical seas, grains are commonly encrusted with
tiny crystals of $CaCO_3$.   These are of two mineralogies, aragonite
and Mg-calcite.   The aragonite has generally the form of needles,
a few microns or tens of microns long (Fig. 1).   The Mg-calcite
occurs mostly as either blades a few tens of microns long or as
rhombs about 4 μm across (Fig. 2).   As a result of coprecipita-
tion from sea-water containing Sr^{2+} and Mg^{2+}, the orthorhombic
aragonite lattice carries about 8000 p.p.m. of Sr^{2+} and the tri-
gonal calcite lattice holds a range of 12-19 mole % of $MgCO_3$, the
value increasing with temperature and, commonly, with rate of
crystal growth.   The Sr^{2+} fits uneasily into the calcite lattice
and Mg^{2+} goes with difficulty into the aragonite lattice.
Calcites bearing Mg^{2+} fall characteristically into two groups,
low-magnesian and high-magnesian, having less than 5 mole % or
11-19 mole % $MgCO_2$.   For brevity these two groups of solid phases
are commonly written as calcite and Mg-calcite.

Although these encrustations (more usually called cement)
have rarely led to wholesale lithification of Holocene sediments,
by intergranular cementation, they seem to be widespread and normal
inside the body chambers of shells, forams and other tests, or in
vacated boreholes or any other sheltered site where agitation and
grain-grain collision cannot destroy the tiny crystals.   Fine
sediment trapped in cavities is rapidly lithified.   Where intergranular cementation of carbonate sand does go on to make rock
layers, it probably always acts, to begin with, some centimetres
under the sea-floor.   On the other hand, cementation of grains
into clusters, called grapestones by Illing (1954), is typically
active at the surface, in shallow lagoons and on peritidal flats
where grains are held in contact by algal mats.   Faecal pellets
of mud-feeding molluscs and crabs may lithify rapidly.   The
permanent textural imprint of algal (Cyanobacteria) stromatolites
owes its preservation to early cementation.   In both grapestone
and algal stromatolites, precipitation is assisted by abstraction
of CO_2 from the water during photosynthesis of blue-green algae.

A micro-environment of cementation that is of particular
significance for later diagenetic history is the blue-green algal
bore.   Studies by Bathurst (1966), Alexandersson (1972) and

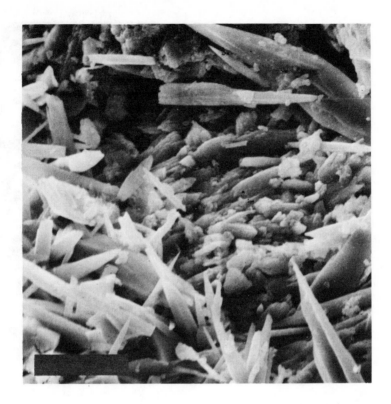

Figure 1. Marine aragonite cement. Holocence.
           Scanning electron microscoope. Bar = 10 μ m.
           Courtesy T. Alexandersson.

Figure 2.   Marine high-magnesian calcite cement.
            Holocene. Scanning electron microscope.
            Bar = 5 µm. Courtesy T. Alexandersson.

Margolis and Rex (1971) have demonstrated that, where blue-green algae bore into carbonate substrates in the photic zone (holes about 6-10 μm diameter), these bore-holes, when vacant following death and decay of the borer, are the sites for precipitation of a fine cement of aragonite or Mg-calcite. The cement crystals are so small that, with the light microscope, the cement looks like micrite. Prolonged boring, dying and cement-filling leads to centripetal replacement of grains by a material resembling micrite: thus micritization. Incomplete replacement leaves a partial alteration in the form of a micrite envelope, the mechanical stability of which in meteoric diagenesis has great importance. In addition to these micrite envelopes formed by replacement, others can be developed by addition of a layer of micritic cement that grows on algal filaments that protrude outward from the grain surface (Kobluk & Risk, 1977).

Precipitation of cement can only proceed in pore-water that is supersaturated with respect to $CaCO_3$. Consequently, this process is limited to the shallow, near-surface, waters of tropical or sub-tropical seas. In the cold waters of the sea floor of the Skaggerak, for example, Alexandersson (1975) has shown that grains are dissolved. Such attack by undersaturated water is probably the norm also on other temperate sea floors, such as the wide shelf of NW Europe. For cementation to occur, some turbulence is essential because some tens to hundreds of thousands of pore-volumes of sea-water must pass through a pore if it is to be filled with cement. This overiding need for a bulk transport system is a recurring theme in the discussion of subsurface diagenesis. Yet too much agitation of the grains will destroy the fragile early crystals. Thus, cements are more abundant in the shelter of intragranular pores, in algal mats or in cavities in reefs. Evidence so far available indicates that the most active cementation has occured where the water was turbulent but the sediment stable, for example where Ginsburg and Schroeder (1973) recorded it in the cup-reefs of Bermuda, or Macintyre (1977) found it on the fore-reef slopes at Galatea near Panama, or James and others (1976) detected it off Belize Cementation was found to be more prolific on the Great Bahama Bank toward the margins or where water is turbulent but the sand is stabilized by algal mats, as over much of the Bahama Bank (Dravis 1979) or again as seen by Harris (1978) on Joulters Cay oolite banks.

It is not clear why the cement in one place is aragonite while elsewhere it is Mg-calcite, nor indeed why, in the same place, the mineralogy differs between adjacent cement generations. On themodynamic grounds, a low-magnesian calcite is the most stable phase of $CaCO_3$ in aqueous solutions at surface temperatures and pressures. In sea-water, dolomite is the thermodynamically expected precipitate, though in fact it does not form in

unmodified sea-water.   It has long been known that, in artificial
preparations of sea-water or in solutions with the characteristic
$Mg^{2+}/Ca^{2+} = 5$ ratio of sea-water, the laboratory precipitation of
$CaCO_3$ gives only aragonite.   It was suggested by Lippman (1960)
that the strongly hydrated Mg^{2+} ion is adsorbed on the calcite
surface, thus inhibiting crystal growth.   Thus the work that
must be done to dehydrate a mixed population of Mg^{2+} and Ca^{2+}
cations, adsorbed on a calcite crystal, exceeds that needed for a
population of purely Ca^{2+} cations.   The difference in change
of Gibbs free energy is such that the growth of aragonite is
preferred, as discussed by Lahann (1978) and Moller and Kubanek
(1976).

Nevertheless, it is now apparent that the growth of calcite
is not prevented by the presence of the Mg^{2+} cation in real sea-
water.   This disparity between the results in artificial sea-
water and natural conditions may well arise from the presence in
sea-water of calcite seed-crystals which were absent in the
earlier experiments.   Berner (1975) has used calcite seeds and
aragonite seeds, in artificial sea-water with and without
dissolved Mg^{2+}.   Calcite growth is slowed kinetically in the
presence of Mg^{2+} but not stopped.

A further kinetic effect is probably involved in the growth
of anomalously high-magnesian calcite cement in sea-water where
the thermodynamically expected precipitate is a low-magnesian
calcite, with between 2-7 mole % $MgCO_3$ as Berner has proposed
(1975).   Recorded Mg-calcite cements have 12-19 mole % MgCO .
Berner (1975) and De Groot and Duyvis (1966) showed that the
Mg^{2+} cations are adsorbed in large quantities on calcite surfaces
in preference to aragonite surfaces, and in greater quantities
than are required for the growth of stable low-magnesian calcite.
Berner has suggested that the excess of adsorbed Mg^{2+} cations
actually become absorbed into the growing calcite lattice,
yielding a metastable high-magnesian calcite which is in "non-
exchange equilibrium" with sea-water.   Berner also follows
Weyl (1967) in claiming that the rate-inhibition of calcite growth
by Mg^{2+} is mainly due, not to the slowness of the dehydration
step, but to the increase of solubility that accompanies enhanced
content of Mg^{2+} in the precipitate.

However that may be, the factors that control cement
mineralogy are far from understood.   Crystallographic matching
of cement with substrate plays some part but recorded sequences
of cement layers in pores show great variation even without the
confines of adjacent chambers in the same test of a foram.   The
tiny size of the spaces, in which cement grows, may be important,
because decay of organic matter in a system not fully open
could, for example, bring about significant changes in pH through
uptake of CO_2 or release of ammonia.   Other possible influences

are breakdown of methane derived from older sediments or the
presence of live or dead tissue.   Adsorbed ions or molecules,
both organic and inorganic, may inhibit selectively the growth of
certain faces or polymorphs.   The variation of crystal morphology
in modern marine aragonite has been examined by Loreau and
Purser (1973) and Loreau (1980).   Aragonite precipitates consist
of three main types: the familiar euhedral needle (Fig. 1) that
forms much cement, the baguette (aiguille emousée, batonnette)
which is well known in Bahamian and Persian Gulf ooids (Fig. 3),
and the nanograin (0.2 - 0.3 μm, Fig. 4) that occurs, for
example, in pelagosite.   The proportions of these crystal forms
in aragonite precipitates are controlled mainly by rate of water-
flow and content of organic matter in the water.

CONCRETIONS AND HARDGROUNDS: HOLOCENE AND EARLIER

The cementation of loose particles into aggregates, to give
concretions, which may coalesce to make beds, has been recorded
from only a few places in Holocene sediments, but is much better
known from older, pre-Pleistocene limestones.   There seem to be
two main overlapping processes that lead to the early growth of
concretions or beds, within a metre or so subsurface: circulatory
exchange with sea water and internal redistribution.

The exchange process depends on abstraction of cations and
bicarbonate from the overlying sea-water.   The process probably
always acts within about the first 50 cm below the sediment
surface, at a critical level where circulation of water allows a
sufficient supply of new ions, by the pumping action of tides and
waves, yet is not so violent as to disrupt the delicate early
growth of intergranular cement.   This cementation is normally a
slow process, commonly requiring hundreds of thousands of years to
make a well-lithified layer.   Beach rocks are exceptional, undergoing lithification in tens of years.   Such fast rates of
cementation may depend on high levels of supersaturation for
$CaCO_3$ accompanying mixing of sea-water with fresh-water
(Schmalz, 1971) or perhaps on degassing of CO_2 (Hanor, 1978).

The redistributory process involves variable degrees of
pore-water migration and diffusion, aided at first by bacterial
fermentation of organic matter and later, perhaps, by abiotic
reactions such as decarboxylation, as revealed by the studies of
Raiswell (1971, 1976), Curtis (1977), Hudston (1978) and Irwin
(1980).   Carbonate may be transferred over long distances, as
compaction waters migrate along paths determined by the various
permeabilities of different horizons.   Locally, over distances
perhaps measurable in centimetres, diffusion plays an important
role, especially during later stages when permeabilities are much
reduced.

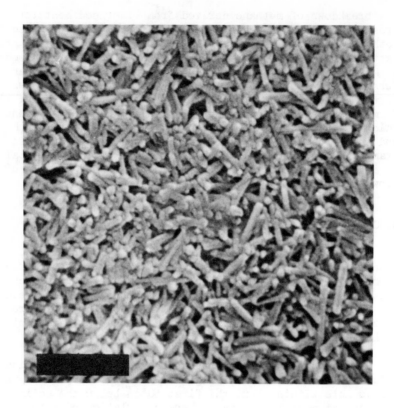

Figure 3.   Marine aragonite baguettes in surface of
            ooid. Holocene. Scanning electron micro-
            scope. Bar = 2 μm. Courtesy J.-P.Loreau.

Figure 4. Marine aragonite nanongrains. Holocene.
Scanning electron microscope. Bar = 2 μm.
Courtesy J.-P. Loreau.

If cemented crusts are later exposed at the sea floor, following erosion of cover, they become hardgrounds. Evidence of such exposure is the presence of bores of bivalves, sponges, algae, etc. and encrustation by oysters, bryozoans and other sedentary organisms. Surfaces may be coated or impregnated with Fe^{2+} or Mn^{2+} salts. Holocene crusts of this kind were described by Shinn (1969) and Taylor and Illing (1969) from near the Qatar Peninsula in the Persian Gulf, by Read (1974) in the Shark Bay region, W. Australia and by Harris (1978) in the oolite of Joulters Cay, Bahamas. Holocene crusts typically have sharp upper surfaces but the base of the crust has an uneven gradational contact with the underlying loose sediment. The Qatar hardgrounds exhibit, in addition, a pattern of expansion ridges of a roughly hexagonal pattern, along which the crusts have been bent upward, overthrust and brecciated. Such deformation leads to a structure which is recognised in older limestones, in section, as tepee (Assereto and Kendall, 1971; Sturani, 1971; Burri et al., 1973; Smith, 1974; Assereto and Folk, 1980). This process is common in Recent halite-cemented supratidal sands (in sabkhas) where the ridge hexagons are about a metre across and generally in peritidal regions. The Qatar ones are about 40 m across. Ordovician deformed crusts were described from Oland in the Baltic by Lindstrom (1963). Some of the most detailed studies of hardgrounds are those by Purser (1969) on cemented Jurassic beachrocks in Burgundy. Some of these were of vadose origin, having the meniscus and pendant cements described by Dunham (1971) and Purser (1969).

On the present deep ocean floor there are crusts, recognisable from submersibles or in dredged samples or in cores (Bathurst, 1975; Milliman, 1966; Schlager and James, 1978). Some of these seem to have evolved by replacement of the original coccolith ooze (low-magnesian calcite) by a microcrystalline crust made of high-magnesian calcite. This process is not understood: the bottom waters are undersaturated for $CaCO_3$ so that reactions may depend on biochemical action in the pore-water of the ooze.

Of particular interest is the observation by Schlager and James (1978), from the submersible Alvin, of lithification of carbonate muds at 1000 - 1800 m in Tongue of the Ocean, Bahamas. These lithified muds form vertical cliffs, but the induration is superficial and overlies soft unaltered mud which is a mixture of aragonite, Mg-calcite and calcite of Pleistocene to Holocene age. The restriction of lithification to cliffs shows that it depended on prolonged exposure of the mud surface to sea-water for over 400,000 years. The original sediment is, in this process, replaced by a low-magnesian calcite (crystals 2-4 μm) with 3.0-5.0 mole % $MgCO_3$, within the range of thermodynamically stable calcite in sea-water.

The cementation of Chalk hardgrounds in the Cretaceous of Europe, by coalescence of nodules, has been discussed by Kennedy and Garrison (1975) and Bromley (1979). Workers have recorded concretions coated with glauconite or bored which must, therefore, have been exhumed and exposed on the sea floor. Raiswell (1971) in his geochemical study of concretions in shales, assumed that the volume of calcite in any part of the concretion can be equated with the porosity at the time of concretionary growth. The high figures obtained indicate that the concretions formed within a metre or so of the sediment surface. It is likely, therefore, that the initiation of concretionary growth is connected with chemical reactions which are part of an early phase of bacterial decay of organic matter trapped in the sediment. Two extremes of hardground evolution thus seem possible, where the $CaCO_3$ is derived by circulation from the overlying sea-water or by diffusional redistribution in a carbonate-rich mud; doubtless the two processes overlap.

The time interval required for growth of hardgrounds before sedimentation continues has been estimated in hundreds of thousands of years.

Not all ancient concretions or hardgrounds are calcite. Many concretions in deltaic facies are rich in siderite or ferroan dolomite and numerous crusts in sabkha are dolomite or dolomitic. The Holocene dolomitic crusts on Andros Is., Bahamas, and on Sugar Loaf Key, Florida Keys (Shinn et al., 1965), are modified aragonite muds and contain from 20% to 80% dolomite.

EARLY DOLOMITIZATION

Although dolomite, theoretically, should be the first mineral to precipitate from sea-water supersaturated for cabonate, in fact kinetic interference leads to the precipitation of aragonite and Mg-calcite from water of the open sea. Since stoichiometric dolomite is a 1:1 cation ordered carbonate with layers of pure Mg^{2+} and pure Ca^{2+}, the high rates of diffusion at the growing crystal face needed to achieve perfect ordering can only be attained above $300°C$ or so. At lower temperatures dolomite is calcium-rich (calcian), commonly around $Ca_{55} Mg_{45} (CO_3)_2$. In cold-waters dolomite grows with difficulty, partly because of the energy obstacle presented by the need to organise two cations into distinct non-equivalent positions, especially where they have similar ionic radii and valencies, and partly because the hydrated Mg^{2+} ion interferes with dolomite growth as it does with that of calcite. No organism precipitates a dolomite skeleton.

All dolomites seem to have grown as a replacement of an

earlier carbonate sediment.   This is not to say that replacement
is texturally precise.   A $CaCO_3$ host is needed to supply at
least half the dolomite volume.   Nevertheless, dolomitization is
a process of dissolution-precipitation in aqueous solution.   The
new solid phase both fills the space previously occupied by the
dissolved $CaCO_3$ and also may occlude any pre-existing porosity
(Gebelein et al., 1980, fig. 11).   Thus, in the strict sense of
Sander (1951), it can be both secondary and primary.   If this
replacement is molecule by molecule, then porosity is increased as
the denser dolomite replaces the less dense carbonate.   Yet
dolomitization can also lead to a reduction in porosity.   The
balance depends on the initial porosity and on the subsequent
supply of carbonate.

Early dolomite is precipitated only where the composition of
the sea-water, within the pores of a carbonate sediment, has been
modified: crystals are small, only a few microns in diameter.
Such modification can arise in a hot, dry climate where intense
evaporation of marine pore-water from a sabkha surface leads to
the precipitation of gypsum and anhydrite and, thus, to levels of
Mg^{2+}/Ca^{2+} greater than the value of 5 on the open shelf.   Sea-
water in the sediment pores can also be modified by mixing with
carbonate-bearing meteoric water.   For example, the solubility
of cation-carbonate is related in a curvilinear way to salinity,
so that mixing even of two under-saturated solutions can bring
about supersaturation (Runnells, 1969).   In the schizohaline
model of Land (1973a and b), which follows the ideas of Hanshaw
et al. (1971), a mixture of fresh-water with 5-30% sea-water
should give a solution which is undersaturated for aragonite, Mg-
calcite and calcite but supersaturated for dolomite - the perfect
condition for dolomitization.   Badiozamani (1973) subsequently
developed this hypothesis and named it the Dorag model.

But more than a chemical change of pore-water is needed.
Magnesium ions must be transported from some source.   Mixing
requires transfer.   Thus various hypotheses for the control of
flow of groundwaters have been put forward.   The seepage
refluxion process of Adams and Rhodes (1960) assumes that brines
with high ratios Mg^{2+}/Ca^{2+}, rendered dense by the evaporation
of overlying sea-water, will move downward into permeable $CaCO_3$
sediments, dolomitizing them on the way.   This model was
extensively tested in a lagoon on the island of Bonaire, in the
Caribbean, but it was found that the downward migration was
largely inhibited by a clay layer.   On the other hand, it seems
to work in the Canary Is. (Muller and Tietz, 1966).

Regarding this evaporitic model for dolomitization, it must
be emphasised that evaporite-carbonate sequences are known in
which the carbonates have not been dolomitized.

A process of evaporative pumping was postulated by Hsu and
Siegenthaler (1969), whereby the levels of adjacent denser marine
and lighter meteoric pore-water in sabkha sediments are controlled
by evaporation in tropical sunlight.   During daylight a continuous
flow of marine pore-water moves landward into the sabkha to replace the fresh-water lost by evaporation from the sediment surface.
Here the sunlight provides the energy for the migration of hypersaline waters with high ratios Mg^{2+}/Ca^{2+} , caused by precipitation of gypsum and anhydrite, as for example in the sabkha of the
Trucial Coast, Persian Gulf.

McKenzie, Hsu and Schneider (1980) have put forward another
model for the Trucial Coast sabkha which requires a three-fold
combination of recharge by flooding sea-water, followed by
capillary evaporation from the sabkha surface and finally a stage
of evaporative pumping.   In this way Mg^{2+}/Ca^{2+} ratios from 7 to
27 are achieved, while aragonite and gypsum and massive quantities
of anhydrite are precipitated.   Constant pumping replenishes
Mg^{2+} , and dolomitization releases Ca^{2+} for more growth of
anhydrite.   This model, instead of relying on landward flow of
sea-water through the porous sediment (a movement that is too
slow), depends on the rapid up and down fluctuation of the water
table, over a vertical distance of about 10 cm, caused by flooding.   The rise of water table up-slope of the flood water,
brings up water with the Mg^{2+}/Ca^{2+} ratios great enough to dolo-
mitize.

Precipitation of evaporites (sulphates) is not essential for
dolomitization.   This is clear from studies of the mounds known
as "hammocks" on the supratidal flats of Andros Is., Bahamas
(Shinn et al., 1965).   Autumnal rain-water accumulates as lenses
under the hammocks where vegetation reduces loss by run-off and
by evaporation, while the elevation is a protection against flooding by sea-water.   The water in tidal inlets is diluted and
fresh-water ponds form on the flats.   The 1-2 m layer of Holocene
aragonite needle-mud on the Pleistocene limestone undergoes some
dissolution.   Dolomitic crusts with micritic crystal size form
peripherally around the hammocks, possibly by schizohaline mixing,
though the conclusions of various workers are far from certain on
this point (Bourrouilh, 1980).

South of Adelaide, Australia, in the ephemeral lakes between
ranges of coastal-plain sand dunes in the Coorong, the water-table
goes up and down seasonally, following periodic recharge with
continentally derived rain-waters (Von de Borsch, 1976).   During
the summer, evaporation exceeds recharge and the water-tables in
the lakes fall, and the lake floors dry out as cracked crusts of
microcrystalline carbonate.   In the winter the water levels rise
and the lake floors change to slurries with the consistency of
yoghurt.   Cores through the first metre or so subsurface reveal

a prograding sequence which has been accumulating over some
6,500 years.   At the base the oldest sediments are marine or
lagoonal, passing upward into a mixture of so-called "protodolomite" and Mg-calcite and finally to magnesite mixed with rare
dolomite.   The "protodolomite" has the basal dolomite reflection
but lacks the order reflections, so it is not a dolomite in the
strict sense.   It is calcian, having up to 8 mole % excess of
calcium.   Whether or not it represents a stage in dolomitization
is not clear.

Chemical reaction of the carbonate muds with Mg-bearing
groundwaters is aided by the annual oscillation of the water-
table.   The ratio Mg^{2+}/Ca^{2+} is low, about 0.6-1.6.   Immediately
inland from the coast, there seems to be a landward-dipping interface between marine pore-water and evaporitically modified, Mg-
bearing, groundwater of continental origin.   A zone of mixing
may enable dolomitization to follow Land's schizohaline pattern.

Among ancient dolomites, Muire et al. (1980) have applied
the tentative Coorong model to a Proterozoic suite in Northern
Australia.   Otherwise, most applications of Holocene studies to
ancient dolomites have relied on the peritidal evaporitic, Abu
Dhabi-type model.   Valuable earlier papers are on the Devonian
Manlius Formation of New York by Laporte (1967), on the
Ordovician of Maryland by Matter (1967), the Mississippian of
eastern Canada by Schenk (1967) and the Carboniferous of Ireland
by West et al. (1968).   In addition to relating the dolomite to
its associated lithofacies, it has been found helpful to appeal
to its $\delta^{18}O$, the presence of inclusions of gypsum or anhydrite,
and the sodium content, as a measure of the role of sea-water
and of evaporite formation in the evolution of the pore-water in
which the dolomite grew.

It has become increasingly clear that there is no unique
process of dolomitization: rather that suitable conditions have
arisen in a variety of circumstances.   Generally speaking, it
seems that coarse, sucrose dolomites are more nearly stoichiometric
and less variable in composition that dolomicrites and dolomitized
carbonate muds.   Evaporite-related dolomite tends to be near
stoichiometric or calcium-depleted and syndepositional in origin.
Non-evaporative, non-stoichiometric, micritic dolomite is syndepositional to early diagenetic.   The sucrose dolomites are
middle to late diagenetic (Lumsden and Chimahusky, 1980).

EARLY DIAGENESIS ISOLATED FROM THE SEA FLOOR

Increasingly it seems that the diagensis of a carbonate
sediment, when it is removed from the influence of the sea and
the immediate sub-sea environment, can follow two paths.   It may

be lifted above sea level, as the reef and associated sediment of
the Caribbean Pleistocene, or it may undergo continuous burial.
The first path leads through meteoric diagensis, the second leads
also through burial diagenesis at enhanced pressures and temperatures. However, before either of these are examined, it is
important first to consider the question of mechanical compaction,
because upon our view of that rests other vital conclusions regarding the mass transport of $CaCO_3$ in solution.

MECHANICAL COMPACTION

If mechanical compaction of carbonate sediments, by grain
breakage and rotation, has generally been insignificant,
particularly in carbonate muds, wherein it might be expected to
have played an important role, then certain conclusions follow.
The most important of these concerns the supply of $CaCO_3$ for
cement. Most carbonate sediments today have initial porosities
around 60-70%. Thus, if porosity is not reduced substantially
by mechanical compaction, it must instead be reduced by cementation with $CaCO_3$ brought mainly from an outside source. This
conclusion in turn leads to serious difficulties regarding supply
and transport.

The effects of mechanical compaction are seen in the
Carboniferous sequence of limestone and shale in North Wales. In
the shales, the crinoid columnals and brachiopod shells are
severely crushed and broken, yet in the adjacent limestones they
remain undeformed. Evidence for mechanical compaction in biomicrites is normally lacking, a point made by Pray (1960),
Bathurst (1975) and Steinen (1978). Mechanical compaction by
breaking and crushing of grains is more obvious in graintstone
(Meyers, 1980), but even here the amount of compaction is small,
a few per cent.

This whole question of mechanical compaction in carbonate
muds received new impetus recently from the publication of a
laboratory study by Shinn et al. (1977). They had subjected wet
cores of Recent carbonate mud, from Florida Bay, to compaction
along the core axis in the laboratory. The resultant textures
showed no breakage of delicate shells. The authors argued that
ancient carbonate muds, also lacking broken shells, could therefore have lost substantial porosity by mechanical compaction.
This certainly put the cat among the pidgeons. However, in 1979,
at a conference in Tulsa organised by the American Association of
Petroleum Geologists (no publication), Shinn gave further details.
Laboratory compaction of Florida Bay muds has yielded porosities
reduced from about 70% down to 40%, rarely to 30%, accompanied by
slight shell breakage. Yet ancient biomicrites commonly have
porosities of less than 5% with no crushed tests. Thus, neither
these interesting experiments, nor those of Bhattacharyya and

Friedman (1979), give grounds so far for abandoning the view that most carbonate muds, and carbonate sediments in general, were cemented and made rigid at a shallow depth subsurface, before severe mechanical compaction could occur. This conclusion implies that substantial quantities of $CaCO_3$ were transported into the sediment from outside to give cement. The sources of this allogenic carbonate, and the supply and drive of water for its transport, remain only partly resolved questions.

DIAGENESIS IN FRESH-WATER AQUIFERS

One approach to post-marine diagenesis is to see what happens when the sea withdraws. Pleistocene limestones that have become exposed to rain and soil water as a result of a fall in sea level, or of an isostatic rise of land, are especially well known in Bermuda, in the Caribbean, and along the Israeli coast of the Mediterranean and elsewhere (summarised in Bathurst, 1975, 1980; also Halley and Harris, 1979; Buchbinder and Friedman, 1980).

Beneath the emergent surface, a fresh-water aquifer either floats on an aquiclude or on denser more saline water, commonly sea-water. Because of the low density of rain-water, the base of the aquifer is depressed so that the ratio of the depth from water-table to sea level is related to the depth from sea level to the base of the aquifer as 1:40. Thus, slight topographic relief of a few metres can yield an aquifer that extends for several hundreds of metres subsurface, as for example in Florida (Back and Hanshaw, 1970). This means that enormous volumes of marine carbonate sediment may be bathed in rain-water. In a limestone sequence containing numerous emersion surfaces, as mapped by Somerville (1979) in the Carboniferous limestones of North Wales, perhaps the entire sequence of sediments has undergone freshwater diagenesis. The fact that, in this succession, there has been substantial loss of thickness by pressure-solution (stratiform stylolites and grain-grain welding) suggests, however, that porosity was not greatly reduced everywhere: perhaps an appropriate outside source of $CaCO_3$ was lacking.

In the freshwater environment, the content of Mg^{2+} is normally low and the precipitation of low-magnesian calcite is no longer inhibited. So, the more soluble phases, aragonite and various Mg-calcites, dissolve and the ions are reprecipitated as calcite (Bathurst, 1975, 1980). Aragonite grains may be dissolved totally to give a mould which may or may not be filled later with calcite cement. It is here that micrite envelopes, along with cement fringes and cemented mud-matrix, preserve the walls of moulds. Or the aragonite may be replaced by calcite neomorphically, as dissolution and precipitation proceed simultaneously across a water film. This latter process generally

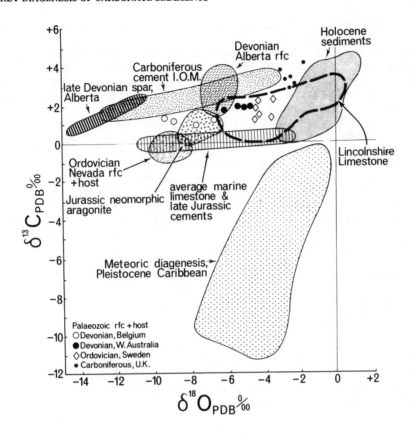

Figure 5. Isotopic compositions of groups of
carbonate sediments and cements. Data
sources are as follows:
Holocene sediments Gross (1964). Meteoric
diagenesis, Pleistocene Caribbean: Gross
(1964).
Average marine limestone and late Jurassic
cements: Hudson (1977).
Jurassic neomorphic aragonite: Tan and Hudson (1974).
Lincolnshire Limestone: Marshall and Ashton (1980).
Carboniferous cement, Isle of Man: Dickson and
Coleman (1980).
Devonian, Alberta, radiaxial fibrous calcite: Walls
et al. (1979).
Late Devonian spar, Alberta: Walls et al. (1979).
Ordovician radiaxial fibrous calcite and host
biomicrite: Ross et al. (1975).
Palaeozoic radiaxial fibrous calcite and host bio-
micrite: Bathurst data.

leaves ghosts of skeletal microstructure in the sparry calcite, where relic organic matter has failed to escape.   The high-magnesian calcite skeletons are also replaced by calcite, without visible loss of detail under the light microscope.   Crystallo-graphic orientation is also preserved as new calcite nucleates on relics of the original Mg-calcite.   These processes redistribute the $CaCO_3$ but do little to reduce total porosity, though the pattern of porosity-permeability is profoundly changed.   The whole process acts in an open system so that the sediment under-goes substantial chemical and isotopic changes. Sr^{2+} is lost from the aragonite and Mg^{2+} from the Mg-calcites, some leaving the system, while some reappears in much reduced quantities in the calcite cements.   Theoretical considerations of partition coefficients, of rock-water ratio and of water chemistry, lead to the conclusion that during meteoric equilibration there will be, in the new calcites, a progressive reduction in the levels of Sr^{2+}, Na and perhaps Mg^{2+}, but an increase in those of Mn^{2+}, Fe^{2+} and Zn^{2+} (Brand and Veizer, 1980).   The CO_2^- anions in the new cements derive their carbon and oxygen isotopes both from dissolution of the original marine skeletons and from the light rain-water.   Thus both $\delta^{18}O$ and $\delta^{13}C$ show a trend toward more negative values (Fig. 5).

At the margins of the fresh-water lens, where fresh-water and sea-water mix, conditions may be ideal for the schizohaline model of dolomitization.   Thus the products of meteoric and schizohaline changes may be closely associated (Land, 1973a and b; Meyers and Lohmann, 1978).

Reduction of the porosity to values less than 5%, as in many ancient limestones, requires influx of additional $CaCO_3$, either from neighbouring regions of karstic dissolution or later during burial.

The time taken for this Pleistocene-type meteoric diagenesis to be completed may commonly be about 80,000-180,000 years, though in phreatic conditions it can be much less.   On the geological time scale this is a short period - a small fraction of a Jurassic ammonite zone.

CARBONATE DIAGENESIS IN A SUBSIDING SEDIMENTARY BASIN

It is impossible to assess the role of early diagenesis in ancient carbonate rocks without relating it to other processes that may have acted at greater depths in the crust.   Hudson (1975) drew attention to the isotope analyses by Keither and Weber (1964) for a wide range of limestones.   These data show a distribution different from that to be expected from carbonate sediments equilibrated in fresh-water aquifer (Fig. 5).   Part

of the lack of fit is due to extreme near-soil values among some
Caribbean data (Allan and Matthews, 1977), but exclusion of these
does not remove the anomaly.   It seems that many ancient limestones became lithified in an environment which caused little
change in $\delta^{13}C$ but led to progressive changes in $\delta^{18}O$ toward
negative values around -15 (Fig. 5).   Since Hudson's pertinent
comment, information confirming this trend has continued to
accumulate.   Interpretation of the progressive negative trend
of $\delta^{18}O$ has drawn workers to interpret this, for the moment, as
evidence for increasing temperature of precipitation of the new
calcite (or dolomite).   Temperatures deduced are around 50 C to
over 80 C, indicating possible depths of burial around 2-3 km.
Jurassic calcite cements show this trend for example (Tan and
Hudson, 1974), also late sparry calcites in Devonian limestones
(Walls et al., 1979).   So do Palaeozoic radiaxial fibrous
calcites and their biomicrite hosts in Palaeozoic mudmounds
(altered Mg-calcite marine cement?), reported by Ross et al.
(1975), Walls et al. (1979) and by Bathurst (unpublished).
Dickson and Coleman (1980) have shown a trend from $\delta^{18}O$ = 3 to
-12 in single, zoned, calcite cement sequences in Carboniferous
limestone.

It is necessary to ask what processes, other than meteoric
diagenesis, can have brought about the changes from a loose marine
sediment, containing aragonite, Mg-calcite and calcite, to a fully
cemented calcite limestone.   We must return here to the ideas
expressed in the Introduction.   What are the available sources of
$CaCO_3$ and $MgCO_3$ and what driving forces controlled the
necessary flow of water, remembering that at least 10,000-50,000
pore volumes of water are needed to bring enough $CaCO_3$ to fill
one pore (summarised in Bathurst, 1975).

Sea-water undoubtedly supplies some ions to make cement and
it is becoming easier to recognised marine cements, both on
textural, mineralogical, chemical and isotopic grounds.   Many
pre-Pleistocene isopachous cement fringes that are now calcite
resemble Holocene marine cements in their textures.   Some are
rich in inclusions and may even contain microdolomites, possible
indication of a Mg-calcite precursor (Lohmann and Meyers, 1977).
Some calcite cements, in Palaeozoic carbonate mud mounds in Arctic
Canada (Davies, 1977), are closely similar in appearance to
Holocene reef cements, for example the Mg-calcite fringes and
botryoidal aragonites.   Evidence of origin is supported by
enrichment, relative to host, in Mg^{2+} and Sr^{2+} respectively,
and $\delta^{13}C$ indicative of marine origin.   Radial-fibrous calcite
fans in the Permian Capitan reef complex also appear to have been
initially aragonite (Mazzullo, 1980).   Conservation of fan-like
rays of aragonite, now altered to calcite, has been described
from tepee fillings in Triassic limestones by Assereto and Folk
(1980) and from the Permian Capitan reef-complex by Loucks and

Folk (1976).   Early submarine cementation has long been recognised
in carbonate mud buildups and their talus (Krebs, 1969; Walls et
al., 1979; Zankl, 1969).   In some buildups syndepositional
fractures in biomicrites were filled with marine cements and
sediments (Pray, 1965).   Karstic (meteoric) features are notably
absent in these examples.   Lindstrom (1963) described Ordovician
hardgrounds which had fragile domed-structures (incipient tepee?)
that must have been of deep-marine (sheltered) origin because
they could never have survived in turbulent coastal waters.
Extensive Jurassic hardgrounds have been recorded in Burgundy by
Purser (1969), with characteristic fringing cements pre-dating a
marine carbonate mud that filtered down later.   Some of these
cements show meniscus and pendant cements indicative of marine
vadose water in beach sands.   Other Chalk and Jurassic hardgrounds, with encrusting and boring biota, have been studied in
detail by Bromley (1979), Hancock (1976) and Fursich (1979).
Analysis of vertical changes in values of δ^{13} C in Jurassic hardgrounds support their marine origin (Marshall and Ashton, 1980).
The cavity-filling cements called radiaxial fibrous calcite
(contra radial-fibrous calcite), described first by Bathurst
(1959, 1975) and interpreted as recrystallized marine cements by
Kendall and Tucker (1973), contain microdolomites (personal
communication from K.C.Lohmann) and may be calcitized Mg-calcites.
Textural evidence shows them to be marine, for example their
alternation with marine internal sediments and their growth prior
to development of marine talus (in Bathurst, 1980).   Isotopic
analyses of these calcilized cements show $\delta^{13}C$ consistent with
a marine origin (Fig. 5).

Nevertheless, many limestones show no evidence (or no
recognised evidence) of having been cemented on or just below the
sea floor.   On the other hand they commonly show evidence of
more than one stage of cementation.   Many grainstones reveal two
cements (at least) one of which preceded compactive grain fracture.
The later cements are approximately equant sparry calcite and it
is unfortunate that no textural criteriae exist for distinguishing,
say, meteoric spar from another of deeper crustal origin.
Certainly the interval of time between cement generations may run
into tens of thousands to tens of millions of years.   Dickson's
painstaking work, over the last decade and more, on cements in
Carboniferous limestones (Dickson, 1966; Dickson and Coleman,
1980) has drawn attention to the importance of zoning and of
cement stratigraphy, a matter developed by Meyers (1974) who has
been able to distinguish cements precipitated from different
pore-waters in Mississippian limestones, on the basis of fine
zonation revealed with cathodoluminescence.   These cements
Meyers has mapped through 100 m of section and across tens of
kilometres of country.   Their relative ages were assessed on
the basis of truncation in clasts at an unconformity or their
burial under filtered-down silts.   Some sparry cements in these

Mississippian rocks may be Pennsylvanian or later in age. Dickson and Coleman (1980) have presented sequences of zoned sparry calcite cements that, on the basis of $\delta^{18}O$ data, perhaps grew during progressive burial to a depth of 3 km. Many limestones containing pressure-welded grains have a sparry calcite cement that must have been delivered after welding had occurred. These developments emphasise the importance of assessing the burial-erosional history of limestones, a task performed by Walls et al. (1979).

If the majority of limestones were not lithified entirely in sea-water or in fresh-water aquifers (on the basis of anomalous isotopic data, Fig. 5) then in what waters were they lithified? Whence and how came the needed $CaCO_3$? Three other main sources of $CaCO_3$ suggest themselves: carbonate released by pressure-solution, carbonate-rich water squeezed from compacting clays and water somehow or other derived from "down there", originating below the limestone in question and moving up through it during compaction in a sedimentary basin. As a background to these problems there are valuable treatments of the burial evolution of sedimentary basins by Magara (1976), Neglia (1979) and two papers by Rumeau and Sourisse (1975, 1976).

Pressure-solution, through the development of grain-grain sutures or stratisform stylolites, rarely brings about a reduction of thickness more than 30%. It is possible to estimate the maximum amount of $CaCO_3$ that could be provided for the cementation of a carbonate sediment during its deformation by pressure-solution. Suppose a certain volume of sediment begins with the quite modest porosity of 60%. If it undergoes a 30% reduction in thickness but all the dissolved $CaCO_3$ is reprecipitated within the pores of the same mass of sediment, then the resultant porosity will be 43%. For lesser intitial porosities of 50% and 30%, the final porosities will be 28% and zero. It is apparent, therefore, that pressure-solution alone cannot change a sediment with a normal starting porosity of 60-70% to a rock with normal porosity less than 5%.

If we look to carbonate-rich waters squeezed from compacting clays, it is clear from calculations by Bjørlykke (1979) and Bathurst (unpublished) that limestone-shale sequences contain far too little shale for this to be a major process by several orders of magnitude.

Appeal to a deeper supply of $CaCO_3$, brought by waters migrating up-basin from deep compacting sediments, provides some help. Yet these waters are strictly limited in quantity and can be used only once. They are not renewable. When it is appreciated that 1 km^3 of carbonate sediment must have imported into it about 0.5 km^3 of calcite cement and that this would need, for its transport at least 200,000 x 0.5 km^3 = 100,000 km^3 of water, the reliance on deeper supplies of $CaCO_3$ in squeezed connate water is seen to require vast sources.

Of course the role of freshwater would not necessarily be limited to a Caribbean-type aquifer in uplifted sediments. Freshwaters can penetrate two kilometres down or more into the crust, given an adequate head, say a mountain range with appropriate precipitation. Such penetration is resisted by the upward trend of compaction waters. Deep fresh-waters have been detected under Florida leaking into the Straits of Florida, and oil companies not uncommonly encounter fresh-water in holes bored through continental shelves. Moreover, some major limestone sequences have clearly been exposed many times above sea level in the course of their accumulation. In this connection the recognition of emersion surfaces, karstic topography and caliches, is a matter of prime importance and significant recent work includes papers by Seminuik (1971), Read (1964), Walkden (1974), Klappa (1979, 1980) and Somerville (1979). Yet the puzzling fact remains that, on isotopic data, most limestones do not appear to have been mainly lithified in cold rain-water. Could it be that hot rain-water at 2-3 km deep, could have played an important role?

CONCLUSION

We are not yet in a position to work out diagenetic budgets for many limestone sequences. Various processes are likely to have acted but it is too early to attempt to draw a balance. Even an assessment of the likely role of submarine cementation on the basis of a Holocene model is difficult, because the few metres of thin Holocene sediment represent only the twinkling of an eye in geological time and we cannot know how much more lithified they might appear in, for example, a mere million years' time.

FURTHER READING

Useful textbooks are by Bricker (1971), Purser (1973, 1980), Fuchtbauer (1974), Bathurst (1974, 1975) and Lucas, Cros and Lang (1976). . There is a review of lithification by Bathurst (1980), a collection of conference papers on the precipitation, deformation and dissolution of carbonates edited by Humbert (1980) and the new SEPM Special Publication 28 on dolomitization (Zenger, Dunham and Ethington, 1980) and Zenger's own review of dolomitization (1979).

REFERENCES

ADAMS, J.E. and RHODES, M.L. 1960. Dolomitization by seepage refluction. Bull.Am.Ass.Petrol.Geol., 44, 1912-1920.

ALEXANDERSSON, T. 1972. Micritization of carbonate particles: processes of precipitation and dissolution in modern shallow marine sediments. Bull.Geol.Insts.Univ.Uppsala, N.S., 3, 201-236.
ALEXANDERSSON, T. 1975. Etch patterns on calcareous sediment grains: petrographic evidence of marine dissolution of carbonate minerals. Science, 189, 47-48.
ALLAN, J.R. and MATTHEWS, R.K. 1977. Carbon and oxygen isotopes as diagenetic and stratigraphic tools: surface and subsurface data, Barbados, West Indies. Geology, 5, 16-20.
ASSERETO, R. and FOLK, R.L. 1980. Diagenetic fabrics of aragonite, calcite, and dolomite in an ancient peritidal-spelean environment: Triassic Calcare Rosso, Lombardia, Italy. J.sedim.Petrol., 50, 371-394.
ASSERETO, R.L. and KENDALL, G.C. St.C. 1971. Megapolygons in Landinian limestones of Triassic of southern Alps: evidence of deformation by penecontemporaneous desiccation and cementation. J.sedim.Petrol., 41, 715-723.
BACK, W. and HANSHAW, B.B. 1970. Comparison of chemical hydrology of the carbonate peninsulas of Florida and Yucatan. J.Hydrol., 10, 330-368.
BADIOZAMANI, K. 1973. The Dorag dolomitization model - application to the Middle Ordovician of Wisconsin. J.sedim.Petrol., 43, 965-984.
BATHURST, R.G.C. 1959. The cavernous structure of some Mississippian Stromatactis reefs in Lancashire, England. J.Geol., 67, 506-521.
BATHURST, R.G.C. 1966. Boring algae, micrite envelopes and lithification of molluscan biosparites. Geol.J., 5, 15-32.
BATHURST, R.G.C. 1974. Marine diagenesis of shallow water calcium carbonate sediments. Ann.Rev.Earth Planet. Sci., 2, 257-274.
BATHURST, R.G.C. 1975. Carbonate Sediments and their Diagenesis, Elsevier, Amsterdam, 658 pp.
BATHURST, R.G.C. 1980. Lithification of carbonate sediments. Sci. Prog.Oxford, 66, 451-471.
BERNER, R.A. 1975. The role of magnesium in the crystal growth of calcite and aragonite from sea water. Geochim.cosmochim.Acta, 39, 489-504.
BHATTACHARYYA, A. and FRIEDMAN, G.M. 1979. Experimental compaction of ooids and lime mud and its implication for lithification during burial. J.sedim.Petrol., 49, 1279-1286.
BJØRLYKKE, K. 1979. Cementation of sandstones: discussion. J.sedim. Petrol., 49, 1358-1359.
BOURROUILH-le JAN, F.G. 1980. Hydrologie des nappes d'eau superficielles de l'ile d'Andros, Bahama. Dolomitisation et diagenese de plaine d'estran en climat tropical humide. Bull. Centre Rech.Explor.-Prod. Elf-Aquitaine, 4, 661-707.
BRAND, U. and VEIZER, J. 1980. Chemical diagenesis of a multicomponent carbonate system - 1: Trace elements. J.sedim.Petrol., 50, 1219-1236.
BRICKER, O.P. (Ed.) 1971. Carbonate Cements. Johns Hopkins Press, Baltimore, Md., 376 pp.

BROMLEY, R.G. 1979. Hardground diagenesis. In Encyclopedia of
    Sedimentology. R.W. Fairbridge and J. Bourgeois (Eds.), Dowden,
    Hutchison and Ross, Stroudsburg, Penn., 397-400.
BUCHBINDER, L.G. and FRIEDMAN, G.M. 1980. Vadose, phreatic, and
    marine diagenesis of Pleistocene-Holocene carbonates in a
    borehole: Mediterranean coast of Israel. J.sedim.Petrol., 50,
    395-407.
BURRI, P., DU DRESNAY, R. and WAGNER, C.W. 1973. Tepee structures
    and associated diagenetic features in intertidal carbonate
    sands (Lower Jurassic, Morocco). Sedim.Geol., 9, 221-228.
CHOQUETTE, P.W. and PRAY, L.C. 1970. Geological nomenclature and
    classification of porosity in sedimentary carbonates. Bull.Am.
    Ass. Petrol.Geol., 54, 207-250.
CURTIS, C.D. 1977. Sedimentary geochemistry: environments and
    processes dominated by involvement of an aqueous phase. Phil.
    Trans.Roy.Soc.Lond., A, 286, 353-372.
DAVIES, R.G. 1977. Former magnesian calcite and aragonite sub-
    marine cement in upper Paleozoic reefs of the Canadian Arctic:
    a summary. Geology, 5, 11-15.
DE GROOT, K. and DUYVIS, E.M. 1966. Crystal form of precipitated
    calcium carbonate as influenced by adsorbed magnesium ions.
    Nature, London, 212, 183-184.
DICKSON, J.A.D. 1966. Carbonate identification and genesis as
    revealed by staining. J.sedim.Petrol., 36, 491-505.
DICKSON, J.A.D. and COLEMAN, M.L. 1980. Changes in carbon and
    oxygen isotope composition during limestone diagenesis.
    Sedimentology, 27, 107-118.
DRAVIS, J. 1979. Rapid and widespread generation of Recent
    oolitic hardgrounds on a high energy Bahamian platform,
    Eleuthera Bank, Bahamas. J.sedim.Petrol., 49, 195-207.
DUNHAM, R.J. 1971. Meniscus cement. In: Carbonate Cements, O.P.
    Bricker (Ed.), Johns Hopkins, Baltimore, Md., 297-300.
FUCHTBAUER, H. 1974. Sediments and Sedimentary Rocks 1.
    Schweizerbart, Stuttgart, 464 pp.
FURSICH, F.T. 1979. Genesis, environments, and ecology of
    Jurassic hardgrounds. N.Jb.Geol.Palaont., 158, 1-63.
GEBELEIN, C.D., STEINEN, R.P., GARRETT, P., HOFFMAN, E.J., QUEEN,
    J.M. and PLUMMER, L.N. 1980. Subsurface dolomitization beneath
    the tidal flats of central west Andros Island, Bahamas. In:
    Concepts and Models of Dolomitization, D.H. Zenger, J.B. Dunham
    and R.L.Ethington (Eds.), Spec.Publ., Soc.Econ.Paleont.Miner.,
    28, 31-49.
GINSBURG, R.N. and SCHROEDER, J.H. 1973. Growth and submarine
    fossilization of algal cup reefs, Bermuda. Sedimentology, 20,
    575-614.
GROSS, MG. 1964. Variations in the O^{18}/O^{16} and $C^{13}/12$ ratios
    of diagenetically altered limestones in the Bermuda Islands.
    J.Geol., 72, 170-194.
HALLEY, R.B. and HARRIS, P.M. 1979. Fresh-water cementation of a
    1,000-year-old oolite. J.sedim.Petrol., 49, 969-987.

HANCOCK, J.M. 1976. The petrology of the Chalk. Proc.geol.Ass.,
86, 499-535.
HANOR, J.S. 1978. Precipitation of beachrock cements: mixing of
marine and meteoric waters vs. CO_2 degassing. J.sedim.Petrol.,
48, 489-501.
HANSHAW, B.B., BACK, W. and DEIKE, R.G. 1971. A geochemical
hypothesis for dolomitization by ground water. Econ.Geol.,
66, 710-724.
HARRIS, P.M. 1978. Holocene marine-cemented sands, Joulters ooid
shoal, Bahamas. Trans.Gulf Coast Ass.geol.Soc., 28, 175-183.
HSU, K.J. and SIEGENTHALER, C. 1969. Preliminary experiments on
hydrodynamic movement induced by evaporation and their bearing
on the dolomite problem. Sedimentology, 12, 11-25.
HUDSON, J.D. 1975. Carbon isotopes and limestone cement. Geology,
3, 19-22.
HUDSON, J.D. 1977. Stable isotopes and limestone lithification.
Jl.geol.Soc.Lond., 133, 637-660.
HUDSON, J.D. 1978. Concretions, isotopes, and the diagenetic
history of the Oxford Clay (Jurassic) of central England.
Sedimentology, 25, 339-369.
HUMBERT, L. (Ed.) 1980. Cristallisation, Deformation, Dissolution
des Carbonates. Report of conference, Inst.Geodynamique, Univ.
Bordeaux 111, 486 pp.
ILLING, L.V. 1954. Bahaman calcareous sands. Bull.Am.Ass. Petrol.
Geol., 38, 1-95.
IRWIN, H. 1980. Early diagenetic carbonate precipitation and pore
fluid migration in the Kimmeridge Clay of Dorset, England.
Sedimentology, 27, 577-591.
JAMES, N.P., GINSBURG, R.N., MARSZALEK, D.S. and CHOQUETTE, P.M.
1976. Facies and fabric specificity of early subsea cements in
shallow Belize (British Honduras) reefs. J.sedim.Petrol., 46,
523-544.
KEITH, M.L. and WEBER, J.N. 1964. Carbon and oxygen isotopic
composition of selected limestones and fossils. Geochim.
cosmochim.Acta. 28, 1787-1816.
KENDALL, A.C. and TUCKER, M.E. 1973. Radiaxial fibrous calcite: a
replacement after acicular carbonate. Sedimentology, 20, 365-389.
KENNEDY, W.J. and GARRISON, R.E. 1975. Morphology and genesis of
nodular chalks and hardgrounds in the Upper Cretaceous of
southern England. Sedimentology, 22, 311-386.
KLAPPA, C.F. 1979. Calcified filaments in Quaternary calcretes:
organo-mineral interactions in the subaerial vadose environment.
J.sedim.Petrol., 49, 955-968.
KLAPPA, C.F. 1980. Rhizoliths in terrestrial carbonates:
classification, recognition, genesis and significance.
Sedimentology, 27, 613-629.
KOBLUK, D.R. and RISK, M.J. 1977. Calcification of exposed
filaments of endolithic algae, micrite envelope formation and
sediment production. J.sedim.Petrol., 47, 517-528.

KREBS, W. 1969. Early void-filling cementation in Devonian fore-
    reef limestones (Germany). Sedimentology, 12, 270-299.
LAHANN, R.W. 1978. A chemical model for calcite crystal growth and
    morphology control. J.sedim.Petrol., 48, 337-343.
LAND, L.S. 1970. Phreatic versus vadose meteoric diagenesis of
    limestones: evidence from a fossil water table. Sedimentology,
    14, 175-185.
LAND, L.S. 1973a. Contemporaneous dolomitization of Middle
    Pleistocene reefs by meteoric water, North Jamaica. Bull.
    Marine Sci. Gulf Caribb., 23, 64-92.
LAND, LS. 1973b. Holocene meteoric dolomitization of Pleistocene
    limestones, North Jamaica. Sedimentology, 20, 441-424.
LAND, L.S. and EPSTEIN, S. 1970. Late Pleistocene diagenesis and
    dolomitization, North Jamaica. Sedimentology, 14, 187-200.
LAPORTE, L.F. 1967. Carbonate deposition near mean sea-level and
    resultant facies mosaic: Manlius Formation (Lower Devonian)
    of New York State. Bull.Am.Ass.Petrol.Geol., 51, 73-101.
LINDSTROM, M. 1963. Sedimentary folds and the development of
    limestone in an Early Ordovician sea. Sedimentology, 2, 243-275.
LIPMANN, F. 1960. Versuche zur Aufklarung der Bildungsbedingungen
    von Kalzit und Aragonit. Fortschr.Mineral., 38, 156-161.
LOHMANN, K.C. and MEYERS, W.J. 1977. Microdolomite inclusions in
    cloudy prismatic calcites: a proposed criterion for former high
    magnesium calcites. J.sedim.Petrol., 47, 1078-1088.
LOREAU, J.-P. 1980. Sediments Aragonitiques et leur Genese.
    Unpubl. thesis Univ. Pierre Marie Curie, Paris, 347 pp.
LOREAU, J.P. and PURSER, B.H. 1973. Distribution and ultra-
    structure of Holocene ooids in the Persian Gulf. In: The
    Persian Gulf - Holocene Carbonate Sedimentation and Diagenesis
    in a Shallow Epicontinental Sea, B.H. Purser (Ed.), Springer
    Verlag, Berlin, 279-328.
LOUCKS, R.G. and FOLK, R.L. 1976. Fanlike rays of former aragonite
    in Permian Capitan Reef pisolite. J.sedim.Petrol., 46, 483-485.
LUCAS, G., CROS, P. and LANG, J. 1976. Les Roches Sedimentaires.
    2. Etude microscopique des roches meubles et consolidees, Doin,
    Paris, 503 pp.
LUMSDEN, D.N. and CHIMAHUSKY, J.S. 1980. Relationship between
    dolomite nonstoichiometry and carbonate facies parameters. In:
    Concepts and Models of Dolomitization, D.H. Zenger, J.B.Dunham
    and R.L. Ethington (Eds.), Spec.Publ., Soc.Econ.Paleont.Miner.,
    28, 123-137.
MACINTYRE, I.G. 1977. Distribution of submarine cements in a
    modern Caribbean fringing reef, Galeta Point, Panama. J.sedim.
    Petrol., 47, 503-516.
MAGARA, K. 1976. Water expulsion from clastic sediments during
    compaction - direction and volumes. Bull.Am.Ass.Petrol.Geol.,
    60, 543-553.
MAGARITZ, M., GAVISH, E., BAKLER, N. and KAFRI, U. 1979. Carbon
    and oxygen isotope composition-indicators of cementation
    environment in Recent, Holocene, and Pleistocene sediments

along the coast of Israel. J.sedim.Petrol., 49, 401-411.

MARGOLIS, S. and REX, R.W. 1971. Endolithic algae and micrite envelope formation in Bahamian oolites as revealed by scanning electron microscopy. Geol.Soc.Am.Bull., 82, 843-852.

MARSHALL, J.D. and ASHTON, M. 1980. Isotopic and trace element evidence for subarine lithification of hardgrounds in the Jurassic of eastern England. Sedimentology, 27, 271-289.

MATTER, A. 1967. Tidal flat deposits in the Ordovician of western Maryland. J.sedim.Petrol., 37, 601-609.

MATTHEWS, R.K. 1974. A process approach to diagenesis of reefs and reef associated limestones. In: Reefs in Time and Space, L.F.Laporte (Ed.), Spec.Publ., Soc.Econ.Paleont. Miner., 18, 234-256.

MAZZULLO, S.J. 1980. Calcite pseudospar replacive of marine acicular aragonite, and implications for aragonite cement diagenesis. J.sedim.Petrol., 50, 409-422.

McKENZIE, J.A., HSU, K.J. and SCHNEIDER, J.F. 1980. Movement of subsurface waters under the sabkha, Abu Dhabi, UAE, and its relation to evaporative dolomite genesis. In: Concepts and Models of Dolomitization, D.H. Zenger, J.B. Dunham and R.L. Ethington (Eds.), Spec.Publ., Soc.Econ.Paleont.Miner., 28, 11-30.

MEYERS, W.J. 1974. Carbonate cement stratigraphy of the Lake Valley Formation (Mississippian) Sacramento Mountains, New Mexico. J.sedim.Petrol., 44, 837-861.

MEYERS, W.J. 1980. Compaction in Mississippian skeletal limestones, southwestern New Mexico. J.sedim.Petrol., 50, 457-474.

MEYERS, W.J. and LOHMANN, K.C. 1978. Microdolomite-rich syntaxial cements: proposed meteoric-marine mixing zone phreatic cements from Mississippian limestones, New Mexico. J.sedim.Petrol., 48, 475-488.

MILLIMAN, J.D. 1966. Submarine lithification of carbonate sediments. Science, 153, 994-997.

MOLLER, P. and KUBANEK, F. 1976. Role of magnesium in nucleation processes of calcite, aragonite and dolomite. N.Jb.Mineral. Abh., 126, 199-220.

MUIR, M., LOCK, D. and VON DER BORSCH, C. 1980. The Coorong model for penecontemporaneous dolomite formation in the Middle Proterozoic McArthur Group. Northern Territory, Australia. In: Concepts and Models of Dolomitization, D.H. Zenger, J.B. Dunham and R.L. Ethington (Eds.), Spec.Publ., Soc.Econ.Paleont. Miner., 28, 51-67.

MULLER, G. and TIETZ, G. 1966. Recent dolomitization of Quaternary biocalcarenites from Fuerteventura (Canary Islands). Contrib. Mineral.Petrol., 13, 89-96.

NEGLIA, S. 1979. Migration of fluids in sedimentary basins. Bull. Am.Ass.Petrol.Geol., 63, 573-597.

PRAY, L.C. 1960. Compaction in calcilutites. Bull.Geol.Soc.Am., 71, 1946 (Abstract).

PRAY, L.C. 1965. Limestone clastic dikes in Mississippian bio-
    herms, New Mexico. Spec.Pap.geol.Soc.Am., 82, 154-155.
PURSER, B.H. 1969. Syn-sedimentary marine lithification of
    Middle Jurassic limestones in the Paris Basin. Sedimentology,
    12, 205-230.
PURSER, B.H. (Ed.). 1973. The Persian Gulf -Holocene Carbonate
    Sedimentation in a Shallow Epicontinental Sea. Springer Verlag,
    Berlin, 471 pp.
PURSER, B.H. 1980. Sedimentation et Diagenese des Carbonates
    Neritiques Recents. Soc.Editions Technip, Paris, 366 pp.
RAISWELL, R. 1971. The growth of Cambrian and Liassic concretions.
    Sedimentology, 17, 147-171.
RAISWELL, R. 1976. The microbiological formation of carbonate
    concretions in the Upper Lias of NE England. Chem.Geol., 18,
    227-244.
READ, J.F. 1974. Calcrete deposits and Quaternary sediments,
    Edel Province, Western Australia. Mem. Am.Ass.Petrol.Geol.,
    22, 250-282.
ROSS, R.J., JAANUSSON, V. and FRIEDMAN, I. 1975. Lithology and
    origin of Middle Ordovician calcareous mudmound at Meikeljohn
    Peak, southern Nevada. U.S. Geol.Surv.Prof.Pap., 871, 48 pp.
RUMEAU, J.-L. and SOURISSE, C. 1975. Essai de classification et
    d'explication genetique d'un ensemble d'eaux de formation.
    Bull.Centre Rech. Pau-SNPR, 9, 215-222.
RUMEAU, J.-L. and SOURISSE, C. 1976. Modes et duree de
    compaction de series-argilo-carbonatees. Bull.Centre Rech.
    Pau-SNPR, 10, 39-52.
RUNNELLS, D.D. 1969. Diagenesis chemical sediments, and the
    mixing of natural waters. J.sedim.Petrol., 39, 1188-1201.
SANDER, B. 1951. Contributions to the study of Depositional
    Fabrics: Rhythmically Deposited Triassic Limestones and
    Dolomites. Am.Ass.Petrol.Geol., Tulsa, Okla., 207 pp.
SCHENK, P.E. 1967. The Macumber Formation of the Maritime
    Provinces, Canada - A Mississippian analogue to Recent strand-
    line carbonates of the Persian Gulf. J.sedim.Petrol., 37,
    365-376.
SCHLAGER, W. and JAMES, N.P. 1978. Low-magnesian calcite lime-
    stones forming at the deep-sea floor, Tongue of the Ocean,
    Bahamas. Sedimentology, 25, 675-702.
SCHMALZ, R.F. 1971. Formation of beach rock at Eniwetok Atoll.
    In: Carbonate Cements, O.P. Bricker (Ed.), Johns Hopkins,
    Baltimore, Md., 17-24.
SEMINIUK, V. 1971. Subaerial leaching in the limestones of the
    Bowan Park Group (Ordovician) of central western New South
    Wales. J.sedim.Petrol., 41, 939-950.
SHINN, E.A. 1969. Submarine lithification of Holocene carbonate
    sediments in the Persian Gulf. Sedimentology, 12, 109-144.
SHINN, E.A., GINSBURG, R.N. and LLOYD, R.M. 1965. Recent supra-
    tidal dolomite from Andros Island, Bahamas. In: Dolomitization
    and Limestone Diagenesis - A Symposium, L.C. Pray and R.C.

Murray (Eds.), Spec.Publ., Soc.Econ.Paleont.Mineral. 13, 112-123.

SHINN, E.A., HALLEY, R.A. and HUDSON, J.H. 1977. Limestone compaction: an enigman. Geology, 5, 21-24.

SMITH, D.B. 1974. Origin of tepees in Upper Permian shelf carbonates of Guadalupe Mountains, New Mexico. Bull.Am.Ass. Petrol.Geol., 58, 63-70.

SOMERVILLE, I. 1979. A cyclicity in the early Brigantian (D2) limestones east of the Clwydian Range, North Wales. Geol.J., 14, 69-86.

STEINEN, R.P. 1978. On the diagenesis of lime mud: scanning electron microscopic observations of subsurface material from Barbados, W.I. J.sedim.Petrol., 48, 1139-1147.

STURANI, C. 1971. Ammonites and stratigraphy of the "Posidonia Alpina" beds of the Venetian Alps. Mem. 1st. Geol.Mineral., Padova, 28, 1-190.

TAN, F.C. and HUDSON, J.D. 1974. Isotopic studies of the palaeoecology and diagenesis of the Great Esturarine Series (Jurassic) of Scotland. Scott.J.Geol., 10, 91-128.

TAYLOR, J.M.C. and ILLING, L.V. 1969. Holocene intertidal calcium carbonate cementation, Qatar, Persian Gulf. Sedimentology, 12, 69-107.

VON DER BORSCH, C.C. 1976. Stratigraphy and formation of Holocene dolomitic carbonate deposits of the Coorong area, South Australia. J.sedim.Petrol., 46, 952-966.

WALKDEN, G.M. 1974. Palaeokarstic surfaces in Upper Visean (Carboniferous) limestones of the Derbyshire Block, England. J.sedim.Petrol., 44, 1232-1247.

WALLS, R.A., MOUNTJOY, E.W. and FRITZ, P. 1979. Isotopic composition and diagenetic history of carbonate cements in Devonian Golden Spike reef, Alberta, Canada. Bull.Geol.Soc. Am., 90, 963-982.

WEST, I.M., BRANDON, A. and SMITH, M. 1968. A tidal flat evaporitic facies in the Visean of Ireland. J.sedim.Petrol., 38, 1079-1093.

WEYL, P.K. 1967. The solution behaviour of carbonate materials in sea water. Studies Tropical Oceanog.Univ.Miami, 5, 178-228.

ZANKL, H. 1969. Structural and textural evidence of early lithification in fine-grained carbonate rock. Sedimentology, 12, 241-256.

ZENGER, D.H. 1979. Dolomites and dolomitization. In: Diagenesis in Sediments and Sedimentary Rocks, G.Larsen and G.V. Chilingar (Eds.), Elsevier, Amsterdam, 423-536.

ZENGER, D.H., DUNHAM, J.B. and ETHINGTON, R.L. (Eds.), 1980. Concepts and Models of Dolomitization. Spec.Publ., Soc.Econ. Paleontol.Mineral. 28.

BURIAL DIAGENESIS IN LIMESTONES

Harold R.Wanless
Marine Geology and Geophysics
Rosenstiel School of Marine and Atmospheric Science
University of Miami
4600 Rickenbacker Causeway
Miami, Florida 33149, USA.

INTRODUCTION

As carbonate sediments become buried, major changes in fabric, composition, rigidity and porosity tend to occur. Character of burial diagenesis is controlled by precursor sediment character, pore fluid, lithostatic and tectonic stresses, evolving aqueous chemical influences, and time for kinetically slow processes to have effect.

Diagenetic changes that take place during burial can be divided into three groups: processes which reduce porosity, processes which maintain porosity and processes which enhance porosity.

Processes which reduce porosity are physical compaction, chemical pressure solution (commonly called chemical compaction), cementation, and growth of authigenic minerals.

Processes which maintain porosity (or otherwise provided rigidity) are: ˙early cementation, increased pore fluid pressure (which reduces lithostatic stress), and early emplacement of hydrocarbons (inhibiting aqueous chemical reactions).

Processes which can enhance porosity during burial are dolomitization, fracturing by tectonic or lithostatic stresses (pervasive or facies selective) and chemical leaching (by acidic waters provided by organic diagenesis within the rock or introduced from elsewhere).

At present, the most controversial aspect of burial diagenesis

A. Parker and B. W. Sellwood (eds.), Sediment Diagenesis, 379–417.
© *1983 by D. Reidel Publishing Company.*

is the timing and relative roles of the porosity reducing processes.
Included is: (a) difficulty in differentiating early cementation
from later cementation, (b) disagreement as to relative importance
of physical compaction and pressure solution, and (c) discord as
to the relationship between pressure solution and cementation and
dolomitization.

As resolution of these controversies is fundamental to understanding burial diagenesis, this paper will focus on, and attempt
to provide, criteria for distinguishing products of pressure solution and solution transfer from other compaction, cementation
and dolomitization.

AN APPROACH TO LIMESTONES UNDER STRESS

Wanless (1979, p.438) offered the following approach for
considering the response of limestones under stress. "It is
useful to consider limestones as packages of crystals which: (a)
will have differing resistance or responsiveness to change and
(b) will or will not contain impurities that would cause them to
behave differently than when pure. Change under stress can
take the form of physical compaction, physical deformation
(plastic or brittle), neomorphism (as micrite conversion to microspar or aggrading pseudospar), or solution. Influencing
impurities can be chemically inert particles (as siliciclastic
sand, silt, and clay), ions in calcite structure (e.g. Mg^{2+},
Folk, 1974) or other ions available to the interstitial fluid
(free ions, adsorbed ions, ions in unstable minerals). Impurities can modify either the physical or chemical response of the
limestone to stress".

"A resistant unit is an entity of grain- to bed-scale having
structural resistance to change. A unit may be resistant because
of crystal size, mineralogy, crystal chemistry, grain/crystal
packing, fabric, permeability and extent of precursor cements.
A structurally resistant unit, stressed excessively, will first
yield at foci of stress, along slightly less resistant internal
surfaces and at boundaries with other units. The bulk of the
unit will remain unchanged. 'Unit', as used here, is comparable
to 'iden' defined by Logan and Semeniuk (1976)".

"A responsive unit is one that does not have internal
structural resistance to stress. This unit, if homogeneous, will
respond to excessive stress, if uniformly applied, by uniform
change throughout the unit: analogues are a sponge squeezed
between two plates or a spring stretched. If the responsive
limestone unit is not homogeneous or if the applied stress is not
uniform, the extent of change will be variable through or along
the unit. Pressure solution in responsive units would not form

seams but rather would be pervasive through a unit (actual sites of dissolution, however, would be at stressed contacts of crystals throughout the unit)".

"Of the many impurities in limestones, two appear to have fundamental influence on resultant response to stress - fine, platy, insoluble minerals and magnesium ions. Very fine-grained insoluble minerals, concentrated by pressure solution, can form surfaces or zones of structural weakness, along which lateral motion (shear) can occur to relieve local stress anomalies. Dolomite may form in limestone units in which significant magnesium is released (or available) during pressure solution (or neomorphism; Folk, 1974). If this occurs, dolomite crystals may eventually become a grain-supporting framework, inhibiting further pressure solution".

"Carbonate rock sequences are composed of stacks of units (layers, zones, beds, laminae, grains) of differing structural resistance and impurity content. Using this general approach, it becomes apparent: (a) that stylolites should not be the only type of solution response to stress and (b) that pressure solution is an integral part of a general system of response to stress that includes physical compaction, deformation, neomorphism, local-source replacement or precipitation (dolomitization), and metasomatism".

Folk (1965), Bathurst (1971) and Rutter (1976) have pointed out that chemical processes of diagenesis (below metamorphic grade) in carbonates occur essentially only in association with an aqueous medium. Diagenesis of sedimentary carbonates during burial, thus, occurs by exchange of ions with and migration of ions through interstitial solutions. These wet diagenetic changes can involve one or more of the following possibilities: (1) molecule for molecule replacement; (2) additions of ions, the precipitated ions being transferred from other areas by interstitial solution; and (3) loss of ions, the dissolved ions being transferred elsewhere by interstitial solutions.

PRESSURE SOLUTION

Pressure solution is the preferential dissolution of mineral material at points of stress. Crystal deformation at stressed contacts will enhance solubility of both crystals. One may preferentially dissolve. Distribution of strain at the surfaces of grains near their points of contact will be a function of grain orientation, size and shape, anisotropy of the crystal lattice and (in polycrystalline grains) the fabric of the crystal mosaic.

Dissolved ions may be reprecipitated locally (Thomson, 1862)

or be carried out of the system.   Pressure solution involves
dissolution but may also be associated with reprecipitation of the
dissolved mineral or of a new mineral at an adjacent non-stressed
site.   Pressure solution is controlled not only by crystal strain
(enhancing mineral solubility) but also the solubility of the
specific crystal, the saturation state of the fluid, the thermodynamics and kinetics of dissolution and reprecipitation, ion
concentration gradients and the mechanism of solute transport
(fluid diffusion or fluid flow).

     Pressure solution has been recognized as the cause of stylo-
lite seams and grain-contact sutures since the era of Thomson and
Sorby (1879, 1908), Stockdale (1922, 1926, 1936, 1943) and
Dunnington (1954).   More recently, Barrett (1964), Wanless (1973),
Garrison and Kennedy (1977) and Mimran (1977) have shown that clay
seams in argillaceous limestones are at least in part a product
of pressure solution.   These smooth to finely undulating seams
look very different from the typical jagged stylolite.

     In 1976, Logan and Semeniuk proposed that burial stresses
initiate pressure solution and fracturing far beyond what had
been previously suggested and cause radical changes in limestone
fabric, form and morphology on grain to formation scale.

     Many scientists still feel that pressure solution can only be
considered a viable probability when jagged stylolites are apparent.   This is wrong.   Figure 1, for example shows two oolites
from the Pennsylvanian Tyro Oolite in southeast Kansas.   Grain
contacts shown by arrows have both been subjected to major
pressure solution.   If, however, the original grain morphology
were not visible, the extensive role of pressure solution could
go unnoticed.   Similarly the clay seams can be easily misinterpreted were it not for morphologic changes in adjacent grains
(Fig.2) or within grains (Fig.3).

     Wanless (1979) proposed that there were three basic styles
of pressure solution (Fig.4) dependent primarily on the resistance
or responsiveness of units and the presence or absence of fine
platy insoluble minerals.   These styles (their resultant morphology; their occurrence) are:

    I) sutured-seam solution (common stylolites and grain con-
        tact sutures; in clean limestones with resistant units);

   II) non-sutured seam solution (microstylolites, micro stylo-
        lite swarms and clay seams, in limestones with a signifi-
        cant amount of disseminated clay, platy silt or carbon-
        aceous material);

  III) non-seam (pervasive) solution (pervasive thinning of a
        unit; in clean responsive limestones).

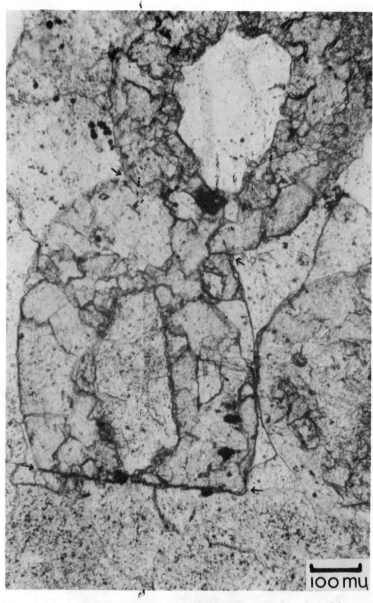

Figure 1.   Thin-section of photomicrograph of oolites at
            base of Eudora-Rock Lake member of Stanton
            Limestone (Pennsylvanian, S.E.Kansas, USA).
            Ovoid ooid has been truncated by pressure
            solution against mollusk grain at bottom and
            another ooid above.   Were the primary grain
            geometry not known, pressure solution could
            be easily overlooked. (Plane-polarized light).

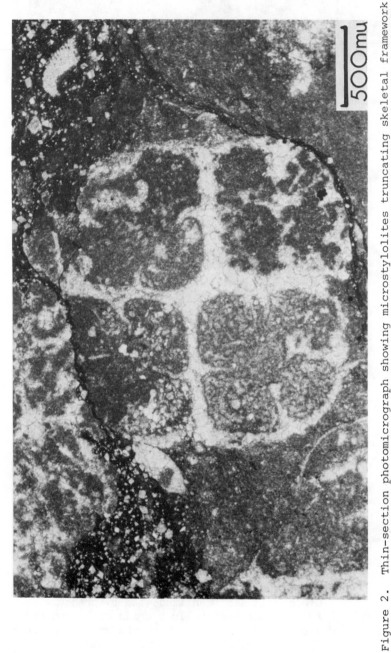

Figure 2. Thin-section photomicrograph showing microstylolites truncating skeletal framework of the coral Tetradium sp. Clays are concentrated along microstylolites and thicker clay seam (at left). Dolomite rhombs are scattered along seams but are not present in adjacent limestone. Ordovician Black River Group, House Creek member, from outcrop on Turin Road west of Lyons Falls, New York, USA. (Plane-polarized light).

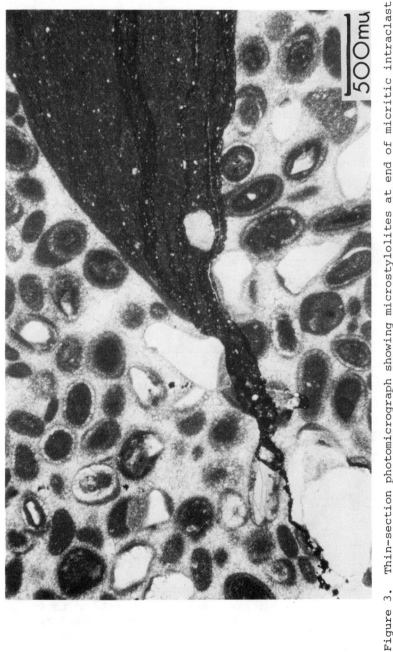

Figure 3. Thin-section photomicrograph showing microstylolites at end of micritic intraclast in quartzose oolitic limestone. Microstylolites concentrate into a clay seam at end of clast. In adjacent clean, resistant oolitic limestone pressure solution changes to sutured seam. Note oolites truncated against both bottom of intraclast and adjacent stylolite. Ordovician Black River Group, Unit A, in outcrop on Turin Road west of Lyons Falls, New York, USA. (Plane-polarized light).

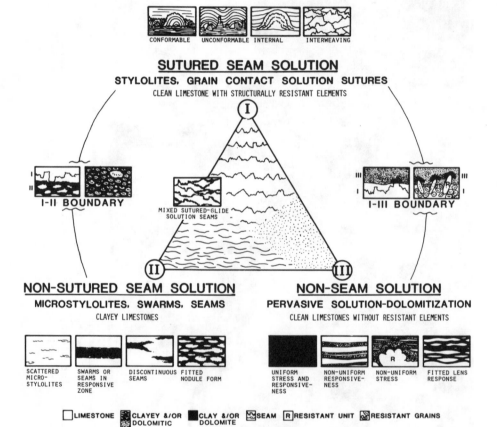

SUTURED SEAM SOLUTION
STYLOLITES, GRAIN CONTACT SOLUTION SUTURES
CLEAN LIMESTONE WITH STRUCTURALLY RESISTANT ELEMENTS

NON-SUTURED SEAM SOLUTION
MICROSTYLOLITES, SWARMS, SEAMS
CLAYEY LIMESTONES

NON-SEAM SOLUTION
PERVASIVE SOLUTION-DOLOMITIZATION
CLEAN LIMESTONES WITHOUT RESISTANT ELEMENTS

Figure 4.   Characteristics of and controls on three
styles of pressure solution proposed by
Wanless (1979). Clean means without significant clay or platy silt.

Wanless (1979), Logan and Semeniuk (1976), Mattes and Montjoy (1980), have also provided evidence that dolomite can form as a by-product of pressure solution.   In the following sections the products of these three styles of pressure solution are summarized and criteria are offered for their recognition.

Sutured-Seam Solution

Stylolites and grain contact sutures are irregular inter-penetrating surfaces of two rock, grain or crystal units. Sutures in cross-section vary from quite smooth to jagged to pillar and socket.   Surface relief is from less than 1mm to decimeters. They can be recognized in thin section as grain or crystal contact sutures and in hand specimen and outcrop as planar to interweaving sutured surfaces.   Sorby (1879, 1908), Stockdale (1922, 1926, 1936, 1943), Dunnington (1954), Bathurst (1971) and Logan and Semeniuk (1976) all concluded that stress-induced dissolution produces stylolites and grain contact sutures.

Lithostatic stress produces stylolites that are horizontal and parallel layering though precursor features (cross laminae) may cause stylolite seams to lie at some angle from horizontal. Rock units that have been under tectonic stress will have sol-ution seams perpendicular to effective stress.   This is commonly at sharp angles to primary layering (Groshong, 1975; Alvarez, Engelder and Lowrie, 1976; Alvarez, Engelder and Geiser, 1978; Droxler and Schaer, 1979).

Sutured solution seams occur at boundaries of structurally resistant units that have little or no platy insoluble material. Clean, resistant units include well crystallized skeletal and non-skeletal grains and crystals, grain-supported beds, and units that have resistance because of large biogenic structures or precursor cementation or neomorphism.

Insoluble minerals concentrate on the stylolite surface. Deep suturing occurs only if units have very little clay or platy mineral content.   Larger crystals, tightly cemented crystal groups and certain crystal orientations will preferentially resist dissolution and cause preferential dissolution of adjacent units (see also Sorby, 1879).   Limestones with abundant silt (not platy) tend to form pillar and socket stylolites rather than jagged seams.

If sutured-seam solution completely dissolves one or more limestone units, the end product should be an irregularly laminated layer of insoluble residue.   Clay and fine platy minerals should be minor constituents.

Logan and Semeniuk (1976) and Mattes and Montjoy (1980)

offer evidence that dolomite has formed along stylolite seams
during or subsequent to pressure solution.

Non-sutured Seam Solution

    Limestones that have a significant amount of clay or platy
silt disseminated through the rock tend not to have sutured seams.
Rather, pressure solution produces fine clay seams, whether the
units are resistant or responsive.   These seams have previously
been described as clay seams (Barrett, 1964), horsetails (Roehl,
1967), wispy laminae (Lucia, 1972, p.170), wavy laminae
(Reinhardt and Hardie, 1976), pseudo-stylolites (Shinn et al.,
1977) and microstylolites (Wanless, 1979).   Wanless (1979) used
microstylolite for a single clay film surface, microstylolite
swarms for anastomosing swarms of microstylolites about limestone
clots or grains, and clay seams for thicker seams of clays and
platy silts containing little calcite. "A microstylolite is a
very thin undulating surface with a relief on the order of only
20 to 40 microns.   Fine silicic-clastic clays and platy silt
occur as films along these surfaces" (Wanless, 1979, p.441).

    Microstylolites are commonly associated with nodular lime-
stones, and it is in these that the origin of microstylolites
(and nodules) from pressure solution can be most clearly seen.

    In nodular limestones, microstylolites tend to occur as
swarms of subhorizontal interconnected surfaces in the shale or
dolomitic shale interbeds, in the shale areas between adjacent
nodules and in a gradational zone 0.5 to 2 mm in thickness at the
top and bottom boundaries of nodules (Fig. 5).   Microstylolite
swarms commonly penetrate into the sides of nodules and fade out
(Fig. 6).

    Five lines of evidence show that clay seams and micro-
stylolites associated with nodules were produced by pressure
solution.

    1.   Limestone nodules contain preserved features of primary
         depositional origin-vague peloids, scattered skeletal
         grains and undistorted burrows.   These features indicate
         that the limestone nodules have been unaffected by major
         diagenetic alteration or volume reduction.   In contrast,
         these attributes are distorted or absent in dolomitic
         shale.   Resistant skeletal units can define nodules
         (Fig. 2 and Jenkyns, 1974).

    2.   Faint silici-clastic laminations, when present, are
         traceable from nodule to nodule through the intervening
         zones of dolomitic shale containing microstylolites.
         These laminae always converge (pinch) in the shale zone

Figure 5.   Bottom: Nodular limestone, Muav Limestone
(Cambrian, Arizona, USA). Lightly etched
slab of limestone composed of faintly
peloidal microsparite nodules (light gray)
and microstylolite swarms and seams of
dolomitic shale (dark, stained by potassium
ferricyanide solution).
Top: Thin-section photomicrograph of top of
nodule showing microstylolite swarms in
transition zone between limestone nodule
(below) and dolomitic shale seam (above).
Clots of microspar isolated by microstylolites become smaller and more isolated
upwards. (Plane-polarized light).

Figure 6. Limestone nodules are dotted on location sketch.
Photomicrograph from ultrathin section (8 microns) shows (a)
microstylolites penetrating nodule edge (upper left), (b)
spar-filled tension fractures in isolated clots and deformed
by glide shear along bounding microstylolites (clear sinuous
fractures in upper left), (c) deformed quartz-silt-filled
burrow tube in dolomitic clay seam between nodules (lower
center). Two large white fractures formed during thin
section preparation. Plane-polarized light; from sample
in Figure 5 (from Wanless, 1979).

between limestone nodules, demonstrating differential
solution thinning in the area between nodules. Thinning
and change in composition (loss of calcium carbonate)
cannot be explained by physical compaction.

3. Clots of limestone are common in dolomitic shale zones
just above and adjacent to nodules (Figs. 5 and 6).
These clots, bounded by microstylolites, appear to have
been separated by microstylolite swarms "eating" into the
edges of the nodules. This fabric cannot be explained
by either depositional structures or physical compaction.

4. Resistant skeletal elements in limestone nodules may be
truncated against microstylolite seams (Figs. 2 and 3).
This, where present is conclusive evidence of pressure
solution along the microstylolite.

5. Thin burrows filled by quartz silt or other vertical
structures may be drastically deformed and shortened in
the microstylolitic zones adjacent to nodules (Fig. 6).
This demonstrates extensive volume reduction. Earlier
listed evidence shows that this volume reduction is a
product of pressure solution.

Limestone clots and nodules commonly have vertical or sinuous spar-filled fractures towards the ends (Fig. 6). These
record tension fracturing caused by preferential thinning between
nodules and shear perpendicular to fractures.

Microstylolite swarms and clay seams do not display fracturing, rather each microstylolite, with its thin film of insoluble
clay and platy silt appears to have been a surface or zone of
lateral shear or flowage distortion.

Nodular bedding. Nodular bedding is here interpreted as
a product of pressure solution along non-sutured (microstylolite)
seams. The unique association of microstylolites with limestones containing a certain amount (at least 10 percent in the
author's experience) of siliciclastic clay and platy silt indicates
that fine platy minerals are important to the formation of microstylolites. Under pressure from over-burden, the limestone beds
are locally subjected to excessive stress. At these stress foci
limestone solution occurs, concentrating platy insoluble material
along a small solution seam. This seam acts as both a pathway
for the removal of dissolved carbonate and a site at which
insoluble clay and silt concentrate. The solution seam (microstylolite) apparently becomes inactive while still very thin,
probably for two reasons. First, the concentration of fine
silt and clay should quickly choke the solution seams as an
effective pathway for the fluid migration required for stylolite
growth. Second, the thin concentration of clay and platy silt

acts as a favourable glide plane along which a component of the locally accentuated stress can most easily be relieved by lateral shear.   These processes produce a new zone of maximum stress either farther along the limestone bed, laterally extending the microstylolite, or at a different horizon, producing a new micro-stylolite.   Continuation of these processes forms the observed 'swarms' of microstylolites.

Thinning of a limestone layer between adjacent nodules by microstylolite solution is gradual (tapered) and less than 30 per-cent in most of the Muav Limestone.   In local zones, especially those containing less silt and clay or dolomite, thinning between nodules is both abrupt and extensive, as much as 80 percent.

Nodules most commonly have a three-dimensionally 'fitted' or closely packed form in which the nodules of one layer fit into the thinned zones of microstylolite swarms or seam of the layers above and below.

Lateral displacement along microstylolite swarm zones can exert significant horizontal stress on limestone nodules and clots. Resultant tension fracturing of the more brittle limestone may have an important influence in decoupling clots from nodules and devel-oping nodular form.

Nodular bedding may or may not be defined by precursor attributes of the limestone.   The nodular facies of the Cambrian Muav Limestone (Arizona, USA) is similar in layering character to and repetitively grades into very thinly interbedded sandstones and shales.   In this case the nodular form and the now contorted dolomitic shale interlayering can be inferred to be a diagenetic alteration from interbedded limestone and limey shale layers.

Microstylolites and nodular or lensoid bedding are also a common product of tectonic stress.   In such cases, microstylolite swarms and clay seams commonly produce nodules that are elongated at a sharp angle to primary layering.   Such nodules usually have a fitted pattern.   This has been termed space solution cleavage (Geiser, 1974; Alvarez, Engelder and Geiser, 1978).   Tectonic examples emphasize two important features of nodular bedding and microstylolites.   First, these are an expected product of pressure solution in argillaceous limestones and need not be defined by primary sedimentary fabric.   Second, pressure solution of clayey limestones can, and does, produce inhomogeneity from a once-homogeneous medium, a concept doubted by previous authors (e.g., Jenkyns, 1974, p.244).

Clay seams. Many silty or clayey limestones contain seams as much as 2 cm in thickness of shale, silty shale or dolomitic shale.   Some of these seams are quite continuous laterally while others fade into adjacent limestone through a zone of micro-

stylolites.   The upper and lower boundaries between clay seams
and limestone are commonly zones of microstylolite swarms.   Clots
of limestone within the clay seams are uncommon to rare.   Clay
seams, because they show gradations in microstylolite zones, are
interpreted to be the end product of microstylolitic dissolution
of clayey limestones.   Clay seams may be controlled in their
distribution by precursor composition, may be due to excessive
stress on a shaly limestone by an adjacent structurally resistant
limestone, or may be essentially random products of pressure
solution.

Nodules can become quite isolated entities with clay seams.
Where this occurs they may take on a more equant form (through
continued pressure solution under more homogeneous stress field)
and may be rotated from their original orientation if the clay
seam undergoes significant shear motion to relieve stress (see
also Jenkyns, 1974).

Microstylolite solution selectively removes finer crystalline
material.   Coarse crystalline calcite grains may thus be selectively preserved in microstylolite swarms or clay seams (Wanless,
1979, Fig. 9a, b; see also Jenkyns, 1974).   Selective preservation
of coarse skeletal material (especially crinoid columnals and
shells) in clay seams can easily be misinterpreted, both by considering the clay seam as a different depositional feature than
the limestone, and by distinguishing the remnant fauna in the
clay seams from that in the limestone.

Dolomite. Dolomite occurs in association with microstylolites
and clay seams in some cases and not in others.   Where present,
such as in the Cambrian Muav Limestone (Arizona, USA), there is
strong evidence that dolomite actively grew during pressure solution.   The nodules contain scattered dolomite rhombs less than
30 microns in size.   They are clear, unzoned and do not stain
for iron with potassium ferricyanide.   Zoned dolomite rhombs,
mostly 30 to 120 microns in size, are abundant in the dolomitic
shale interbeds and along microstylolites marginal to and penetrating nodules.   These larger rhombs, stain for iron in the
outer zones, are cloudy in the outer portion (presumably from
included clays) and overgrow platy silt grains.   The dolomite
rhombs associated with microstylolites are so different and so
abundant that they cannot be explained as a simple residue from
solution of the precursor limestone.   Rather, conditions during
pressure solution must have provided preferential conditions for
the outer zones of dolomite growth.

Dolomite is uncommon in nodular limestones and chalks that
are basinal in origin and well removed from the influence of
shallow marine sediments.

Non-seam Solution

Non-seam or pervasive pressure solution was proposed by the author in 1979 to explain the unusual sedimentary structures and delicate interweaving of limestone and dolomite fabrics so common in finely crystalline limestones having little clay content. Since that time non-seam solution has taken on yet broader application.

Non-seam pressure solution occurs in responsive limestone units which have little clay or platy silt. The unit will respond to stress by dissolution throughout rather than along selected surfaces. Actual pressure solution of course takes place at stressed crystal contacts within the unit. This style of pressure solution is not dissimilar to pressure welding in fine sandstones in which pressure solution occurs at grain contacts throughout a unit reducing the volume but not producing preferential surfaces or seams of dissolution.

Non-seam pressure solution can result in complete dissolution of a unit. It has only been recognizable by the author, however, where pre-existing grains are concentrated as an insoluble residue or where dolomite has been precipitated within the unit as a "reactate" mineral during pressure solution of the limestone; Logan and Semeniuk (1976, p. 6) defined "reactate" as a mineral formed in association with and because of pressure solution conditions.

Non-seam solution is difficult to confidently ascertain except where there are lateral changes in thickness and composition of a limestone unit. Five of the following examples are taken from Wanless (1979), the sixth is the result of more recent studies. These are presented in detail because the concept of non-seam solution is new and the implications to burial diagenesis are profound.

Example 1: A mudcracked zone in the New Market Limestone (St. Paul Group, Ordovician) of western Maryland contains "lenses" of clean micritic limestone centered around large mudcracks (Fig. 7a, b). Between these limestone lenses, the layers "gradationally" thin and become laminated micritic dolomite. Numerous fine laminae can be traced laterally from one micritic lens to another through the intervening micritic dolomite. Dolomite content of limestone laminae increases laterally to about 60 percent in association with a gradual decrease in to about 21 percent of that in the adjacent limestone (Fig. 7b, d). Certain limestone laminae or zones extend farther than others before thinning and becoming dolomitic. The relative persistence of a certain limestone laminae remains similar as traced from micrite lens to lens (Fig. 7b).

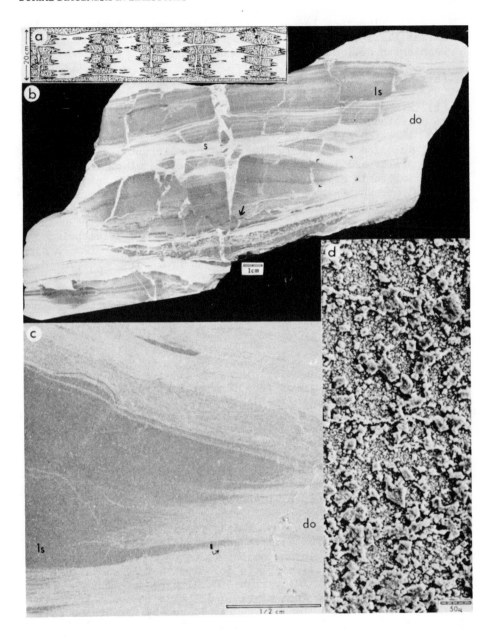

See legend on p. 396.

Figure 7. Limestone lenses about desiccation cracks in New
  Market Limestone, pasture southwest of St. Paul's Church,
  Washington Co., Maryland (from Wanless, 1979).

  (a) Sketch of outcrop on which dotted areas are limestone and
       clear areas are dolomitic.
  (b) Etched slab of three limestone lenses (dark) and dolomitic
       desiccation crack. Away from cracks laminae in micritic
       limestone lenses gradually thin and become micritic
       dolomite (do). Dolomite seams (s) disrupt desiccation
       pillar and laterally thicken and become micritic dolomite.
       Note thin stylolite (arrow) in lower part of central lens.
  (c) Close-up of area boxed in (b) shows laminae laterally thin-
       ning and becoming dolomitic (light). At bottom left and
       lower central, lateral change is through a zone of swarms
       of very fine dolomite seam (no clay). Negative print
       from acetate peel).
  (d) Scanning electron photomicrograph from boxed area in (c)
       shows intergrown rhombs in dolomitic micrite (bottom)
       scattered small dolomite rhombs in micritic laminae (in
       area of lateral transition) and two thin dolomite-rich
       seams in micritic zone (center and near top).

      In some cases, the transition from micrite to micritic dol-
omite is through a swarm of very fine microstylolite-like seams
containing dolomite but essentially no clay (Fig. 7c). Most
commonly, however, the change is simply a gradual increase in
the abundance and size of dolomite concomittant with bed thinning. Dolomite rhombs in the laminated dolomitic areas are
zoned, 10 to 30 microns in size and commonly slightly intergrown
(Fig. 7d). Dolomite rhombs in limestone are widely scattered
and less than 10 microns in size. This example contains very
little siliclastic sand, silt, or clay.

      At certain levels, the mudcrack "pillar" and adjacent
micritic lens are disrupted by a subhorizontal seam of dolomite
in the form of 10-30 micron rhombs. Seams can be traced
laterally into laminated micritic dolomite that is similar to
that described above. A decrease in thickness of about 40 percent occurs in passing from micritic dolomite to dolomite seam.
The micritic limestone lenses are rarely cut by very thin, stylolite seams with suture relief of 5 mm (arrow in Fig. 7b).

      This feature, although containing elements that appear to
be primary in origin, is clearly not a primary structure per se.
As the lateral change in layer thickness is associated with a
corresponding change in composition, simply physical compaction
has not been important. Rather, the dolomitic zones have formed

from differential solution thinning from limestone with associated
growth and concentration of dolomite.   The area in the vicinity
of the mudcrack tended to remain preferentially resistant to
pressure solution.   The areas between were responsive to pressure
solution which occurred not along a single seam but pervasively.
Certain laminae were more responsive to pressure solution than
others.   In this example, the solution of a 10 mm thickness of
limestone has produced a 60% dolomitic lamina 2.1 mm in thickness.
One hundred volumes of limestone has produced about 13 volumes of
dolomite.

Structural resistance to pressure solution was provided by
some attribute of the mudcracks or the adjacent limestone, such
as grain size, early cements, or permeability difference.   In
the same way that a shell may provide structural resistance that
now defines a limestone nodule, here the mudcrack zones acted as
structural pillars defining a limestone lens.

Example 2: The "ribbon limestones" so characteristic of the
cleaner Cambro-Ordovician limestones consist of laterally continuous dolomitic laminae or layers and rather lensoid limestones
(Fig. 8a; see also illustrations in Matter, 1967; Reinhardt and
Hardie, 1976; and Sando, 1957).   In outcrop, the dolomitic layers
vary from featureless to laminated. Lenses commonly contain
apparent cross-laminations (of quartz) and in some cases look
like starved ripples.   On careful examination, however, there
is clear evidence that the lensoid form is produced by differential solution of previously more continuous limestone layers.
Peels or etched slabs of ribbon limestones show four related sets
of features demonstrating that the limestone lenses are not starved
ripples or other primary structures.

(a) As quartz sand laminae approach the base or the top of
     a limestone bed, there is a corresponding increase in
     the amount of insoluble residue along the base (or top)
     of the bed (arrows in Fig. 8c).   This quartz con-
     centrate extends as a thin layer between limestone
     lenses (Q in Figs. 8b and c).

(b) At the end of a limestone lens all the laminae in the
     limestone converge and the lens shows convex outward
     surfaces (Fig. 8d).

(c) Associated with this thinning is an increase in the
     amount of dolomite and size of the rhombs.

As with the first example, these limestone lenses have been
created by pervasive solution-dolomitization of parts of
previously more continuous limestone beds.   The limestone lenses
are most commonly arranged in a fitted pattern much like that of

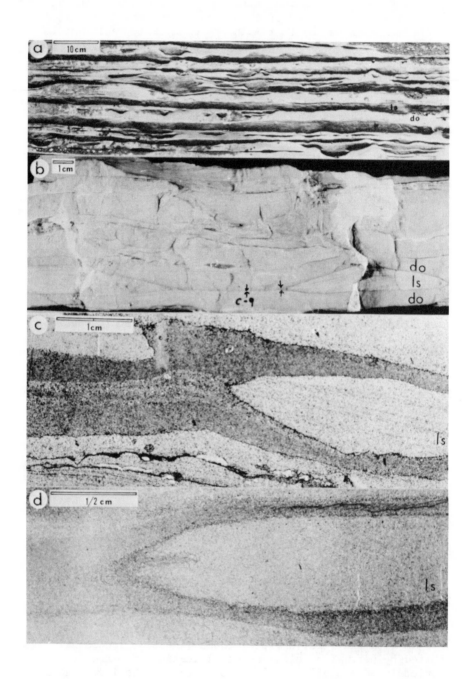

Figure 8. Ribbon limestone (from Wanless, 1979).

(a) Field photograph of typical ribbon limestone with continuous
     dolomite layers (light) and discontinuous limestone layers.
     New Market Limestone; location as in Figure 7.
(b) Quartz and laminae in limestone lenses at base of sample
     pinch as lenses thin, and quartz is concentrated in thin
     dolomitic lamina (arrows) between lenses. Weathered
     slab of Conococheague Limestone (Cambrian), Corwell's
     pasture 1 km south of Clear Spring, Washington Co.,
     Maryland, USA.
(c) As quartz sand laminae in limestone lens intersect base of
     lens, quartz content of laminated dolomite just beneath
     increases. Then quartzose dolomite laminae persist from
     lens to lens. Positive print from acetate peel;
     Conococheague Limestone, Downs pasture, 6 km north of
     Clear Spring, Washington Co., Maryland, USA.
(d) Limestone lens laterally tapering and becoming dolomitic.
     Resulting distinctive dolomite lamina can be traced from
     lens to lens in hand specimen. Positive print from
     acetate peel; Conococheague Limestone, Downs pasture.

nodular limestones. There is generally no apparent cause for
the pattern of limestone lenses; instead, the fitted lensoid
limestone pattern seems to be a "spaced" product of pressure
solution. This kind of pressure solution is less capable of
lateral shear motion than that in clayey nodular limestones, and
these lenses are more tapered and pointed than are nodules.

In the example in Figure 8b, c, the more continuous dolomitic layers do not contain quartz sand whereas the limestone
layers do. It can thus be concluded that this ribbon limestone
originally consisted of coarser and finer limestone layers - the
finer-grained layers being responsive to complete pervasive
solution-dolomitization at spaced intervals. That the continuous
dolomitic layers are the result of pervasive solution-dolomitization is examined further in Example 4.

Example 3: The lower part of the sample used for Example 1
contains a double limestone bed with dolomitic burrow fillings
(Figs. 7b and 9). Laterally this unit thins and becomes dolomitic, the dolomite burrow fillings taking on the appearance of
dolomite clasts in a micritic dolomite matrix (Fig. 9). As with
the previous examples, the lateral change can only be explained
by a solution-dolomitization process. Here, however, is evidence that a rock evolves from stage to another. First, as
seen in the limestone zone, the burrow fillings are preferentially
dolomitized. Each burrow tube visible is connected by a very

thin solution seam cutting across the limestone (Fig. 9). Dolomite rhombs, coarser than in the surrounding limestone, occur
along these seams. There was thus a first stage in which the
burrow fills were preferentially responsive to pressure solution
and dolomitization. A part of this response was seam solution
at stress foci in the contained limestone. Subsequently, parts
of the remaining limestone bed responded by pervasive solution-
dolomitization in the manner described in Example 2.

Example 5: Where an outside influence causes a stress on a
dolomitic zone, the degree of pressure solution- dolomitization
will be greater. Resistant intraclasts or stromatolitic lumps
(Fig. 10a, d) protruding into a micritic dolomite zone cause the
dolomitic laminae above to pinch (thin) and become more dolomitic
(see also Bathurst, 1971, Fig. 317 . The stromatolite and intraclasts acted as resistant structural units; the dolomitic layers
are from limestone units that were responsive, each to a different
degree.

From these examples, it is concluded that pervasive pressure
solution dolomitization is an important cause of some finely
crystalline stratigraphic dolomite. A solution-dolomitization
origin is mot apparent where a unit undergoes lateral change from
limestone to dolomite (Examples 1-3) or where solution-dolomitization is associated with limestone anomalies (Examples 4, 5).
There is, however, clear evidence through the Paleozoic ribbon
limestones (Sando, 1957; Reinhardt and Hardie, 1976), the more
confused lumpy limestones (Matter, 1967), and finely interlaminated
(cryptalgal) limestones and dolomites (Gebelein and Hoffman, 1973)
of western Maryland and Virigina that pressure solution-dolomitization has formed the dolomitic zones and, as a result, slightly
to severely deformed the primary sedimentary structures.

Example 6: Ordovician thinly layered slope deposits from
western Newfoundland and Virginia contain elongate lenses in which
there are laminae of quartzose, peloidal and radiolarian-sponge
spicule packstone and wackestone. Laminae thin away from these
lenses but are each traceable for the extent of exposure in outcrop or hand specimen.

Layers in the lenses look to have a primary depositional
fabric (Figs. 11a and b). As a laminae thins away from the lens
fine carbonate is lost, peloids and most other carbonate grains
become partially or completely pressure solved against other
grains (Fig.11c). Radiolaria become crushed and spicules become
reoriented from essentially random to parallel to layering.

Layers with abundant quartz or more resistant peloid layers
reduce in volume about 50%. Layers with few resistant units
reduce in volume as much as 80%. Laminae are commonly traceable

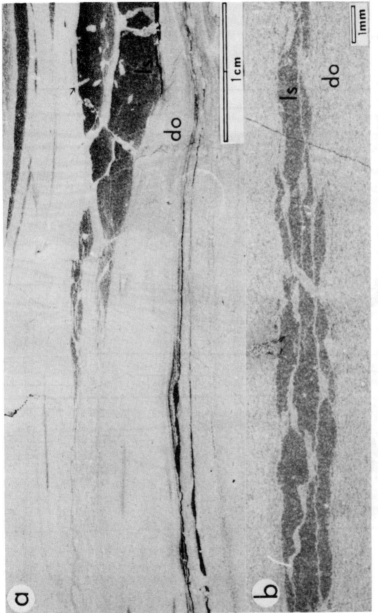

Figure 9.   (a) Edge of limestone lens (dark) with dolomitic burrow tube fillings (arrows).
Limestone layers thin laterally and become increasingly dolomitic, and the burrow
fillings take on the appearance of dolomite intraclasts. Etched slab; close up of
sample illustrated in Figure 7b.
         (b) Close up of 9a showing thin solution seams (dolomitic) connecting dolomitized
burrow fillings. Negative print from acetate peel (from Wanless, 1979).

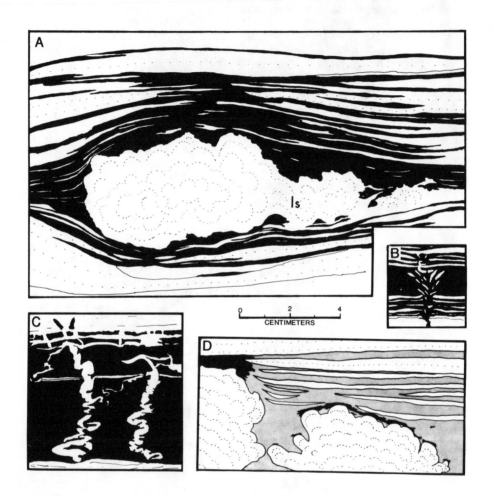

Figure 10. Sketches from field photographs of causes and results
of non-seam pressure solution in ribbon limestones. White is
limestone, shaded is dolomitic and black is dolomite.
Conococheague Limestone (Cambrian), Ashton, Maryland, USA (from
Wanless, 1979).

(a) Resistant limestone stromatolite caused local solution-
     dolomitization of surrounding limestone layers.
(b) Limestone desiccation crack filling is broken and tele-
     scoped by volume reduction during solution-dolomitization
     of the associated layer.
(c) Limestone desiccation crack fillings folded and faulted
     by volume reduction during solution-dolomitization of
     associated layers.
(d) Partial to intense solution-dolomitization in stress foci
     adjacent to resistant limestone stromatolites.

from lens to lens with the apparent change in amount of insoluble
quartz, spicules or other diagnostic grain.

As with Example 1, the thicknened layers associated with the
limestone lenses are remnants of once continuous layers that have
been largely lost through pervasive pressure solution. Further
many layers in these rocks are quite comparable to the thinned
portions of layers (e.g., bottom three layers in Fig. 11) but
have no thicker remnant exposed. It is very probable that each
of these layers is a remnant, thinned by pressure solution of a
once thicker carbonate rich layer.

Resultant Characteristics

I. Suture seam solution forms distinct dissolution surfaces
on the boundaries of units. The surrounding limestone has not
responded, but may have received pore filling cement (Bathurst,
1971, p. 770). Stylolite surfaces are sutured. Shortening
parallel to the direction of maximum stress has created sutured
interpenetrating limestone surfaces along which no lateral shear
can take place without extensive brittle fracture.

II. Limestones containing clay or fine platy silt respond to
pressure solution by forming many small interconnected solution
seams. Once formed, the microstylolite seams are potential
shear surfaces which, in conjunction with continued pressure
solution (forming more microstylolite surfaces), provide a com-
pound mechanism for response to stress. Solution surfaces,
though wavy, are not sutured. Individual seams are only a few
microns in thickness. With continued solution, many microstylolite seams form and may become an interconnected maze in

Figure 11. Pervasive solution of Ordovician distal slope limestone. F and M Quarry, north of Hillsville, Virginia, USA.
  (a) Composite thin-section photomicrograph showing thinning
      of three layers away from a limestone lens. Top and
      bottom layers are packstone of quartz (white) and
      peloids (dark); middle layer is peloidal wackestone.
      Packstone layers thin to 50 percent volume; wacke-
      stone, to 20 percent.
  (b) and (c) Close up of boxed areas. In (c) peloid are
      partially pressure solved against quartz grain.

which small limestone clots are isolated.

If solution of limestone is carried to completion, the end
product is a seam, layer or lens of laminated clay, shale, platy
silt or dolomitic shale. Grains of more coarsely crystalline
clacite tend to survive non-sutured pressure solution and may be
concentrated in such seams.

Pressure solution can form microstylolites (1) along
boundary zones of resistant clayey limestone units, (2) along
specific zones in a resistant unit (much as suture stylolites
do), or (3) throughout more responsive clayey limestone units.
In the third case, incomplete dissolution will produce a rock
comprising myriads of small limestone clots in a swarm of microstylolites.

III. Non-seam (pervasive) pressure solution is only recognized where it is associated with the formation and concentration
of dolomite. Pervasive pressure solution-dolomitization occurs
in limestones that have very low silt and clay content and have
little structural resistance during response to applied stress.
Pervasive pressure solution-dolomitization causes a reduction
in thickness of affected units. This is closely associated
with an increase in abundance of dolomite rhombs (by growth),
not as an insoluble concentrate. Pervasive solution-dolomitization is most apparent where there is evidence of some lateral
variability in intensity of the process (Examples 1-3 and 6) or
some evident distortion about associated limestone features that
remained resistant (Examples 4, 5).

Partial pervasive solution-dolomitization will result in
an essentially randomly mixed fabric of micrite or microspar
and dolomite. Complete pressure solution-dolomitization yields
a vaguely laminated dolomitic zone that contains little silic-
clastic silt or clay.

DISTINGUISHING THE ROLE OF PRESSURE SOLUTION

Dolimitization

Dolomite can form during many stages in the evolution of
a carbonate sediment.   Mattes and Montjoy (1980), for example,
make a valiant effort to distinguish various stages of dolomitization in the Devonian Miette build up of the southern Canadian
Rockies.   There are many ancient dolomites that are partially or
wholly formed by penecontemporaneous (Illing, et al., 1965; Shinn,
et al., 1965) or pre-stress diagenetic processes (Deffeyes, et al.,
1965; Folk, 1973).   In the variety of ancient limestones examined
by the author, however, pressure solution dolomitization is an
important influence on dolomite distribution.

Dolomite generated during pressure solution has certain
characteristics that together serve as distinguished criteria.
The following paragraphs provide characteristics to distinguish
first precursor dolomite and then postpressure solution dolomite
from dolomite formed during pressure solution.

Dolomite that is simply concentrated as an insoluble residue
should be similar in size and character to the dolomite in
adjacent limestones that have not undergone pressure solution.
If the zone that has undergone pressure solution can be traced
laterally into a zone that has not undergone pressure solution,
there should be conservation in the amount of dolomite crystals if
they are an insoluble residue.   If dolomite crystals are clearly
different in size, character, and abundance then that in equivalent adjacent limestone, they must have formed during or after
pressure solution.   These can easily be confused.

Dolomite crystals that formed during pressure solution will
tend to look like they formed after pressure solution.   The
dolomite crystal is an expanding component; calcite is a yielding
component.   Clays accumulating along a microstylolite may deform
around dolomite crystals, be partially overgrown by the dolomite
or be incorporated within the dolomite.   Each of these petrographic characteristics could easily be misinterpreted as recording dolomite growth after pressure solution.

Characteristics that show that dolomite formed during pressure
solution of limestone rather than after are subtle.   The following taken together should be sufficient:

1) Larger dolomite rhombs are restricted to areas of volume
reduction by pressure solution.   These rhombs contain
characteristics not found in other dolomite rhombs elsewhere
in the limestone (size, zonation, iron content, clay inclusions).

2) Dolomite rhombs form a grain-supported framework but are at most only slightly intergrown. This relation occurs because, with pressure solution of limestone, the growing euhedral dolomite rhombs eventually come into contact with one another. As the unit becomes grain supported by dolomite, limestone pressure solution ceases and that local source for magnesium is terminated. If dolomite grew after volume reduction (pressure solution or physical compaction), such a dolomite fabric would be much less probable; intergrowths should be more common.

3) Magnesium should be a conservative element in rocks that have had dolomite as a local by-product of pressure solution.

Source of Ions

The extensiveness of pressure solution suggested by this study indicates that much of the magnesium for dolomitization and calcium carbonate for pore filling cementation can be quite local in origin. This supports some recent interpretations by Oldershaw and Scoffin (1967) and Lloyd (1977) that carbonate diagenesis (cementation) can be a largely closed-system process.

Referring back to Example 1 of non-seam pressure solution-dolomitization, an apparent 100 volumes of limestone evolved, under pressure solution-dolomitization, to 13 volumes of dolomite. Assuming that the original limestone had initial porosity of 50 percent, it would have had to contain an average of about 14 mole percent $MgCO_3$ if all the magnesium were derived from the zone of pressure solution. The amount of magnesium could have been less if some was supplied by local interstitial waters, organic material, or from nearby limestone areas, or by penecontemporaneous dolomite. Because of pressure solution, however, the source for magnesium need no longer be a mysterious problem requiring special outside influences to cause widespread dolomitization in limestones of many depositional origins.

Why do certain limestones not contain dolomite as a by-product of pressure solution? Nodular limestones containing little or no dolomite along the microstylolite seams tend to have formed in shelf (portions of Chambersburg Limestone) to deeper water (post-Paleozoic, e.g. red nodular limestones of Mediterranean Triassic) environments. These environments should have contained low to negligible amounts of magnesium calcite.

The source of calcium carbonate for porosity filling cementation and addition during neomorphism may now be considered from a new point of view in light of the large volumes of calcium carbonate released by the three styles of pressure solution recognised here.

Volume Reduction

Mimran (1977), Schlanger and Douglas (1974) and others have documented that significant physical compaction can occur in fine-grained carbonates during early burial.   As outlined below, however, physical compaction during early burial cannot have produced the nodular-clay seam, "ribbon limestone" and other diagenetic fabrics described in this paper.

First, the published experiments of Shinn et al. (1977) on squeezing lime muds from Florida Bay cannot be used to explain the origin of clay seams or microstylolites.   The laminae they produced are simply squashed seagrass blades and roots.   Clays disseminated in a carbonate mud would not become organized into anastomosing microstylolite-like seams simply by physical compaction.   These are a product of pressure solution of carbonate.

Anastomosing swarms of microstylolites in the vertical and lateral transition zone between a limestone nodule and adjacent clay seam are evidence for pressure solution.   Physical compaction neither changes the composition nor concentrates disseminated clays.

Rock fabric modification by non-sutured seam pressure solution is a mixture of pressure solution and mechanical flowage (shear) along the microstylolites.   It is easy to recognize the physical deformation and ignore the intimately associated pressure solution.

Features formed by non-seam (pervasive) pressure solution have commonly been misinterpreted as primary structures, features produced by differential compaction and/or preferential early cementation.   Physical compaction alone cannot cause a lateral thinning and change in composition.   A coincidence of lateral thinning of a limestone layer and change in composition (actual increase in dolomite content) is excellent evidence for differential pressure solution - presuming no lateral change was originally present.   This becomes unequivocal where there is the same amount of insoluble residue (quartz, silt, resistant carbonate grains) in both the thicker limestone and the adjacent thinned zone (Figures 8b and c and 11).   The transitional zone of thinning also commonly contains small seams, also a strong demonstration that volume reduction is a product of pressure solution.

Physical changes during burial stress can include soft sediment water loss compaction, soft sediment flowage, shear and deformation, fracturing of isolated units (e.g. shells) by localized stresses in loose sediment, fracturing (tensional and shear) of structurally resistant units by localized or regional stresses

on or within a unit.   Physical compaction cannot reorganise the
distribution of grains within a unit, nor can it change the
character of boundaries between units.   Physical flowage and
shear can cause some modification of unit boundaries and distribution of grains.

NEOMORPHISM AND CEMENTATION

Increasing temperature and stress during burial can also
induce neomorphic changes, local dissolution and reprecipitation,
and cementation in carbonate sediments.   Increasing strain on
fine crystals, the presence of carbonate minerals that are unstable
at higher temperatures, and pore fluid migration, enhance neomorphism, local dissolutions and reprecipitation, and cementation
during burial.

Neomorphism, "the transformation between one mineral and
itself or a polymorph" (Folk, 1965, p. 21) includes wet polymorphic transformation of aragonite to calcite, wet recrystallization, and aggrading neomorphism (Bathurst, 1971, p. 476).   Neomorphism theoretically does not involve addition or loss of ions.
However, as neomorphic diagenesis in carbonates occurs via an
aqueous medium, the opportunity exists to add or remove material.
In fact, it is generally impossible to distinguish purely neomorphic processes from those which have involved some addition or
loss of material (Bathurst, 1971, p. 476).

Neomorphism, local solution and reprecipitation, and cementation can cause several important changes during burial.

Aragonite becomes unstable with increased temperature of
burial and will tend to dissolve or to neomorphose to calcite.
Dravis (1979) found that portions of the Austin Chalk
(Cretaceous, Texas, USA) had contained aragonite skeletal material
and matrix which changed to calcite during early burial by neomorphism and local dissolution and reprecipitation.   A minor
amount of lithification associated with this diagenesis strength-
ened the chalk sufficiently to inhibit subsequent pressure solution to depths of over 2,000 m.   In contrast, adjacent units
which did not originally contain aragonite have been subjected to
major modification and porosity reduction by pressure solution.

Fine-grained carbonate muds (of mixed mineralogy) tend to
become lithified and converted (stablized) to finely crystalline
low magnesium calcite (micrite).   Conversion of carbonate mud
to crystalline micrite can occur as a near surface diagenetic
change (driven by mineral solubility and pore water chemistry)
or during burial (driven in addition by temperature and strain
deformation of crystals).   Wet neomorphism to crystalline

micrite tends to result in slight increase in crystal size (0.5
to 4 microns), creation of a closely packed crystalline framework
and major reduction in porosity. Porosity reduction must result
from addition of material (cementation), reduction in rock volume
(crystal contact pressure solution and adjacent pore-filling reprecipitation) or both. As neomorphism proceeds and crystal
surfaces come more and more into contact, solution transfer occurs
less through intercrystalline pores and more and more in the thin
solution film at intracrystalline boundaries (a disorded layer
between crystals a few unit cells in thickness; Bathurst, 1971,
p. 498).

Micrite may neomorphose to microspar (5 to 50 micron-sized
crystals of calcite). This is commonly thought to be a neomorphic response to increasing stress similar to grain growth in
metals but occurring in a wet medium (Folk, 1965). Bathurst
(1971, p. 502), however, has suggested that the earlier lithification history, together with rate of earlier neomorphism and
porosity evolution might control neomorphic crystal size.

Folk (1974) has pointed out that magnesium "poisons" the
surface lattice positions of micrite-sized calcite. Magnesium,
if present, would be largely released during aggrading neomorphism from micrite to microspar. Limestone nodules in the
Muav Limestone (Cambrian, Arizona, USA) are dominantly microspar.
Small dolomite rhombs (less than 30 microns in size) are present
scattered through the microspar. Some of these crystals are
present within microspar crystals; others occur at triple junc-
tions of the microspar. These may very well have formed during
aggrading neomorphism from micrite to microspar.

Yet coarser aggrading neomorphism is common in zones of some
limestones that have received foci of stress. In Figure 12,
aggrading neomorphic pseudospar has formed in the strain zone
adjacent to a burrow tube. Pseudospar crystals are made up of
numerous slightly disoriented smaller crystals. Crystal size
of pseudospar decreases gradationally into adjacent limestone.
In the figured example, scattered dolomite rhombs as much as 80
microns in size occur within pseudospar crystals; they are not
present in the adjacent limestone.

Cementation, syntaxial precipitation onto pre-existing
crystals or nucleation and growth of new crystals, can occur at
a variety of stages in burial diagenesis. Previous sections
have presented evidence that rock strengthening and/or pore filling cementation can be associated with pressure solution and neomorphism. In addition, carbonate cements can be provided during
burial by supersaturated pore waters moving in from elsewhere.
Bathurst gives an excellent discussion of origins of cement (1971),
but in the author's estimation underestimates the role of local

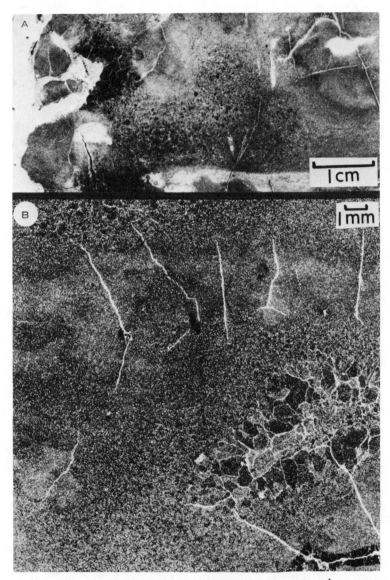

Figure 12. Pseudospar in Cambrian Muav Limestone (Arizona, USA).
  (a) Patch of dark aggrading neomorphic pseudospar at stress
       focus on outside of responsive burrow tube.   Polished
       slab.
  (b) Negative print from acetate peel of above sample showing
       pseudospar containing included dolomite and fine silici-
       clastic grains.   Fracturing is commonly observed in asso-
       ciation with aggradational neomorphic alteration.

pressure solution origin.   Oldershaw and Scoffin (1967) and
Mattes and Montjoy (1980) respectively suggest that late stage
ferroan calcite and dolomite cements are derived from an influx
of fluids from adjacent shaley limestones during dewatering
associated with pressure solution.

RESPONSE EVOLUTION

Pressure solution constantly modifies the fabric of the
associated limestone and provides a continuing supply of calcium
carbonate that can be locally reprecipitated (DeBoer, 1977).   Neomorphism and cementation continually modify the strength of the
limestone.

While a thorough evaluation is not yet possible, this study
provides several observations which suggest that some cementation
is associated with pressure solution.   Layers that have resisted
pressure solution generally have primary fabrics and do not
appear to have undergone significant compaction.   Porosity of
these units is much reduced, however, from what must have
initially been present.   Most of the porosity loss must have
been caused by addition of calcite cement during the time that
solution and neomorphic processes were active (see also Bathurst,
1971, p. 417, 504).   Such zones of preferential cementation,
whether primary structures or randomly-spaced features between
zones of preferential solution (nodules), may become increasingly
resistant structural units.   The small stylolite in Figure 7b,
for example, suggests that this limestone lens evolved from a
responsive to a resistant unit.

A limestone unit's response to stress can change through
time in several ways.   First, a limestone unit may evolve from
inactive to active.   Certain units will respond to increasing
stress before others or in preference to others.   Second, pervasive pressure solution-dolomitization will cause a unit to
become increasingly resistant as the dolomite content increases
(at a reduced unit volume).   Third, cementation in silty limestone nodules during pressure solution of an adjacent part of
the unit may decrease the ratio of clay to calcite in the nodule
sufficiently to cause it to act more resistant and respond to
stress by sutured seam solution.   Jenkyns (1974) also notes
that stylolite seams may form between adjoining nodules.
Cementation in responsive limestone lenses may cause them to
increase in strength and subsequently respond as resistant units
(see the fine sutured stylolite in limestone lenses of Figure 7b).
Fourth, limestone may become less responsive because (a) shear
motion along seams can more easily relieve stress and/or (b)
permeability is sufficiently reduced to inhibit effective fluid
migration.

THE END PRODUCT

Evidence in this paper and from others (e.g. Oldershaw and Scoffin, 1967; Lloyd, 1977) suggests that some carbonate diagenesis can be a closed-system process, we should ask whether the three styles of pressure solution can commonly continue to completion (total limestone solution). Do the processes of pressure solution gradually reduce the porosity and permeability of a unit to a point where it cannot effectively eliminate material in solution? Will pressure solution in a limestone trend towards a final stable form, the extent and form bing controlled by limestone attributes defined earlier as well as by pressure and temperature conditions and water chemistry? Bathurst (1971, p.500-501) also expresses this concern. Perhaps the adjustment of fabric and the production of dolomite continue well after exchange or any other mass transfer with the external system has ceased.

SECONDARY POROSITY AND PERMEABILITY ENHANCEMENT

There are three important processes which either can directly enhance porosity or permeability or can provide conditions favorable for subsequent enhancement. These are dolomitization, fracturing, and release of CO_2 during deeper burial.

Dolomitization

Pre-existing dolomite and dolomite produced during burial can enhance porosity or more commonly provide a rock conducive to porosity enhancement by other processes.

DeBeaumont (in Van Tuyl, 1916) noted that, if dolomite were produced by the recrystallization of calcium carbonate but without addition of new carbonate, a 12.3 percent decrease in mineral volume would occur. Were this recrystallization to occur without collapse, a significant increase in porosity could occur. Such a process has not been found to be effective.

Dolomite, once formed, is more resistant to both pressure solution (Dunnington, 1967) and acid leaching than is calcium carbonate. Thus, if dolomite becomes grain contact or crystal framework supporting, it will inhibit further pressure solution of the contained limestone. Subsequent acidic waters can then dissolve or partially dissolve the remaining limestone without further loss of porosity.

Well-dolomitized carbonate units are somewhat more brittle than limestone. Dolomitic units or zones, may, thus, preferentially fracture.

Fracturing

Four attributes of lithostatic stress or tectonic stress on a carbonate rock enhance the probability that fracturing will occur. Rapid increase in stress, low confining pressure, inherent brittleness of the rock unit, and slowed pathways for fluid migration can promote brittle fracture and shear.

Rapid loading, as by glaciers, deltas or basalt flows, can induce brittle fracture or grain breakage. Similarly, rapid application of tectonic stress will tend to cause brittle fracture of lithified limestones and dolomites. Fracturing will be enhanced if confining pressure is small (provided by intergranular voids, adjacent non-rigid units, or small burial depth in the case of tectonic stress). Groshong (1975), Alvarez, Engelder and Lowrie (1976), and Drexler and Schaer (1979) document extensional fracturing and compressional pressure solution associated with tectonic folding. Deformation is accomplished by compressional stylolites and solution cleavage, extensional fractures and glide along shear planes. These surfaces divide the rock into rigid units, within which little change (strain) occurs.

Fracturing may only slightly increase the porosity of a limestone or dolomite. This small porosity increase, however, generally produces a dramatic increase in permeability.

Acid Waters

Secondary porosity can be created in buried limestones by passage of acidic waters through the pore spaces. Acidic waters can be derived from dewatering of shales or from CO_2 generation during organic diagenesis. Towards the end of kerogen diagenesis, the first stage in degradation to oil and gas, large volumes of CO_2 are released. Kerogen at this stage is immature, and little hydrocarbon material has been released (Tissot and Welte, 1978). The CO_2 lowers the pH of the local aqueous system. This can cause dissolution in both local limestone and limestone through which the fluid migrates. Porosity and permeability developed in this stage just prior to hydrocarbon release can improve reservoir quality of the containing limestone and enhance migration pathways to other reservoir sites. Porosity created can be diminished or lost by subsequent pressure solution and cementation.

ACKNOWLEDGEMENTS

Portions of this paper were adapted from or taken from Wanless (1979). Research was supported by Grant No. EAR 78-63670 from the National Science Foundation and grants 1308-69 and 1340-70 from the Geological Society of America.

REFERENCES

ALVAREZ, W., ENGELDER, T. and GEISER, P.A., 1978. Classification
    of solution cleavage in pelagic limestones. Geology, 6, 263-
    266.
ALVAREZ, W., ENGELDER, T. and LOWRIE, W., 1976. Formation of
    spaced cleavage and folds in brittle limestone by dissolution.
    Geology, 4, 648-701.
BARRETT, P.J. 1964. Residual seams and cementation in Oligocene
    shell calcarenites, Te Kuiti Group. Jour.Sed.Petrol., 34, 524-
    531.
BATHURST, R.G.C. 1971. Carbonate Sediments and Their Diagenesis,
    Dev. in Sedimentology 12: Elsevier, 620p.
DeBOER, R.B. 1977. On the thermodynamics of pressure solution-
    interaction between chemical and mechanical forces. Geochim.et
    Cosmochim. Acta, 41, 249-256.
DEFEYES, K.S., LUCIA, F.J. and WEYL, P.K., 1965. Dolomitization
    of Recent and Plio-Pleistocene sediments by marine evaporite
    waters on Bonaire, Netherlands Antilles. Soc.Econ.Paleontolo-
    gists Mineralogists Spec.Pub. No.13, 71-88.
DRAVIS, J. 1979. The sedimentology and diagenesis of the Upper
    Cretaceous Austin Chalk Formation, South Texas and Northern New
    Mexico. Unpubl. PhD thesis, Rice University, Houston, Texas,
    513p.
DROXLER, A., and SCHAER, J.P. 1979. Deformation cataclastique
    plastique lors du plissment sous faible couverture, de strates
    calcaires. Ecloque Geol.Helv., 72, 551-570.
DUNNINGTON, H.V. 1954. Stylolite development postdates rock
    induration. Jour.Sed.Petrology, 24, 27-49.
FOLK, R.L. 1965. Some aspects of recrystallization in ancient
    limestones. Soc.Econ.Paleontologists Mineralogists Spec.Pub.
    No.13, 14-48.
FOLK, R.L. 1973. Carbonate petrography in the post-Sorbian age,
    in Ginsburg, R.N., ed. Evolving Concepts in Sedimentology:
    Baltimore, Maryland, The Johns Hopkins University Press, p.118-
    158.
FOLK, R.L. 1974. The natural history of crystalline calcium
    carbonate: effect of magnesium content and salinity. Jour.
    Sed.Petrology, 44, 40-53.
GARRISON, R.E., and KENNEDY, W.J. 1977. Origin of solution seams
    and flaser structure in Upper Cretaceous chalks of southern
    England. Sed.Geology, 19, 107-137.
GEBELEIN, C.D. and HOFFMAN, P.H. 1973. Algal origin of dolomitic
    laminations in stromatolitic limestone. Jour.Sed. Petrology,
    43, 603-613.

GEISER, P.A. 1974. Cleavage in some sedimentary rocks of the
  Central Valley and Ridge province, Maryland. Geol.Soc.America
  Bull., 85, 1399-1412.
GROSHONG, R.H. 1975. Strain, fractures, and pressure solution in
  natural single-layer folds. Geol.Soc.America Bull., 86, 1363-
  1376.
ILLING, L.V., WELLS, A.J. and TAYLOR, J.C.M. 1965. Penecontempo-
  rary dolomite in the Persian Gulf. Soc.Econ. Paleontologists
  Mineralogists Spec.Pub. No.13, 89-111.
JENKYNS, H.C. 1974. Origin of red nodular limestones (Ammonitico
  Rosso, Knollenkalke) in the Mediterranean Jurassic: a diagenetic
  model. Spec.Pubs. Int.Assoc.Sedimentologists, 1, 249-271.
LLOYD, R.M. 1977. Porosity reduction by chemical compaction-
  stable isotope model. Am.Assoc.Petroleum Geologists Bull.,
  61, 809.
LOGAN, B.W., and SEMENIUK, V. 1976. Dynamic metamorphism; processes
  and products in Devonian carbonate rocks, Canning Basin, Western
  Australia. Geol.Soc.Australia Spec.Pub. No. 6, 138.
LUCIA, F.J. 1972. Recognition of evaporite-carbonate shoreline
  sedimentation, in Rigby, J.K., and Hamblin, W.K., eds.
  Recognition of ancient sedimentary environments. Soc.Econ.
  Paleontologists Mineralogists Spec.Pub. No. 16, 160-191.
MATTER, A. 1967. Tidal flat deposits in the Ordovician of Western
  Maryland. Jour.Sed.Petrology, 37, 601-609.
MATTES, B.W. and MONTJOY, E.W. 1980. Aerial dolomitization of the
  Upper Devonian Miette buildup, Jasper National Park, Alberta.
  Soc.Econ.Paleontologists Mineralogists Spec.Pub. No.28, 259-
  297.
MIMRAN, Y. 1977. Chalk deformations and large scale migration of
  calcium carbonate. Sedimentology, 24, 333-360.
OLDERSHAW, A.E. and SCOFFIN, T.P. 1967. The source of ferroan
  and non-ferroan calcite cements in the Halkin and Wenlock
  Limestones. Jour.Geology, 5, 309-320.
REINHARDT, J. and HARDIE, L.A. 1976. Selected examples of carbonate
  sedimentation, Lower Paleozoic of Maryland. Maryland Geol.
  Survey, Guidebook 5, 53p.
ROEHL, P.O. 1967. Stony Mountain (Ordovician) and Interlake
  (Silurian) facies analogs of Recent low-energy marine and sub-
  aerial carbonates, Bahamas. Am.Assoc.Petroleum Geologists, 51,
  1979-2032.
RUTTER, E.H. 1976. The kinetics of rock deformation by pressure
  solution. Phil.Trans.R. Soc.Lond., A, 283, 203-219.
SANDO, W.J. 1957. Beekmantown Group (Lower Ordovician) of Maryland. Geol.Soc.America Mem. 68, 161p.
SCHLANGER, S.O. and DOUGLAS, R.G. 1974. The pelagic ooze-chalk-
  limestone transition and its implications for marine strati-
  graphy.Spec.Pubs.Int.Assoc. Sedimentologists, 1, 117-148.
SHINN, E.A., GINSBURG, R.N. and LLOYD, R.M. 1965. Recent supratidal dolomite from Andros Island, Bahamas. Soc.Econ.
  Paleontologists Mineralogists Spec.Pub. No.13, 112-123.

SHINN, E.A., HALLEY, R.B., HUDSTON, J.H. and LIDZ, B.H. 1977.
   Limestone compaction: an enigma. Geology, 5, 21-24.
SORBY, H.C. 1879. On the structure and origin of limestones.
   Quart. Jour.Geol.Soc.London, 35, 56-95.
SORBY, H.C. 1908. On the application of quantitative methods to
   the study of the structure and history of rocks. Quart.Jour.
   Geol.Soc.London, 64, 171-233.
STOCKDALE, P.B. 1922. Stylolites: their nature and origin.
   Indiana Univ.Studies, 9, 1-97.
STOCKDALE, P.B. 1926. The stratigraphic significance of solution
   in rocks. Jour.Geology, 34, 399-414.
STOCKDALE, P.B. 1936. Rare stylolites. Am.Jour.Sci., 32, 129-133.
STOCKDALE, P.B. 1943. Stylolites: primary or secondary? Jour.Sed.
   Petrology, 13, 3-12.
THOMSON, J. 1862. On crystallization and liquefaction, as
   influenced by stresses tending to change of form in crystals.
   Phil.Mag., 24, 395-401.
VAN TUYL, F.M. 1916. The origin of dolomite. Iowa Geol.Surv.
   Rept., 25, 251-422.
TISSOT, B.P. and WELTE, D.H. 1978. Petroleum Formation and
   Occurrence. Springer-Verlag, 538p.
WANLESS, H.R. 1973. Microstylolites, bedding and dolomitization.
   Am.Assoc.Petroleum Geologists, 57, 811.
WANLESS, H.R. 1979. Limestone response to stress: pressure
   solution and dolomitization. Jour.Sed.Petrology. 49, 437-462.

INDEX

419